LLM for SoC Design and Security

Farimah Farahmandi • Jingbo Zhou • Upoma Das •
Sujan Kumar Saha • Mark Tehranipoor
Editors

LLM for SoC Design and Security

Springer

Editors
Farimah Farahmandi
University of Florida
Gainesville, FL, USA

Jingbo Zhou
University of Florida
Gainesville, FL, USA

Upoma Das
University of Florida
Gainesville, FL, USA

Sujan Kumar Saha
University of Florida
Gainesville, FL, USA

Mark Tehranipoor
University of Florida
Gainesville, FL, USA

ISBN 978-3-032-34115-0 ISBN 978-3-032-34116-7 (eBook)
https://doi.org/10.1007/978-3-032-34116-7

This Springer imprint is published by the registered company Springer Nature Switzerland AG
The registered company address is: Gewerbestrasse 11, 6330 Cham, Switzerland

Acknowledgments This book is the result of the collective efforts of many people, and we are sincerely grateful to everyone who contributed to its completion.

First and foremost, we would like to thank all of the chapter authors for sharing their expertise, ideas, and time. Their dedication and thoughtful contributions form the foundation of this book and reflect the exciting progress being made at the intersection of hardware security, System-on-Chip (SoC) design, verification, and Artificial Intelligence.

Our sincere thanks go to the editorial and production team at Springer Nature for their guidance, patience, and professionalism throughout the publication process. Their support made the journey from manuscript to book both smooth and enjoyable.

We would also like to acknowledge our colleagues, collaborators, and the research communities at our respective universities and laboratories. The discussions, collaborations, and shared enthusiasm for research have greatly influenced many of the ideas presented in this book.

The remarkable progress in Large Language Models is reshaping many aspects of engineering, and hardware design and security are no exception. We hope this book provides students, researchers, and practitioners with both a useful reference and a source of inspiration for exploring the exciting opportunities that lie ahead at the intersection of AI and trustworthy hardware systems.

Finally, we would like to thank our families, friends, mentors, colleagues, and students for their constant encouragement, patience, and support throughout the preparation of this book. Their belief in our work and understanding during the many hours devoted to this project made its completion possible.

Contents

Chapter 1
Introduction to SoC Design and Security

Amisha Srivastava, Samit S. Miftah, and Kanad Basu

System-on-chip (SoC) architectures have become the backbone of modern computing, enabling compact, high-performance platforms across domains such as mobile, automotive, and edge AI. As SoCs evolve to integrate heterogeneous components like CPUs, GPUs, accelerators, and security modules, the design and verification process has grown increasingly complex. Traditional simulation-based and formal verification methods often fall short in handling the concurrency, state-space explosion, and intricate interactions in such systems. Moreover, hardware security has emerged as a first-class concern, with vulnerabilities such as side-channel leaks and hardware Trojans demanding specialized verification strategies beyond functional correctness. This chapter introduces the architectural foundations of SoCs, outlines the challenges in ensuring their correctness and security, and motivates the need for intelligent, automated solutions. It highlights the gap between conventional verification tools and the demands of next-generation SoCs, setting the stage for the use of large language models (LLMs) in this domain. By bridging the semantic gap between specifications, RTL code, and security policies, LLMs offer a promising avenue for improving design productivity, reducing manual effort, and enhancing presilicon security analysis. This chapter lays the groundwork for understanding how LLMs can be effectively integrated into SoC design and security workflows, as explored in the subsequent chapters.

A. Srivastava
The University of Texas at Dallas, Richardson, TX, USA

S. S. Miftah · K. Basu (✉)
Rensselaer Polytechnic Institute, Troy, NY, USA
e-mail: basuk@rpi.edu

F. Farahmandi et al. (eds.), *LLM for SoC Design and Security*,
https://doi.org/10.1007/978-3-032-34116-7_1

1.1 The Evolution of System-on-Chip (SoC) Designs

System-on-chip (SoC) architectures have progressed dramatically over the past two decades, moving from simple microcontroller-based devices to highly integrated, heterogeneous computing platforms. Early SoCs were designed primarily for embedded control tasks, typically combining a CPU core with basic peripherals such as timers, UARTs, and GPIO controllers. These systems were optimized for cost and energy efficiency, targeting industrial controllers, consumer electronics, and early mobile devices. The rise of mobile computing and multimedia-rich applications in the 2000s accelerated SoC complexity. Designers began integrating digital signal processors (DSPs) for audio and video processing, graphics processing units (GPUs) for rendering, and wireless communication modules for connectivity. Standardized on-chip interconnects further enabled scalable reuse of third-party intellectual property (IP) blocks, fostering rapid feature integration on a single silicon die. In recent years, the explosion of artificial intelligence (AI) and machine learning (ML) workloads has further reshaped SoC design. Dedicated neural processing units (NPUs) and AI accelerators now provide massive parallelism for matrix computations, while cryptographic engines and trusted execution environments deliver hardware-level security. These additions have made SoCs indispensable in domains such as autonomous vehicles, edge AI, smart devices, and secure communications. Technology scaling has also played a pivotal role. Progress from planar CMOS to FinFET and, more recently, gate-all-around (GAA) devices has enabled higher transistor densities and improved power-performance efficiency. Emerging techniques such as 2.5D packaging, 3D stacking, and chiplet-based integration allow heterogeneous dies, potentially fabricated at different technology nodes, to be combined into a unified system. This trend blurs the distinction between traditional SoCs and system-in-package (SiP) solutions, pushing the boundaries of integration. A representative SoC is shown in Fig. 1.1, illustrating how contemporary designs bring together memory hierarchies, cryptographic engines, digital signal processors (DSPs), and communication IPs on a single chip. While such integration enables unprecedented functionality, it also introduces exponentially greater design complexity, verification overhead [22, 87], and security risks [65, 69]: themes that form the focus of this book.

1.2 Modern SoC Architecture and Functional Blocks

While the evolution of SoC design explains the trajectory of integration, examining the functional composition of modern SoCs highlights why verification and security have become such formidable challenges. Contemporary SoCs are no longer loose collections of modules but tightly interconnected ecosystems where the correctness and trustworthiness of the whole system depend on intricate interactions between diverse components. At their core, most designs feature *general-purpose CPUs*,

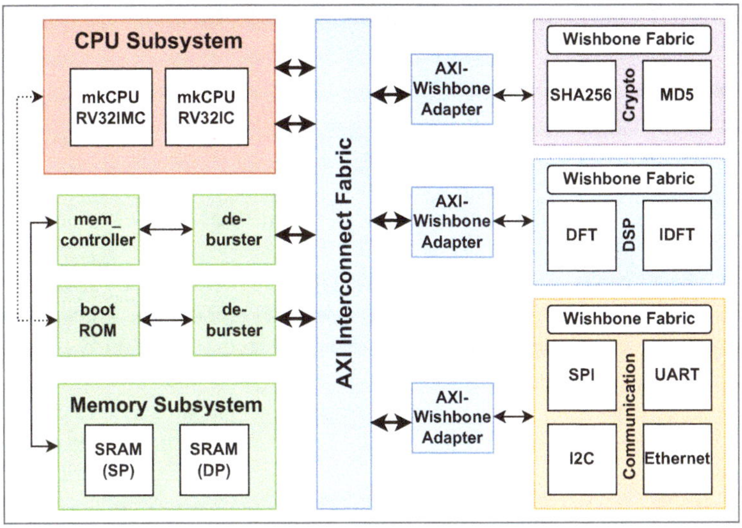

Fig. 1.1 A representative SoC architecture incorporating memory, cryptographic modules, DSPs, and communication IPs

which provide programmability and control, complemented by *GPUs and DSPs* that accelerate parallel workloads such as graphics and signal processing. Increasingly, domain-specific accelerators ranging from AI/ML engines to cryptographic cores are added to meet performance and security requirements. A complex memory hierarchy, spanning on-chip caches, memory controllers, and off-chip DRAM, ensures data availability, while high-speed I/O controllers provide connectivity to sensors, networks, and storage devices. Security is now a first-class design requirement, with primitives such as trusted execution environments, secure boot engines, and hardware random number generators embedded directly into silicon. These blocks operate in tandem with networks-on-chip (NoCs), which have replaced traditional buses as the communication backbone. Unlike simpler interconnects, NoCs must provide concurrency, quality-of-service guarantees, and isolation across heterogeneous blocks. This functional diversity amplifies verification challenges: ensuring protocol compliance, maintaining data integrity across modules, and defending against subtle timing or covert-channel vulnerabilities. As will be discussed in the following sections, the convergence of functional correctness and security assurance demands methodologies that can reason across both structural and behavioral layers of design. Understanding the building blocks and their interdependencies is thus a prerequisite for leveraging large language models (LLMs) to transform SoC verification and security.

1.3 Design Complexity and Verification Bottlenecks

As illustrated in Fig. 1.2, the development of an SoC proceeds through multiple stages, beginning with design exploration and architecture planning, followed by microarchitectural specification, RTL implementation, verification, fabrication, and postsilicon validation. Verification is positioned as a critical stage that ensures correctness before costly fabrication, yet it often dominates the overall design effort due to its complexity.

The design of modern SoC architectures has grown exponentially in complexity, driven by the integration of multiple heterogeneous components, such as general-purpose processors, AI accelerators, cryptographic engines, memory controllers, and communication interfaces onto a single silicon die. These components must interact seamlessly under strict power, performance, and area (PPA) constraints, often with real-time and safety-critical requirements. As a result, ensuring functional correctness and reliability has become a major bottleneck in the SoC development cycle. Traditional simulation-based verification methods rely on stimulus generation and testbenches to validate expected behavior. However, as designs scale, the number of possible input sequences, timing interactions, and corner cases grows combinatorially. This results in the well-known *state-space explosion* problem, where exhaustive simulation becomes computationally infeasible [89]. Consequently, even well-verified designs may harbor latent bugs that only manifest under rare or adversarial conditions.

Formal verification techniques attempt to overcome this by mathematically proving correctness properties over all possible states [59]. While powerful, these methods face their own limitations. Formal verification techniques often struggle to scale to large designs: The symbolic representations required become unmanageable in the presence of deep pipelines, multiple asynchronous clock domains, or complex cross-domain interactions. As a result, many formal tools are limited to small or isolated IP blocks rather than full SoC-level validation. Moreover, the growing use of third-party IP blocks further complicates the verification landscape. IP integration introduces variability in design quality, verification methodology, and documentation, making it difficult to apply uniform verification strategies across the SoC. Interfacing bugs, configuration mismatches, and protocol violations at IP

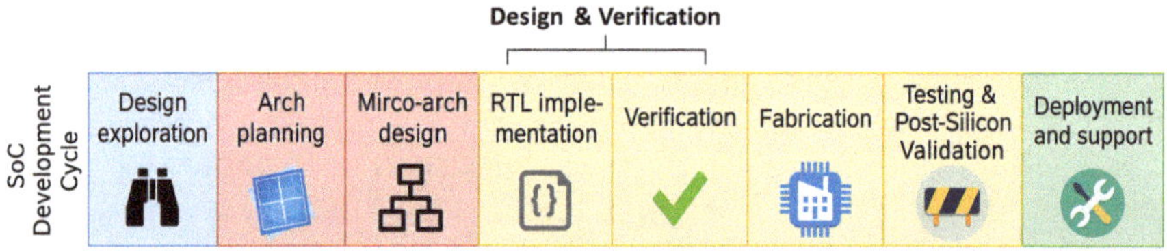

Fig. 1.2 Lifecycle of SoC functional blocks and their interactions. Verification is a central stage that bridges design and fabrication, but it increasingly dominates the development timeline

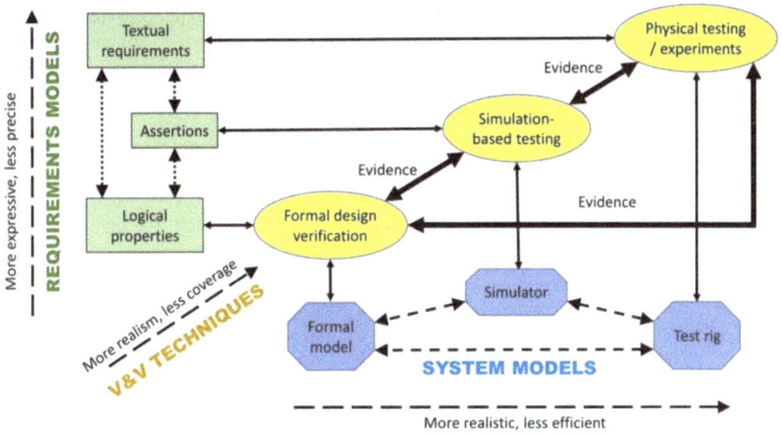

Fig. 1.3 Verification and validation techniques linking requirements, system models, and evidence generation through formal, simulation-based, and physical testing

boundaries are common sources of functional errors that elude standard simulation campaigns.

Importantly, Fig. 1.3 also highlights the tradeoff between expressiveness and efficiency in verification and validation. Requirements models (e.g., logical properties, and assertions) provide strong coverage but limited realism, while system models (e.g., simulators and test rigs) capture real behavior but are less exhaustive. The interplay between these models underscores the need for scalable, automated solutions. *Large language models (LLMs)* offer a promising direction, as they can parse RTL, generate assertions, detect ambiguities, and assist in verification planning—bridging the gap across this spectrum with minimal human intervention.

In response to these challenges, hybrid methodologies are emerging, such as simulation-guided formal techniques, coverage-driven verification, and ML-assisted regression filtering in order to improve coverage and reduce debug time. However, these approaches still require significant manual effort to write assertions, interpret counterexamples, and configure test environments. Overall, the verification bottleneck remains a critical obstacle in the SoC design flow. There is a pressing need for scalable, automated, and intelligent solutions that can reason about both the structure and semantics of hardware designs. The rise of *LLMs* offers a novel opportunity to address this gap, as they are capable of understanding RTL, generating assertions, identifying functional ambiguities, and assisting in verification planning, all while scaling with minimal human intervention.

While the challenges discussed thus far largely pertain to functional verification ensuring that a design performs its intended operations under specified conditions, modern SoCs must also contend with a broader set of adversarial threats that cannot be captured through functional correctness alone. These include subtle vulnerabilities such as power/EM side-channel leakage [6, 7, 65], fault injection attacks [70, 74, 75, 81], hardware trojans [80], and timing-based information disclosure [94]. Such issues demand a fundamentally different class of verification known as security verification, which focuses on protecting the system against malicious exploitation rather than validating expected behavior. As we will explore in the next section, functional and security verifications differ in goals, methodologies, and assumptions and must be addressed together to ensure truly trustworthy SoC designs [22, 26].

1.4 Functional vs. Security Verification

Functional verification focuses on validating that a hardware design correctly implements its specification under normal operating conditions. This process primarily seeks to ensure that, given a defined set of inputs, the system produces the expected outputs. Common approaches include simulation-based verification using constrained-random testbenches, coverage-driven verification to maximize exercised states, and formal verification to mathematically prove specific functional properties. Functional correctness is crucial for ensuring that a chip can meet performance, timing, and power requirements without logical errors.

However, functional verification alone is insufficient to guarantee the trustworthiness of modern SoCs. Security verification extends beyond intended behaviors to examine how a design might fail or be exploited under adversarial conditions. The goal is not only to validate correctness but also to uncover potential vulnerabilities that could compromise confidentiality, integrity, or availability. For example, even if a cryptographic core produces the correct ciphertext functionally, it may still leak secret keys through power side-channel emissions. Similarly, a processor may pass all functional tests yet be vulnerable to privilege escalation, speculative execution attacks, or hardware Trojans embedded in third-party IP blocks.

Security verification therefore requires a different mindset and complementary toolsets. While functional verification assumes cooperative environments and predictable workloads, security verification must account for malicious actors, rare corner cases, and unintended interactions between modules. Techniques such as formal information flow tracking (IFT) [4, 34], side-channel leakage analysis [5, 7, 24], fuzzing [15], symbolic execution [48, 95], and penetration testing [2, 73] are commonly employed to identify weaknesses that functional tests overlook. These approaches are often more adversary-driven, aiming to break the design rather than validate it.

The distinction between functional and security verification is summarized in Fig. 1.4, which illustrates how security must be integrated throughout the SoC development lifecycle. Beginning with the production concept, designers establish threat models and security architectures (e.g., SGX, TrustZone) that guide secure implementation choices. Subsequent stages such as design reviews, formal analysis, and simulation-based validation help identify weaknesses before fabrication. Security testing then evaluates the design against both functional correctness and adversarial scenarios, while postdeployment monitoring and incident response

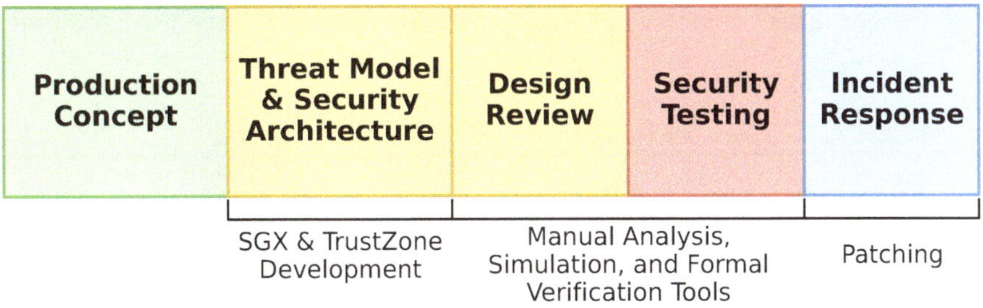

Fig. 1.4 Security lifecycle in SoC development, spanning from production concept and threat modeling to design review, security testing, and incident response

mechanisms, such as patching, provide resilience against newly discovered threats. This end-to-end perspective demonstrates that functional and security verification are not parallel tracks but complementary processes that must evolve together to ensure robust and trustworthy SoCs.

1.5 Common Security Threats in SoC Designs

In this section, we examine the range of vulnerabilities that frequently appear in SoCs and can be exploited either through software alone or through coordinated hardware-software weaknesses. It is known that approximately 43% of such vulnerabilities arise from this combined exposure and can be grouped into seven categories: privilege misuse, buffer overflows, resource mismanagement, information leakage, numeric faults, cryptographic errors, and code injection [21]. Detecting these issues is particularly challenging, since attackers often leverage rare execution paths and corner cases that escape traditional functional verification. This highlights the importance of systematic security verification across the design lifecycle [26, 60]. To guide this process, it is essential to define a clear threat model. In this section, we classify threats into four categories that broadly capture hardware security concerns: system-level design vulnerabilities, hardware trojans, side-channel attacks, and access violations. Each of these threat models will be discussed in detail in the following subsections.

1.5.1 System-Level Design Vulnerabilities

Flaws at the system architecture level can introduce severe security risks, even when all intended functionalities are present [41, 42, 88]. A common example is the storage or transmission of sensitive data in plaintext within memory, which exposes it to unauthorized reads. Another case arises in mailbox subsystems, where

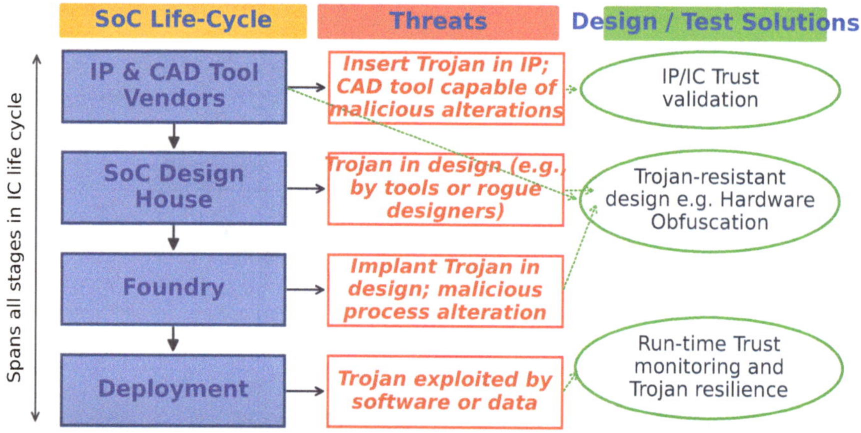

Fig. 1.5 Hardware Trojan threats across the IC lifecycle, spanning IP vendors, design houses, foundries, and deployment, along with representative countermeasures [8]

incoming data from the system infrastructure is verified and forwarded to the SoC. If corrupted data is received and the mailbox lacks an appropriate feedback or error-reporting mechanism, the corrupted data may propagate silently, complicating diagnosis and undermining system reliability. Additional vulnerabilities emerge from simplistic signaling schemes, such as reliance on single-bit flags for critical operations [31, 84]. For instance, an interrupt controller may employ individual flags for tasks such as validity, error, and parity checking. Attackers can exploit this fragility through bit-flip attacks [16, 17, 37], manipulating these control bits to falsely indicate correct behavior. A more resilient design would use multibit encodings or flag arrays to represent state information. Such encodings can distinguish between multiple error conditions and define specific valid patterns, thereby raising the difficulty of tampering and enhancing the robustness of the system against adversarial manipulation.

1.5.2 *Hardware Trojans*

Hardware Trojans (HTs) are deliberate and malicious modifications to integrated circuits (ICs) that are designed to compromise the security, reliability, or trustworthiness of a system [9, 12]. Unlike random design bugs, which usually manifest during functional testing, HTs are engineered to remain hidden during standard validation. They are typically composed of two main components: a *trigger* and a *payload*. The trigger monitors for rare or unlikely events, such as specific input patterns, counter values, or environmental conditions, while the payload carries out the attack once activated. Payloads can range from subtle information leakage to catastrophic denial of service, making HTs one of the most serious threats in modern SoCs [32, 82].

Lifecycle Perspective of Trojan Insertion: As shown in Fig. 1.5, HTs can infiltrate the IC supply chain at *any stage of the design lifecycle*:

- **IP and CAD Tool Vendors:** Third-party IPs or compromised CAD tools may contain preinserted Trojans. Since IP reuse is widespread in SoC design, such malicious inclusions can propagate to multiple products.
- **SoC Design House:** Rogue designers or inadequately secured EDA flows can directly embed Trojans at the RTL or gate level. These modifications often bypass internal checks due to their stealthy nature.
- **Foundry:** Trojans can be implanted during fabrication through malicious process alterations or mask tampering, making detection extremely difficult at the postsilicon stage.
- **Deployment:** Even after a device is shipped, Trojans can be activated and exploited in the field using adversarial inputs, software triggers, or environmental manipulation (ex. temperature, voltage).

This lifecycle perspective highlights why Trojans represent a unique challenge: They exploit the globalization of hardware supply chains, the dependence on third-party IP, and the sheer complexity of SoC design. Defenses must therefore combine proactive design techniques (ex. obfuscation, redundancy) with postdeployment trust monitoring to ensure long-term resilience.

Stealthy Insertion Methods: Beyond the stage of introduction, Trojans are also characterized by the stealth mechanisms used to evade detection during conventional verification. Two common insertion strategies include:

- **Exploiting Rare Nodes and Branches:** Adversaries target internal nodes or execution branches that are rarely activated under normal workloads [13]. By embedding Trojan triggers in these seldom-used states, attackers ensure that functional verification, which relies on constrained-random or directed tests, fails to exercise the malicious condition. The Trojan remains dormant until the adversary deliberately applies the rare input sequence or condition, at which point the payload activates.
- **Gate Misplacement or Replacement:** Another subtle insertion technique involves altering the gate-level netlist by misplacing or replacing gates with malicious variants [38, 63]. These small perturbations may mimic benign design choices, yet they can create vulnerabilities such as unauthorized state transitions, denial of service, or bit-flip behavior. Because these changes minimally affect area, power, or timing, they are often invisible to standard design reviews and physical verification checks, making them particularly dangerous [83, 85].

In summary, hardware Trojans represent a multidimensional threat: They can enter the design at different points in the lifecycle (from IP to deployment) and can be inserted using stealthy circuit-level strategies that evade standard verification. Countering HTs therefore requires a combination of supply-chain trust validation, Trojan-resistant design (ex. hardware obfuscation), and run-time monitoring mechanisms that can detect anomalous activity in-field.

1.5.3 Side-Channel Attacks

Side-channel attacks exploit indirect information that is unintentionally leaked during the normal operation of a system rather than relying on direct access to the system's internal states or data [99]. These attacks focus on observing and analyzing various physical phenomena, such as power consumption, electromagnetic emissions, timing variations, and even acoustic signals. By analyzing these phenomena, attackers can infer sensitive information about the operations being performed within an SoC or cryptographic device. In the context of cryptography, a side channel is any observable characteristic of a system that can be measured during its operation, aside from the actual data output. For instance, when a cryptographic algorithm is executed, the power consumption of the device running the algorithm may vary depending on the data being processed. Although this power consumption is not part of the algorithm's intended output, it can still provide clues about the data or the operations performed. Listed below are some types of popular side-channel attacks.

1.5.3.1 Microarchitectural Side-Channel Attacks

Microarchitectural side-channel attacks exploit unintended information leakage arising from the internal structures and optimizations of modern processors. These attacks generally fall into two main categories: *timing-based attacks*, which exploit variations in execution time caused by microarchitectural events, and *cache-based attacks*, which take advantage of shared cache behavior to reveal memory access patterns. Together, these attack classes demonstrate how performance-enhancing features of CPUs can inadvertently undermine system security:

- **Timing Attacks:** Timing attacks exploit variations in the time it takes to execute certain operations, particularly in cryptographic algorithms. These timing variations can arise from different execution paths or optimizations that depend on secret data, such as cryptographic keys. By carefully measuring the execution time of specific operations, an attacker can infer the secret parameters involved in the computation. For example, in a chosen ciphertext attack, an adversary may measure the time taken to decrypt different ciphertexts and use statistical methods to deduce the secret key. These attacks are particularly effective when the algorithm's execution time varies based on the data being processed.
- **Cache-Based Attacks:** Cache-based attacks exploit the behavior of the CPU cache, which is a small, fast memory used to store frequently accessed data. Since the cache is shared among different processes, an attacker can infer information about another process by observing changes in the cache state. Common types of cache-based attacks include:
  - **Flush + Reload:** In a Flush + Reload attack, the attacker begins by flushing (removing) a specific cache line, forcing any data in that line to be fetched

from the main memory the next time it is accessed. The attacker then waits and reloads the same cache line, measuring the time it takes to access the data [91]. If the access is fast, it indicates that another process has loaded the data back into the cache, suggesting that the data was recently used. This allows the attacker to determine whether a specific memory location was accessed by the victim process, potentially revealing information about the victim's activities, such as cryptographic key usage in algorithms like AES.

– **Prime + Probe:** In this attack, the attacker takes advantage of the cache's shared nature in modern CPUs [44]. As shown in Fig. 1.6, the attack involves two key phases: the *prime* phase, where the attacker fills the cache set with their own data, and the *probe* phase, where the attacker reaccesses these lines after the victim executes. Cache hits and misses observed during probing indicate which lines were replaced by the victim, allowing the attacker to reconstruct the victim's memory access pattern. This leakage can be exploited to extract sensitive information such as cryptographic keys or plaintext. Prime + Probe is particularly concerning in multitenant environments, like cloud servers, where processes or virtual machines share hardware resources.

1.5.3.2 Speculative Execution Attacks

Speculative execution is a performance optimization used in modern CPUs, where instructions are executed ahead of time based on predicted outcomes of operations

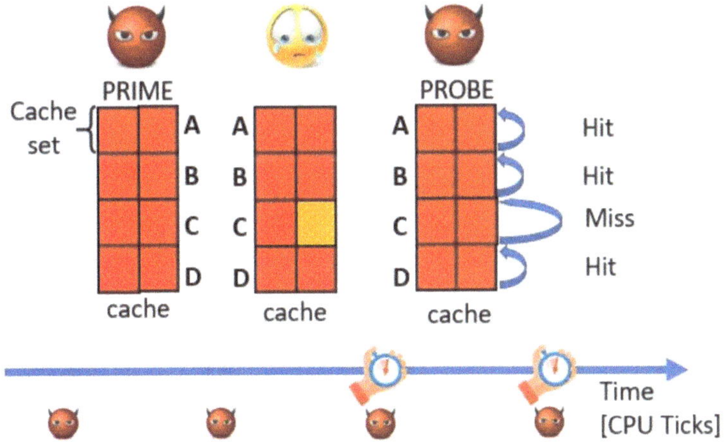

Fig. 1.6 Illustration of the Prime + Probe cache side-channel attack. The attacker first *primes* the cache set with their own data. When the victim process executes, some of these cache lines may be evicted. During the *probe* phase, the attacker measures access times to detect which lines were evicted, thereby inferring the victim's memory access patterns

such as conditional branches [47]. By speculating, the processor keeps its pipeline busy, achieving higher throughput and efficiency. However, if speculation involves sensitive data, even though the speculative instructions may later be rolled back, they can still leave measurable microarchitectural traces (ex. cache state changes). Attackers exploit these transient effects to leak information that should remain inaccessible:

- **Meltdown:** Meltdown exploits a flaw in out-of-order execution, the mechanism that allows CPUs to execute instructions as soon as their operands are ready, without waiting for previous instructions to complete [43]. This attack leverages the fact that speculative instructions can transiently bypass memory protection checks and access privileged kernel memory. Although the CPU discards the illegal results once the violation is detected, side effects—such as changes in the cache—remain observable. By carefully measuring these side effects, an attacker can reconstruct the contents of otherwise protected memory. Meltdown is therefore particularly severe, as it compromises the isolation between user applications and the operating system kernel, exposing highly sensitive data.
- **Spectre:** Spectre targets speculative execution through manipulation of the CPU's branch prediction logic [35]. Branch predictors guess the direction of conditional branches to maintain an uninterrupted instruction pipeline [1, 19]. By deliberately training the predictor with crafted inputs, an attacker can cause the CPU to mispredict and speculatively execute instructions that access secret data. While the speculative path is later discarded, the memory accesses leave microarchitectural side effects, such as modified cache states. By timing memory accesses in the *probe* phase, attackers can infer which paths were taken and, consequently, the secret data involved. Spectre is especially insidious because it generalizes across CPUs and software environments, breaking isolation across process, kernel, and even virtualization boundaries, making it one of the most far-reaching speculative execution attacks discovered to date.

1.5.3.3 Power Side-Channel Attacks

Power side-channel attacks exploit the relationship between a device's power consumption and the operations it performs. By observing the power usage of cryptographic hardware during execution, attackers can uncover sensitive information, such as encryption keys, that would otherwise remain hidden. These attacks are especially effective against hardware-based implementations like smart cards and embedded processors [23]. Figure 1.7 illustrates the typical flow of a power side-channel attack, where the adversary collects power traces and analyzes them to reconstruct secret data.

Power analysis attacks are generally divided into three major categories:

- **Simple Power Analysis (SPA):** SPA involves visually examining the recorded power traces to detect patterns corresponding to executed instructions or processed data values [46]. Because different instructions often consume distinct

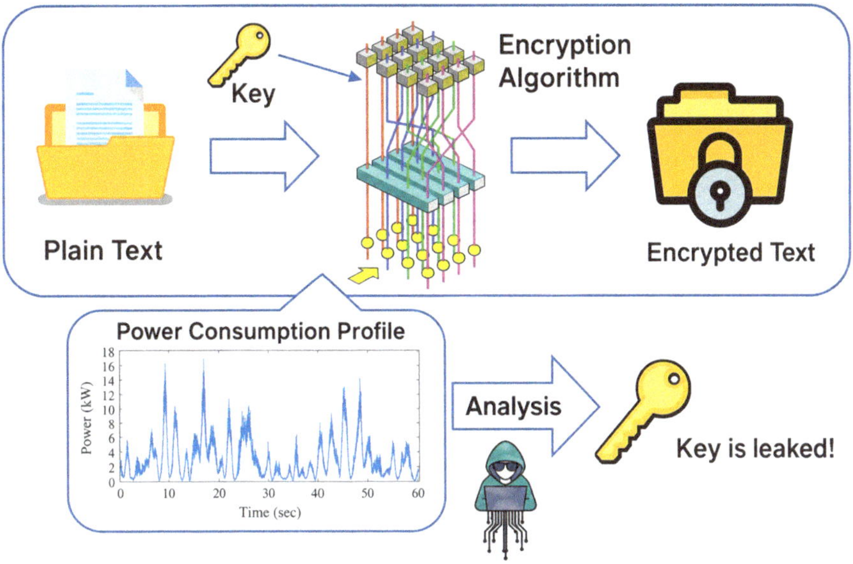

Fig. 1.7 Overview of a power side-channel attack. The attacker records the power consumption profile of a security-critical hardware device and analyzes it to recover secret values such as encryption keys [76]

amounts of power, SPA can reveal instruction sequences and even specific operations. While it requires detailed knowledge of the implementation, SPA can directly expose sensitive algorithmic behavior from raw traces.

- **Differential Power Analysis (DPA):** DPA is a statistical technique that does not rely on detailed knowledge of the hardware design [36]. Instead, it collects many power traces from cryptographic operations and applies statistical methods to uncover correlations between power variations and secret-dependent computations. By analyzing these differences, an attacker can separate the influence of the secret key from the rest of the operations and successfully recover it. DPA has proven effective against widely used algorithms, including those in elliptic curve cryptography.
- **Correlation Power Analysis (CPA):** CPA extends DPA by using mathematical models of power consumption to make hypotheses about the secret key and then correlating the predicted consumption with the measured traces [11]. The correct key hypothesis produces the strongest correlation, enabling the attacker to identify the secret key with high accuracy. Because it leverages correlation techniques, CPA can exploit even very small differences in power usage, making it one of the most powerful and precise forms of power analysis.

In summary, SPA represents the most straightforward form of power analysis, relying on direct inspection of traces, while DPA introduces statistical methods to extract secrets without requiring in-depth design knowledge. CPA further enhances this process by employing correlation models, making it more robust and precise than traditional DPA. Together, these techniques illustrate how increasingly sophisticated statistical and modeling approaches allow attackers to exploit subtle power consumption variations and compromise cryptographic implementations.

1.5.4 Electromagnetic (EM) Side-Channel Attacks

Electromagnetic attacks exploit the EM radiation emitted by integrated circuits (ICs) as they switch internal states during computation. These emanations, which are unintentionally radiated due to current fluctuations and high-frequency transitions, can be captured using EM probes connected to low-noise amplifiers (LNAs) and oscilloscopes (see Fig. 1.8). The captured waveforms, referred to as EM side-channel traces, reveal patterns closely correlated with the underlying computations, including sensitive cryptographic operations.

Similar to power analysis attacks, EM side-channel analysis is divided into two categories:

- **Simple Electromagnetic Analysis (SEMA):** Relies on direct inspection of EM traces to identify operations such as key-dependent branches or specific instructions.
- **Differential Electromagnetic Analysis (DEMA):** Uses statistical methods over multiple traces to correlate EM leakages with secret data, analogous to differential power analysis (DPA).

A unique strength of EM attacks lies in their ability to bypass power-based countermeasures. Even if power lines are filtered, masked, or shielded, electromagnetic emanations can still leak exploitable information through near-field

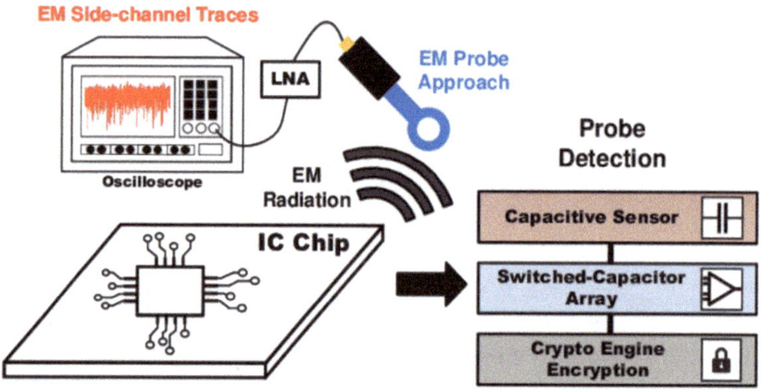

Fig. 1.8 Illustration of an electromagnetic (EM) side-channel attack. An EM probe collects radiation emitted by the IC chip. The captured traces are analyzed through capacitive sensors and switched-capacitor arrays, ultimately revealing sensitive cryptographic operations [72]

probes. Probe detection countermeasures, such as capacitive sensors or switched-capacitor arrays that monitor unusual EM activity, are being explored to protect cryptographic engines. Nevertheless, EM attacks remain a serious threat since they can compromise cryptographic devices in scenarios where traditional power analysis is effectively mitigated [62].

1.5.5 Privilege Escalation Attacks

Privilege escalation represents one of the most severe threats in modern hardware and system security. It occurs when an attacker is able to bypass intended access restrictions and acquire elevated privileges, granting unauthorized control over sensitive system resources. Such an escalation undermines both confidentiality and integrity, enabling adversaries to read, modify, or execute operations beyond their intended permissions. Attackers rely on a wide range of techniques to achieve privilege escalation. At the hardware level, phenomena such as *RowHammer* [71] can induce bit flips in DRAM cells, corrupting adjacent memory regions and enabling unauthorized access. Similarly, microarchitectural side channels such as *Spectre* and *Meltdown* [35, 42] exploit branch predictors, speculative execution, and exception handling mechanisms to extract secret data from protected memory spaces. These vulnerabilities highlight that even functionally correct designs can harbor subtle flaws that expose critical state information.

Beyond these well-studied attacks, security risks also stem from auxiliary design features. Design-for-test (DFT) and design-for-debug (DFD) infrastructures, if insufficiently secured, can inadvertently reveal sensitive internal signals during functional operation. Likewise, classical software vulnerabilities such as buffer overflows or integer overflows can extend into hardware-software interfaces, enabling overwriting of protected memory regions and unauthorized privilege escalation. Given the severity of these threats, robust mitigation strategies are essential. Security verification must consider not only functional correctness but also all potential access paths to privileged states, spanning caches, registers, on-chip memory, and external storage. Protecting these assets requires careful validation of control logic, memory hierarchies, and debug/test features. Ultimately, privilege escalation underscores the need for holistic hardware security validation, ensuring that even rare execution scenarios and auxiliary interfaces cannot be abused by adversaries.

The breadth of vulnerabilities in SoC designs makes it evident that security cannot be treated as a peripheral concern but must be embedded into the verification process itself. Traditional methods, though effective at catching functional flaws, struggle to scale against the adversarial, stealthy, and cross-layer nature of modern threats. This gap creates a compelling case for intelligent, automated solutions that can reason about design semantics, adapt to diverse attack surfaces, and accelerate vulnerability detection. LLMs, with their ability to parse specifications, analyze RTL, generate security-aware assertions, and hold promise as a transformative force in this domain. In the following section, we examine how existing techniques address these challenges today, setting the stage for how LLMs can augment and eventually redefine the future of SoC security verification.

1.6 State of the Art in Security Verification

Security verification in the hardware design flow has emerged as a critical research domain, motivated by the increasing reliance on hardware platforms in security-sensitive applications. Unlike traditional functional verification, which emphasizes correctness with respect to specifications, security verification ensures resilience against adversarial threats and compliance with explicit security guarantees [22]. Modern security verification builds on a broad set of techniques. Formal methods remain central: Theorem-proving frameworks, such as Coq or Isabelle/HOL, allow rigorous reasoning about security properties by constructing mathematical proofs, while algorithmic approaches, such as model checking and satisfiability modulo theories (SMT) solvers, systematically explore possible states and execution traces. Symbolic execution extends this paradigm by treating program inputs symbolically, thereby enabling the exploration of multiple paths simultaneously.

In parallel, information flow tracking (IFT) has become an important mechanism for enforcing confidentiality and integrity properties. Static IFT frameworks check policies at design time, while dynamic approaches add runtime monitoring to detect violations during execution. Additionally, researchers increasingly recognize the importance of hardware-software coverification, given that many vulnerabilities manifest only when both layers are considered together. Collectively, these techniques form the state of the art in security verification, each contributing unique strengths and limitations [22].

1.6.1 Challenges in Modern Security Verification

Despite significant progress, several challenges persist in verifying the security of modern hardware platforms [22]. The increasing complexity of system-on-chip (SoC) designs, coupled with diverse threat models and evolving attack surfaces, has made it difficult for existing frameworks to provide strong security guarantees. These challenges span across microarchitectural effects, hardware-software interactions, specification clarity, and the scalability of verification techniques:

- **Cross-Modular Effects and State-Space Explosion:** Modern SoC designs integrate numerous heterogeneous modules, and vulnerabilities often emerge from subtle interactions across module boundaries. Even a minor flaw in one block may combine with behaviors in others to produce exploitable weaknesses. Capturing such cross-modular effects requires multimodule information flow analysis, but this quickly leads to state-space explosion, where the exponential number of states renders exhaustive exploration computationally infeasible.

- **Timing-Related Vulnerabilities:** Security verification frameworks are traditionally tailored for functional correctness and rarely account for timing behaviors. This leaves designs vulnerable to timing-based side channels, where adversaries infer sensitive information from differences in execution latency, memory access times, or fault handling behaviors. Detecting and reasoning about such vulnerabilities remains difficult, as current verification models lack the expressiveness to capture fine-grained temporal properties.
- **Cache-State Gaps:** Caches and buffers are integral to modern microarchitectures, yet they remain a persistent blind spot in security verification. Most frameworks focus on register-level correctness, ignoring cache states such as eviction policies, privilege-level transitions, and buffer contents. These structures have been repeatedly exploited in high-profile side-channel attacks, underscoring the need for verification approaches that explicitly model nonregister states.
- **Hardware-Software Interactions:** Hardware vulnerabilities often become severe only in the presence of specific firmware or operating system configurations. Examples include insecure debug logic, firmware-controlled cryptographic key storage, or overlapping peripheral address ranges. The inability of current frameworks to jointly analyze hardware and software behaviors limits their ability to capture such cross-layer vulnerabilities.
- **Specification Ambiguity:** Security properties are frequently underspecified or conflated with functional correctness requirements. This ambiguity leaves corner cases undefined and threat models incomplete, allowing adversaries to exploit unspecified behaviors. Moreover, when vulnerabilities are discovered postsilicon, mitigations such as microcode updates or firmware patches typically address symptoms rather than root causes, since RTL-level bugs cannot be fundamentally corrected without costly redesigns.
- **Scalability and Usability:** Formal verification approaches face inherent tradeoffs between rigor and scalability. Theorem proving provides strong guarantees but requires substantial manual effort and expertise. Algorithmic methods such as model checking and symbolic execution are more automated but often succumb to state explosion or require oversimplified abstractions that risk missing critical vulnerabilities. As designs grow in complexity, the scalability and usability of current methods remain pressing concerns.

1.6.2 State-of-the-Art Methods to Address These Challenges

In response to these challenges, researchers have developed a diverse set of methods, each targeting specific limitations of current verification frameworks [22]. In Fig. 1.9, we show a collection of different tools that are applied on different levels of hardware and software of a computer system:

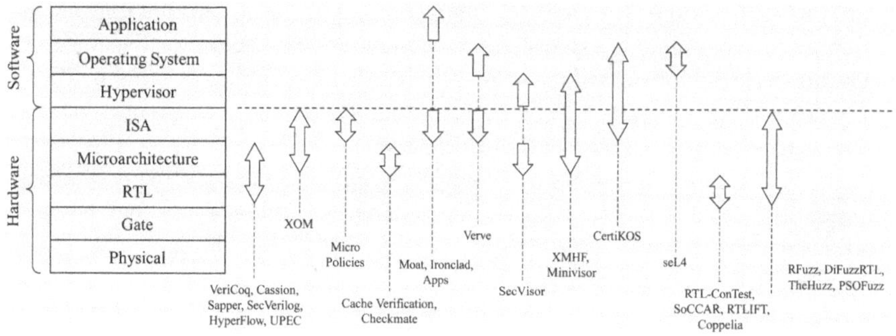

Fig. 1.9 Tools applied on various hardware and software levels

- **Addressing Cross-Modular Effects:** To mitigate the state-space explosion problem, *RTL-ConTest* combines symbolic execution with concrete simulation, enabling the systematic validation of security properties at the RTL level of complex SoCs [48]. By generating control-flow graphs and targeted test cases, RTL-ConTest has been shown to uncover issues such as insecure debug logic and address overlaps between SoC peripherals. Similarly, *unique program execution checking (UPEC)* automatically derives computational models from RTL and identifies covert channels without requiring manual modeling [61]. Both approaches illustrate scalable strategies to capture cross-modular vulnerabilities.
- **Modeling Timing-Related Vulnerabilities:** Approaches such as *SecVerilog* extend hardware description languages with type systems that enforce timing-sensitive noninterference [92]. Designers can annotate variables with security labels, and dependent types enable reasoning about resource sharing across security domains. These mechanisms allow execution timing to be explicitly considered as a source of leakage, reducing the likelihood of covert channels that exploit temporal differences.

  To address vulnerabilities arising from asynchronous resets, the tool *SoCCAR* was developed. This tool is designed specifically to identify weaknesses originating from asynchronous resets in SoC designs, which are often overlooked by formal verification methods or concealed within rare corner cases. As a result, traditional simulation-based verification tools are unable to detect them effectively [49, 56].
- **Capturing Cache-State Behaviors:** Formal cache modeling has been explored using tools such as the *Murφ* model checker [18, 93]. By explicitly representing cache hits, misses, and transitions, these methods enable quantitative analysis of attacker-victim interference. Applied to a range of cache architectures, Murφ-based analysis has revealed that most designs fail to satisfy noninterference properties, highlighting the widespread vulnerability of caches to side-channel exploitation.
- **Integrating Hardware-Software Verification:** Frameworks such as *Coppelia* extend symbolic execution to hardware/software coverification [95]. Coppelia

translates RTL into LLVM intermediate representation, integrates software-level execution, and generates concrete exploits when violations are detected. By validating hardware vulnerabilities in realistic software contexts, Coppelia demonstrates how joint analysis can bridge the gap between hardware-only and software-only verification.

- **Clarifying Specifications and Preserving Guarantees:** To resolve specification ambiguity, interface-based approaches such as *deep specifications* in CertiKOS provide rigorous definitions for abstraction layers [25]. Verification proceeds modularly, ensuring that each layer adheres to its formal specification and that higher-level properties are preserved. In addition, verified compilers such as *CompCert* ensure that verified source-level properties are faithfully preserved through compilation, reducing reliance on unverified tool chains [40].
- **Improving Scalability and Usability:** Efforts to lower the barrier of entry and improve scalability include tools such as *VeriCoq*, which translates Verilog into Coq, enabling formal reasoning without extensive manual modeling [10]. Likewise, *RTLIFT* automatically instruments RTL with information flow tracking logic, allowing conventional verification tools to analyze security properties efficiently [3]. These approaches enhance automation and reusability, though challenges in property expressiveness and industrial-scale adoption remain.
- **Fuzzing as an Emerging Verification Method:** An emerging and highly promising verification method in hardware security is *fuzzing*. Fuzzing is an automated, coverage-guided testing approach that systematically generates and mutates input programs to drive hardware into diverse and often unexpected states, thereby uncovering vulnerabilities that traditional verification methods often miss (see Fig. 1.10). Unlike purely formal methods, which face scalability challenges, fuzzing scales naturally to large SoCs and processor designs while requiring minimal manual effort.

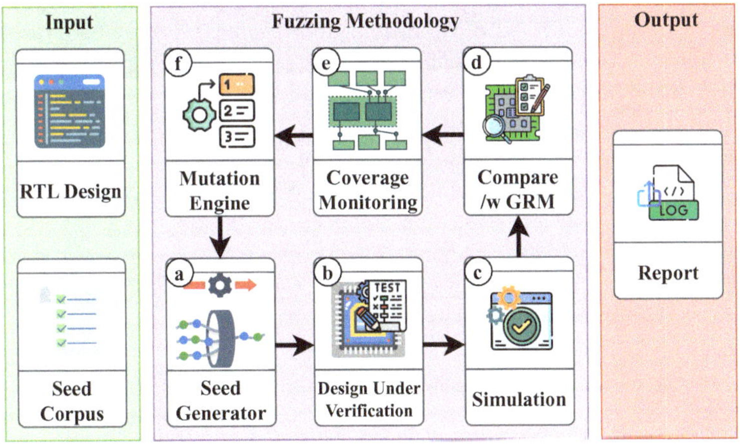

Fig. 1.10 Workflow of general fuzzing methodology

The first application of coverage-guided mutational fuzzing to RTL verification was proposed in *RFUZZ*. By introducing FPGA acceleration and novel reset mechanisms, RFUZZ demonstrated that fuzzing could be applied effectively at the RTL level, addressing state-space explosion by automating coverage-directed test generation and reducing the manual burden of property specification [39].

Subsequent efforts built on this foundation to tackle additional challenges. *DifuzzRTL* introduced a differential testing approach, comparing RTL execution against golden ISA models to expose semantic mismatches. This method addresses the problem of specification ambiguity by removing the need for complete formal specifications, while also uncovering vulnerabilities that emerge only through architectural inconsistencies [29].

TheHuzz advanced this line of work by combining symbolic execution with fuzzing. By systematically exploring deeper execution paths and generating inputs tailored to corner cases, TheHuzz addressed the challenge of path explosion and ensured that vulnerabilities dependent on rare execution conditions could be reached [33].

PSOFuzz incorporated particle swarm optimization (PSO) into the fuzzing workflow. This addition enabled adaptive guidance of mutation operators and seed generation, overcoming the stagnation problem in traditional fuzzers. In doing so, PSOFuzz addressed scalability challenges by ensuring that coverage continues to expand even in highly complex designs [14].

Beyond processor pipelines, *fuzzing hardware like software* broadened the applicability of fuzzing by leveraging mature software fuzzers on translated RTL models. This approach addressed usability concerns by eliminating the need for design-specific harnesses, reducing verification overhead, and making fuzzing more accessible for diverse hardware designs [86].

More recently, UVM-based fuzzing frameworks such as *InterConFuzz* and *SymbFuzz* have demonstrated how fuzzing can be deeply integrated into industry-standard verification workflows [55, 57]. Both frameworks leverage UVM to ensure compatibility with commercial tool chains, but they target different verification challenges. InterConFuzz focuses on network-on-chip (NoC) architectures, where cross-modular interactions and concurrency pose unique vulnerabilities. By integrating symbolic execution and checkpoint-guided exploration, it systematically addresses the challenges of verifying communication-heavy subsystems. In contrast, SymbFuzz applies symbolic execution-guided fuzzing to processor and IP-level verification. By redefining coverage in terms of control-register interactions and introducing checkpoint and partial-reset strategies, it addresses specification ambiguity, scalability bottlenecks, and coverage stagnation in complex processor designs.

Taken together, these advances illustrate how fuzzing has evolved into a versatile and comprehensive verification paradigm. From RFUZZ's foundational framework to recent UVM-based extensions, fuzzing systematically tackles state-space explosion, specification ambiguity, cross-modular effects, and scalability. As modern SoCs grow in complexity and integrate diverse components, fuzzing provides an adaptive, automated, and industrially viable methodology that complements and extends traditional verification. This positions fuzzing as a key driver in improving both the efficiency and effectiveness of hardware security verification in future system-on-chip designs.

Together, these advances illustrate how fuzzing directly addresses key challenges in modern hardware verification. It mitigates state-space explosion through guided exploration, uncovers timing and cross-layer vulnerabilities, scales efficiently to industrial designs, and automates coverage closure. As demonstrated by RFUZZ, DifuzzRTL, TheHuzz, PSOFuzz, and hardware-software fuzzing pipelines, fuzzing is rapidly evolving into a cornerstone of hardware security verification, offering both depth and breadth in bug discovery while reducing verification cost and complexity.

1.7 AI/ML in Hardware Verification: The New Frontier

Recent advances in AI and ML have opened new opportunities for design automation, particularly in hardware verification. Traditional ML techniques have already demonstrated value in several areas. For example, supervised learning models trained on past bug repositories can predict modules most likely to contain defects, thereby guiding verification resources toward high-risk regions of the design. Clustering and anomaly-detection methods have been employed to identify unusual simulation traces or to highlight coverage gaps that may indicate untested or corner-case behaviors. In hardware verification, such techniques have been applied to group similar failure modes and detect missing behaviors in large SoC security verification workflows. Related approaches are also widely explored in malware and intrusion detection, where clustering and anomaly detection on execution traces or system logs are used to uncover malicious activity [54, 77]. Reinforcement learning (RL) has further emerged as a powerful strategy for guiding test stimulus generation. By adaptively exploring the design's state space, RL enables coverage of rare events and hard-to-reach states more efficiently than random or constrained-random approaches [96].

These contributions illustrate how ML can complement existing simulation and formal verification flows by automating repetitive tasks and accelerating coverage closure. However, despite these successes, conventional ML techniques face important limitations. They typically require large, high-quality training datasets that may not exist for every design or IP block. Models often generalize poorly across projects, as patterns learned from one SoC may not transfer directly to another with different architectures or verification methodologies. Moreover, conventional ML is largely statistical, it can highlight correlations or anomalies, but it lacks the ability to interpret specifications, reason about system semantics, or generate human-readable verification artifacts. As a result, ML alone cannot bridge the gap between natural language requirements, RTL implementations, and formal security policies. This is where LLMs represent a transformative step forward. Unlike traditional ML, LLMs are trained on vast corpora of code, documentation, and natural language, allowing them to operate simultaneously in the symbolic space of hardware description languages (e.g., Verilog/SystemVerilog) and the narrative space of human-readable specifications [68, 78, 79]. LLMs can explain RTL code, generate SystemVerilog assertions, propose UVM testbench scaffolds, or extract threat models directly from specifications [50, 51, 58]. By combining pattern recognition with semantic understanding, they offer a pathway to automate verification tasks that previously required deep domain expertise. Thus, LLMs extend the reach of AI/ML in hardware verification from data-driven assistance to knowledge-driven reasoning, setting the stage for their integration into SoC design and security workflows [26, 27, 30, 64, 67].

1.8 Why Use LLMs for SoC Design and Security?

The preceding sections highlighted how the trajectory of SoC design has produced both opportunity and complexity. Modern SoCs integrate CPUs, GPUs, accelerators, cryptographic engines, and high-speed interconnects into tightly coupled systems. This integration has made verification the dominant bottleneck in the development cycle, with functional correctness and security posing orthogonal but equally critical challenges (Sect. 1.4). The classification of common threats further reinforced that security vulnerabilities often emerge at the boundaries between functional correctness and adversarial exploitation. Against this backdrop, LLMs offer a promising new paradigm for addressing the dual demands of scalability and trustworthiness in SoC design.

The scale of the verification challenge is underscored by the 2022 Wilson Research Group Functional Verification Study (Fig. 1.11). Verification engineers devote nearly half of their effort to debugging (47%), with substantial additional time spent developing testbenches (15–17%), creating tests and simulations (19–21%), and planning strategies (11–13%). These phases are not only labor-intensive but also repetitive, requiring deep expertise in hardware description languages, simulation environments, and verification methodologies. Importantly, many of these tasks—test generation, coverage analysis, debugging assistance—are precisely the types of structured yet language-driven reasoning problems where LLMs excel [26, 64].

LLMs provide a new abstraction layer that directly complements the limitations of simulation-based and formal verification. Unlike traditional EDA tools, which rely on rigid rule-based engines, LLMs can bridge informal specifications and machine-readable code. They can parse RTL and natural language documentation, generate SystemVerilog assertions, propose UVM testbench scaffolds, and assist in root-cause localization when simulation mismatches occur. In doing so, they shift the burden from manual, repetitive tasks toward automated, semantically guided workflows. For example, during debugging, LLMs can suggest likely failure points based on log analysis; during test generation, they can synthesize corner-case stimuli from textual requirements; and during security validation, they can propose attack

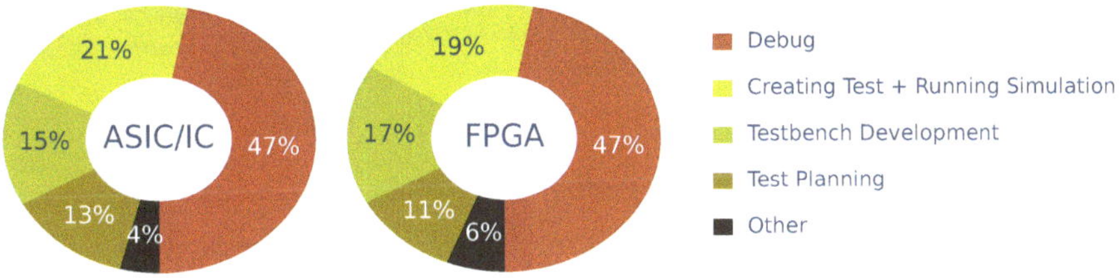

Fig. 1.11 Breakdown of the average time spent on different verification steps by verification engineers in 2022, redrawn from the 2022 Wilson Research Group Functional Verification Study [90]. Debugging dominates the workload, highlighting opportunities for LLMs to reduce manual effort

scenarios and generate targeted checks for privilege escalation or side-channel leakage.

Existing frameworks already hint at this potential. AssertLLM [20] explores assertion generation from specifications but highlights the difficulty of aligning general-purpose LLMs with formal semantics. LLM4DV [98] demonstrates automatic test stimulus generation yet struggles to capture the nuanced interactions of hardware components. SCAR [76] applies LLMs to generate masked hardware implementations for side-channel resilience, while UVLLM [28] integrates LLMs into UVM workflows for automated testbench creation. These efforts illustrate both the promise and current shortcomings: Issues of syntactic correctness, incomplete reasoning about HDL semantics, and a lack of structured training datasets remain barriers. Benchmark efforts such as VERT, MG-Verilog, VerilogEval, and Vul-FSM are steps toward addressing these gaps, providing domain-specific corpora that better align LLMs with hardware design semantics [45, 52, 53, 68, 97]. Crucially, the application of LLMs is not limited to functional verification. Security verification, long treated as an afterthought, demands adversary-driven reasoning that conventional regression suites cannot provide. LLMs can assist in building threat models from design documentation, suggesting possible Trojan insertion points, identifying cache- or timing-based covert channels, and generating security-focused assertions that probe beyond expected functionality [26, 64, 66]. By enabling such adversarial perspectives at scale, LLMs position themselves as enablers of trustworthy hardware design. In summary, the integration of LLMs into SoC design and verification is not merely an incremental productivity improvement but a paradigm shift. They offer a unifying framework capable of spanning the spectrum from specification to functional correctness to adversarial validation.

As SoCs become more heterogeneous and security-critical, the inability of traditional verification flows to keep pace highlights the need for intelligent, automated, and scalable approaches. LLMs embody this next step, combining human-centric interaction with machine-scale reasoning. The remainder of this book builds on this foundation, exploring how LLMs can reshape the landscape of SoC design and security, and outlining methodologies to harness their potential for the next generation of trustworthy computing systems.

1.9 Book Organization and Reader Roadmap

This book explores the intersection of LLMs and SoC workflows, focusing on their application to design, verification, and security. The upcoming chapters delve into verification techniques, LLM architectures, prompt engineering, case studies, and future outlooks. Whether you are a hardware engineer, a security researcher, or an AI practitioner, this book aims to provide a unified view of how LLMs can revolutionize the SoC development lifecycle.

References

1. Onur Acıiçmez, Çetin Kaya Koç, and Jean-Pierre Seifert. Predicting secret keys via branch prediction. In *Topics in Cryptology–CT-RSA 2007: The Cryptographers' Track at the RSA Conference 2007, San Francisco, CA, USA, February 5–9, 2007. Proceedings*, pages 225–242. Springer, 2006.
2. Hasan Al-Shaikh, Arash Vafaei, Mridha Md Mashahedur Rahman, Kimia Zamiri Azar, Fahim Rahman, Farimah Farahmandi, and Mark Tehranipoor. SHarPen: SoC security verification by hardware penetration test. In *Proceedings of the 28th Asia and South Pacific Design Automation Conference*, pages 579–584, 2023.
3. Armaiti Ardeshiricham, Wei Hu, Joshua Marxen, and Ryan Kastner. Register transfer level information flow tracking for provably secure hardware design. In *Design, Automation & Test in Europe Conference & Exhibition (DATE), 2017*, pages 1691–1696. IEEE, 2017.
4. Avinash Ayalasomayajula, Nusrat Farzana, Debjit Pal, and Farimah Farahmandi. Prioritizing information flow violations: Generation of ranked security assertions for hardware designs. In *2024 IEEE International Symposium on Hardware Oriented Security and Trust (HOST)*, pages 128–138. IEEE, 2024.
5. Md Kawser Bepary, Arunabho Basu, Sajeed Mohammad, Rakibul Hassan, Farimah Farahmandi, and Mark Tehranipoor. SPY-PMU: Side-channel profiling of your performance monitoring unit to leak remote user activity. *Cryptology ePrint Archive*, 2025.
6. Md Kawser Bepary, Tao Zhang, Farimah Farahmandi, and Mark Tehranipoor. PreSCAN: A comprehensive review of pre-silicon physical side-channel vulnerability assessment methodologies. *Chips*, 3(4):311–333, 2024.
7. Md Kawser Bepary, Tao Zhang, Jingbo Zhou, Fahim Rahman, Farimah Farahmandi, and Mark Tehranipoor. Towards efficient gate-level electromagnetic side-channel leakage modeling and vulnerability assessment. *Journal of Hardware and Systems Security*, 9(1):24–37, 2025.
8. Swarup Bhunia, Atul Prasad Deb Nath, and Mark M. Tehranipoor. *Introduction*. Springer International Publishing, 2018.
9. Swarup Bhunia, Michael S Hsiao, Mainak Banga, and Seetharam Narasimhan. Hardware Trojan attacks: Threat analysis and countermeasures. *Proceedings of the IEEE*, 102(8):1229–1247, 2014.
10. Mohammad-Mahdi Bidmeshki and Yiorgos Makris. VeriCoq: A Verilog-to-Coq converter for proof-carrying hardware automation. In *2015 IEEE International Symposium on Circuits and Systems (ISCAS)*, pages 29–32. IEEE, 2015.
11. Eric Brier, Christophe Clavier, and Francis Olivier. Correlation power analysis with a leakage model. In *Cryptographic Hardware and Embedded Systems-CHES 2004: 6th International Workshop Cambridge, MA, USA, August 11–13, 2004. Proceedings 6*, pages 16–29. Springer, 2004.
12. Rajat Subhra Chakraborty, Seetharam Narasimhan, and Swarup Bhunia. Hardware Trojan: Threats and emerging solutions. In *2009 IEEE International high level design validation and test workshop*, pages 166–171. IEEE, 2009.

13. Rajat Subhra Chakraborty, Francis Wolff, Somnath Paul, Christos Papachristou, and Swarup Bhunia. MERO: A statistical approach for hardware Trojan detection. In *International Workshop on Cryptographic Hardware and Embedded Systems*, pages 396–410. Springer, 2009.
14. Chen Chen, Vasudev Gohil, Rahul Kande, Ahmad-Reza Sadeghi, and Jeyavijayan Rajendran. Psofuzz: Fuzzing processors with particle swarm optimization. In *2023 IEEE/ACM International Conference on Computer Aided Design (ICCAD)*, pages 1–9. IEEE, 2023.
15. https://google.github.io/clusterfuzz/. Accessed: 04/03/2024.
16. Sanjay Das, Shamik Kundu, and Kanad Basu. Bottlenecks in secure adoption of deep neural networks in safety-critical applications. In *2023 IEEE 66th International Midwest Symposium on Circuits and Systems (MWSCAS)*, pages 801–805. IEEE, 2023.
17. Sanjay Das, Shamik Kundu, and Kanad Basu. Explainability to the rescue: A pattern-based approach for detecting adversarial attacks. In *2024 IEEE International Symposium on Hardware Oriented Security and Trust (HOST)*, pages 160–170. IEEE, 2024.
18. David L Dill. The Mur ϕ verification system. In *International Conference on Computer Aided Verification*, pages 390–393. Springer, 1996.
19. Dmitry Evtyushkin, Ryan Riley, Nael CSE Abu-Ghazaleh, ECE, and Dmitry Ponomarev. Branchscope: A new side-channel attack on directional branch predictor. *ACM SIGPLAN Notices*, 53(2):693–707, 2018.
20. W Fang, M Li, M Li, Z Yan, S Liu, H Zhang, and Z Xie. AssertLLM: Generating and evaluating hardware verification assertions from design specifications via multi-LLMs. arxiv 2024. *arXiv preprint arXiv:2402.00386*.
21. Farimah Farahmandi, Yuanwen Huang, Prabhat Mishra, Farimah Farahmandi, Yuanwen Huang, and Prabhat Mishra. System-on-chip security vulnerabilities. *System-on-Chip Security: Validation and Verification*, pages 1–13, 2020.
22. Farimah Farahmandi, M Sazadur Rahman, Sree Ranjani Rajendran, and Mark Tehranipoor. *CAD for hardware security*. Springer, 2023.
23. Hasindu Gamaarachchi and Harsha Ganegoda. Power analysis based side channel attack. *arXiv preprint arXiv:1801.00932*, 2018.
24. Navyata Gattu et al. Power side channel attack analysis and detection. In *Proceedings of the 39th International Conference on Computer-Aided Design*, 2020.
25. Ronghui Gu, Zhong Shao, Hao Chen, Xiongnan Newman Wu, Jieung Kim, Vilhelm Sjöberg, and David Costanzo. {CertiKOS}: An extensible architecture for building certified concurrent {OS} kernels. In *12th USENIX Symposium on Operating Systems Design and Implementation (OSDI 16)*, pages 653–669, 2016.
26. Khan Thamid Hasan, Md Ajoad Hasan, Nashmin Alam, Md Touhidul Islam, Upoma Das, and Farimah Farahmandi. AI-assisted hardware security verification: A survey and AI accelerator case study. *arXiv preprint arXiv:2604.01572*, 2026.
27. Md Ajoad Hasan, Dipayan Saha, Khan Thamid Hasan, Nashmin Alam, Azim Uddin, Sujan Kumar Saha, Mark Tehranipoor, and Farimah Farahmandi. LAsset: An LLM-assisted security asset identification framework for system-on-chip (SoC) verification. *arXiv preprint arXiv:2601.02624*, 2026.
28. Yuchen Hu, Junhao Ye, Ke Xu, Jialin Sun, Shiyue Zhang, Xinyao Jiao, Dingrong Pan, Jie Zhou, Ning Wang, Weiwei Shan, et al. UVLLM: An automated universal RTL verification framework using LLMs. *arXiv preprint arXiv:2411.16238*, 2024.
29. Jaewon Hur, Suhwan Song, Dongup Kwon, Eunjin Baek, Jangwoo Kim, and Byoungyoung Lee. DIFUZZRTL: Differential fuzz testing to find CPU bugs. In *2021 IEEE Symposium on Security and Privacy (SP)*, pages 1286–1303. IEEE, 2021.
30. Zahin Ibnat, Paul E Calzada, Dipayan Saha, Hasan Al-Shaikh, Sujan Kumar Saha, Jingbo Zhou, Farimah Farahmandi, and Mark Tehranipoor. Trusting the machine: How secure is LLM-generated RTL code? In *2025 ACM/IEEE 7th Symposium on Machine Learning for CAD (MLCAD)*, pages 1–8. IEEE, 2025.
31. Fehmi Jaafar, Gabriela Nicolescu, and Christian Richard. A systematic approach for privilege escalation prevention. In *2016 IEEE International Conference on Software Quality, Reliability and Security Companion (QRS-C)*, pages 101–108. IEEE, 2016.

32. Yier Jin, Nathan Kupp, and Yiorgos Makris. Experiences in hardware Trojan design and implementation. In *2009 IEEE international workshop on hardware-oriented security and trust*, pages 50–57. IEEE, 2009.
33. Rahul Kande, Addison Crump, Garrett Persyn, Patrick Jauernig, Ahmad-Reza Sadeghi, Aakash Tyagi, and Jeyavijayan Rajendran. {TheHuzz}: Instruction fuzzing of processors using {Golden-Reference} models for finding {Software-Exploitable} vulnerabilities. In *31st USENIX Security Symposium (USENIX Security 22)*, pages 3219–3236, 2022.
34. Rasheed Kibria, Farimah Farahmandi, and Mark Tehranipoor. IFV: Information flow verification at the pre-silicon stage utilizing static-formal methodology. *Cryptology ePrint Archive*, 2026.
35. Paul Kocher, Jann Horn, Anders Fogh, Daniel Genkin, Daniel Gruss, Werner Haas, Mike Hamburg, Moritz Lipp, Stefan Mangard, Thomas Prescher, et al. Spectre attacks: Exploiting speculative execution. *Communications of the ACM*, 63(7):93–101, 2020.
36. Paul Kocher, Joshua Jaffe, and Benjamin Jun. Differential power analysis. In *Advances in Cryptology—CRYPTO'99: 19th Annual International Cryptology Conference Santa Barbara, California, USA, August 15–19, 1999 Proceedings 19*, pages 388–397. Springer, 1999.
37. Shamik Kundu, Sanjay Das, Sayar Karmakar, Arnab Raha, Souvik Kundu, Yiorgos Makris, and Kanad Basu. Bit-by-bit: Investigating the vulnerabilities of binary neural networks to adversarial bit flipping. *Transactions on Machine Learning Research*, 2023.
38. Pran Kurup and Taher Abbasi. *Logic synthesis using Synopsys®*. Springer Science & Business Media, 1997.
39. Kevin Laeufer, Jack Koenig, Donggyu Kim, Jonathan Bachrach, and Koushik Sen. RFUZZ: Coverage-directed fuzz testing of RTL on FPGAs. In *2018 IEEE/ACM International Conference on Computer-Aided Design (ICCAD)*, pages 1–8. IEEE, 2018.
40. Xavier Leroy, Sandrine Blazy, Daniel Kästner, Bernhard Schommer, Markus Pister, and Christian Ferdinand. CompCert-a formally verified optimizing compiler. In *ERTS 2016: Embedded Real Time Software and Systems, 8th European Congress*, 2016.
41. Congmiao Li and Jean-Luc Gaudiot. Online detection of spectre attacks using microarchitectural traces from performance counters. In *2018 30th International Symposium on Computer Architecture and High Performance Computing (SBAC-PAD)*, pages 25–28. IEEE, 2018.
42. Moritz Lipp, Michael Schwarz, Daniel Gruss, Thomas Prescher, Werner Haas, Anders Fogh, Jann Horn, Stefan Mangard, Paul Kocher, Daniel Genkin, Yuval Yarom, and Mike Hamburg. Meltdown: Reading kernel memory from user space. In *27th USENIX Security Symposium (USENIX Security 18)*, pages 973–990, Baltimore, MD, August 2018. USENIX Association.
43. Moritz Lipp, Michael Schwarz, Daniel Gruss, Thomas Prescher, Werner Haas, Stefan Mangard, Paul Kocher, Daniel Genkin, Yuval Yarom, and Mike Hamburg. Meltdown. *arXiv preprint arXiv:1801.01207*, 2018.
44. Fangfei Liu, Yuval Yarom, Qian Ge, Gernot Heiser, and Ruby B Lee. Last-level cache side-channel attacks are practical. In *2015 IEEE symposium on security and privacy*, pages 605–622. IEEE, 2015.
45. Mingjie Liu, Nathaniel Pinckney, Brucek Khailany, and Haoxing Ren. VerilogEval: Evaluating large language models for Verilog code generation. In *2023 IEEE/ACM International Conference on Computer Aided Design (ICCAD)*, pages 1–8. IEEE, 2023.
46. Stefan Mangard. A simple power-analysis (SPA) attack on implementations of the AES key expansion. In *Information –ICISC 2002: 5th International Conference Seoul, Korea, November 28–29, 2002 Revised Papers 5*, pages 343–358. Springer, 2003.
47. Ross Mcilroy, Jaroslav Sevcik, Tobias Tebbi, Ben L Titzer, and Toon Verwaest. Spectre is here to stay: An analysis of side-channels and speculative execution. *arXiv preprint arXiv:1902.05178*, 2019.
48. Xingyu Meng, Shamik Kundu, Arun K Kanuparthi, and Kanad Basu. RTL-contest: Concolic testing on RTL for detecting security vulnerabilities. *IEEE Transactions on Computer-Aided Design of Integrated Circuits and Systems*, 41(3):466–477, 2021.
49. Xingyu Meng, Kshitij Raj, Atul Prasad Deb Nath, Kanad Basu, and Sandip Ray. SoC-CAR: Detecting system-on-chip security violations under asynchronous resets. In *2021 58th ACM/IEEE Design Automation Conference (DAC)*, pages 625–630. IEEE, 2021.

50. Xingyu Meng, Amisha Srivastava, Ayush Arunachalam, Avik Ray, Pedro Henrique Silva, Rafail Psiakis, Yiorgos Makris, and Kanad Basu. Unlocking hardware security assurance: The potential of LLMs. *arXiv preprint arXiv:2308.11042*, 2023.
51. Xingyu Meng, Amisha Srivastava, Ayush Arunachalam, Avik Ray, Pedro Henrique Silva, Rafail Psiakis, Yiorgos Makris, and Kanad Basu. NSPG: Natural language processing-based security property generator for hardware security assurance. In *Proceedings of the 61st ACM/IEEE Design Automation Conference*, pages 1–6, 2024.
52. Anand Menon, Samit Miftah, Shamik Kundu, Souvik Kundu, Amisha Srivastava, Arnab Raha, Gaberiel Sonnenschien, Suvadeep Banerjee, Deepak Mathaikutty, and Kanad Basu. Enhancing large language models for hardware verification: A novel systemverilog assertion dataset. *ACM Transactions on Design Automation of Electronic Systems*, 2025.
53. Anand Menon, Samit Shahnawaz Miftah, Amisha Srivastava, Shamik Kundu, Shovik Kundu, Arnab Raha, Suvadeep Banerjee, Deepak Mathaikutty, and Kanad Basu. Openassert: Towards secure assertion generation using large language models. In *2025 IEEE 43rd VLSI Test Symposium (VTS)*, pages 1–5. IEEE, 2025.
54. Anand Menon, Amisha Srivastava, Shamik Kundu, and Kanad Basu. Application profiling using register-instruction hardware performance counters. In *2023 IEEE Computer Society Annual Symposium on VLSI (ISVLSI)*, pages 1–6. IEEE, 2023.
55. Samit S. Miftah, Hyunmin Kim, and Kanad Basu. Interconfuzz: A fuzzing-based comprehensive noc verification framework. In *2025 62nd IEEE/ACM Design Automation Conference (DAC)*. ACM, 2025.
56. Samit S Miftah, Kshitij Raj, Xingyu Meng, Sandip Ray, and Kanad Basu. System-on-chip information flow validation under asynchronous resets. *IEEE Transactions on Computer-Aided Design of Integrated Circuits and Systems*, 2024.
57. Samit S. Miftah, Amisha Srivastava, Hyunmin Kim, Shiyi Wei, and Kanad Basu. Symbfuzz: Symbolic execution guided hardware fuzzing. In *2025 58th IEEE/ACM International Symposium on Microarchitecture (MICRO)*. ACM, 2025.
58. Samit Shahnawaz Miftah, Amisha Srivastava, Hyunmin Kim, and Kanad Basu. Assert-o: Context-based assertion optimization using llms. In *Proceedings of the Great Lakes Symposium on VLSI 2024*, pages 233–239, 2024.
59. Radek Pelánek. Fighting state space explosion: Review and evaluation. In *International Workshop on Formal Methods for Industrial Critical Systems*, pages 37–52. Springer, 2008.
60. Tanvir Rahman, Shuvagata Saha, Sujan Kumar Saha, Farimah Farahmandi, and Mark Tehranipoor. EmFIA: A novel emulation-based fault injection vulnerability assessment framework at RTL level. In *2025 IFIP/IEEE 33rd International Conference on Very Large Scale Integration (VLSI-SoC)*, pages 1–5. IEEE, 2025.
61. Mohammad Rahmani Fadiheh. *Unique Program Execution Checking: A Novel Approach for Formal Security Analysis of Hardware*. PhD thesis, Technische Universität Kaiserslautern, 2022.
62. Pankaj Rohatgi. Electromagnetic attacks and countermeasures. *Cryptographic Engineering*, pages 407–430, 2009.
63. Richard L Rudell. *Logic synthesis for VLSI design*. University of California, Berkeley, 1989.
64. Dipayan Saha, Hasan Al Shaikh, Shams Tarek, and Farimah Farahmandi. Special session: ThreatLens: LLM-guided threat modeling and test plan generation for hardware security verification. In *2025 IEEE 43rd VLSI Test Symposium (VTS)*, pages 1–5. IEEE, 2025.
65. Dipayan Saha and Farimah Farahmandi. DL-SCADS: Deep learning-based post-silicon side-channel analysis using decomposed signal. In *2024 58th Asilomar Conference on Signals, Systems, and Computers*, pages 1787–1791. IEEE, 2024.
66. Dipayan Saha, Shams Tarek, Hasan Al Shaikh, Khan Thamid Hasan, Pavan Sai Nalluri, Md Ajoad Hasan, Nashmin Alam, Jingbo Zhou, Sujan Kumar Saha, Mark Tehranipoor, et al. SV-LLM: An agentic approach for SoC security verification using large language models. *arXiv preprint arXiv:2506.20415*, 2025.
67. Dipayan Saha, Shams Tarek, Katayoon Yahyaei, Sujan Kumar Saha, Jingbo Zhou, Mark Tehranipoor, and Farimah Farahmandi. LLM for SoC security: A paradigm shift. *IEEe Access*, 12:155498–155521, 2024.

68. Dipayan Saha, Katayoon Yahyaei, Sujan Kumar Saha, Mark Tehranipoor, and Farimah Farahmandi. Empowering hardware security with LLM: The development of a vulnerable hardware database. In *2024 IEEE International Symposium on Hardware Oriented Security and Trust (HOST)*, pages 233–243. IEEE, 2024.
69. Pantha Protim Sarker, Upoma Das, Nitin Varshney, Shang Shi, Akshay Kulkarni, Farimah Farahmandi, and Mark Tehranipoor. When everyday devices become weapons: A closer look at the pager and walkie-talkie attacks. *arXiv preprint arXiv:2501.17405*, 2025.
70. Pantha Protim Sarker, Tianze Kan, Jingchen Liang, Ozgur Tuncer, Bo He, Zelin Lu, Sudarshan Mallu, Lang Lin, Norman Chang, Riku Hasegawa, et al. ML-EMFI: A machine learning-driven pre-silicon electromagnetic fault injection security evaluation for robust IC design. In *2025 IEEE International Symposium on Hardware Oriented Security and Trust (HOST)*, pages 57–66. IEEE, 2025.
71. Mark Seaborn and Thomas Dullien. Exploiting the DRAM rowhammer bug to gain kernel privileges. *Black Hat*, 15(71):2, 2015.
72. Dong-Hyun Seo, Archisman Ghosh, Debayan Das, Mayukh Nath, Santosh Ghosh, and Shreyas Sen. PG-CAS: Pro-active EM-SCA probe detection using switched-capacitor-based patterned-ground co-planar capacitive asymmetry sensing. *IEEE Open Journal of Circuits and Systems*, 4:271–282, 2023.
73. Sugandh Shah and Babu M Mehtre. An overview of vulnerability assessment and penetration testing techniques. *Journal of Computer Virology and Hacking Techniques*, 11:27–49, 2015.
74. Amit Mazumder Shuvo, Md Latifur Rahman, Jingbo Zhou, Farimah Farahmandi, and Mark Tehranipoor. ReFID: A system-aware remote fault-injection attack detection & mitigation for secure heterogeneous system. In *2025 IEEE International Symposium on Hardware Oriented Security and Trust (HOST)*, pages 46–56. IEEE, 2025.
75. Amit Mazumder Shuvo, Tao Zhang, Farimah Farahmandi, and Mark Tehranipoor. Flat: Layout-aware and security property-assisted timing fault-injection attack assessment. *IEEE Transactions on Very Large Scale Integration (VLSI) Systems*, 32(6):1150–1163, 2024.
76. Amisha Srivastava, Sanjay Das, Navnil Choudhury, Rafail Psiakis, Pedro Henrique Silva, Debjit Pal, and Kanad Basu. Scar: Power side-channel analysis at RTL level. *IEEE Transactions on Very Large Scale Integration (VLSI) Systems*, 32(6):1110–1123, 2024.
77. Amisha Srivastava, Sneha Thakur, Abraham Peedikayil Kuruvila, Poras T Balsara, and Kanad Basu. Hardware-based detection of malicious firmware modification in microgrids. In *2024 37th International Conference on VLSI Design and 2024 23rd International Conference on Embedded Systems (VLSID)*, pages 186–191. IEEE, 2024.
78. Shams Tarek, Dipayan Saha, Khan Thamid Hasan, Sujan Kumar Saha, Mark Tehranipoor, and Farimah Farahmandi. Assertain: Automated security assertion generation using large language models. *arXiv preprint arXiv:2604.01583*, 2026.
79. Shams Tarek, Dipayan Saha, Sujan Kumar Saha, and Farimah Farahmandi. Bugwhisperer: Fine-tuning LLMs for SoC hardware vulnerability detection. In *2025 IEEE 43rd VLSI Test Symposium (VTS)*, pages 1–5. IEEE, 2025.
80. Mark Tehranipoor, N Nalla Anandakumar, and Farimah Farahmandi. Hardware Trojan insertion. In *Hardware Security Training, Hands-on!*, pages 73–92. Springer, 2023.
81. Mark M Tehranipoor, Farimah Farahmandi, and Huanyu Wang. Security property-driven vulnerability assessments of ICS against fault-injection attacks, January 21 2025. US Patent 12,204,685.
82. Mohammad Tehranipoor and Farinaz Koushanfar. A survey of hardware Trojan taxonomy and detection. *IEEE design & test of computers*, 27(1):10–25, 2010.
83. Eleonora Testa, Mathias Soeken, Heinz Riener, Luca Amaru, and Giovanni De Micheli. A logic synthesis toolbox for reducing the multiplicative complexity in logic networks. In *2020 Design, Automation & Test in Europe Conference & Exhibition (DATE)*, pages 568–573. IEEE, 2020.
84. Niek Timmers and Cristofaro Mune. Escalating privileges in Linux using voltage fault injection. In *2017 Workshop on Fault Diagnosis and Tolerance in Cryptography (FDTC)*, pages 1–8. IEEE, 2017.

85. Kris Tiri and Ingrid Verbauwhede. Secure logic synthesis. In *International Conference on Field Programmable Logic and Applications*, pages 1052–1056. Springer, 2004.
86. Timothy Trippel, Kang G. Shin, Alex Chernyakhovsky, Garret Kelly, Dominic Rizzo, and Matthew Hicks. Fuzzing hardware like software. In *31st USENIX Security Symposium (USENIX Security 22)*, pages 3237–3254, Boston, MA, August 2022. USENIX Association.
87. Azim Uddin, Sujan Kumar Saha, Farimah Farahmandi, and Mark Tehranipoor. Case study: Fault-injection vulnerability assessment at RTL level. In *2024 IEEE Physical Assurance and Inspection of Electronics (PAINE)*, pages 1–7. IEEE, 2024.
88. Michael Vai, Ben Nahill, Josh Kramer, Michael Geis, Dan Utin, David Whelihan, and Roger Khazan. Secure architecture for embedded systems. In *2015 IEEE High Performance Extreme Computing Conference (HPEC)*, pages 1–5. IEEE, 2015.
89. Antti Valmari. The state explosion problem. In *Advanced Course on Petri Nets*, pages 429–528. Springer, 1996.
90. Nan Wu, Yingjie Li, Hang Yang, Hanqiu Chen, Steve Dai, Cong Hao, Cunxi Yu, and Yuan Xie. Survey of machine learning for software-assisted hardware design verification: Past, present, and prospect. *ACM Transactions on Design Automation of Electronic Systems*, 29(4):1–42, 2024.
91. Yuval Yarom and Katrina Falkner. {FLUSH+ RELOAD}: A high resolution, low noise, l3 cache {Side-Channel} attack. In *23rd USENIX security symposium (USENIX security 14)*, pages 719–732, 2014.
92. Drew Zagieboylo, Charles Sherk, Andrew C Myers, and G Edward Suh. Specverilog: Adapting information flow control for secure speculation. In *Proceedings of the 2023 ACM SIGSAC Conference on Computer and Communications Security*, pages 2068–2082, 2023.
93. An Qi Zhang, Andrés Goens, Nicolai Oswald, Tobias Grosser, Daniel Sorin, and Vijay Nagarajan. PipeGen: Automated transformation of a single-core pipeline into a multicore pipeline for a given memory consistency model. In *Proceedings of the 2024 International Conference on Parallel Architectures and Compilation Techniques*, pages 1–13, 2024.
94. Jiliang Zhang, Congcong Chen, Jinhua Cui, and Keqin Li. Timing side-channel attacks and countermeasures in CPU microarchitectures. *ACM Computing Surveys*, 56(7):1–40, 2024.
95. Rui Zhang, Calvin Deutschbein, Peng Huang, and Cynthia Sturton. End-to-end automated exploit generation for validating the security of processor designs. In *2018 51st Annual IEEE/ACM International Symposium on Microarchitecture (MICRO)*, pages 815–827. IEEE, 2018.
96. Tao Zhang, Jungmin Park, Mark Tehranipoor, and Farimah Farahmandi. PSC-TG: RTL power side-channel leakage assessment with test pattern generation. In *2021 58th ACM/IEEE Design Automation Conference (DAC)*, pages 709–714. IEEE, 2021.
97. Yongan Zhang, Zhongzhi Yu, Yonggan Fu, Cheng Wan, and Yingyan Celine Lin. MG-Verilog: Multi-grained dataset towards enhanced LLM-assisted Verilog generation. In *2024 IEEE LLM Aided Design Workshop (LAD)*, pages 1–5. IEEE, 2024.
98. Zixi Zhang, Balint Szekely, Pedro Gimenes, Greg Chadwick, Hugo McNally, Jianyi Cheng, Robert Mullins, and Yiren Zhao. Llm4dv: Using large language models for hardware test stimuli generation. In *2025 IEEE 33rd Annual International Symposium on Field-Programmable Custom Computing Machines (FCCM)*, pages 133–137. IEEE, 2025.
99. YongBin Zhou and DengGuo Feng. Side-channel attacks: Ten years after its publication and the impacts on cryptographic module security testing. *Cryptology ePrint Archive*, 2005.

Chapter 2
Introduction to Large Language Models (LLMs)

Ishraq Tashdid, Kimia Tasnia, and Sazadur Rahman

Building on the motivation from the previous chapter, this chapter moves from first principles to practical use for system-on-chip (SoC) teams. We begin by explaining what large language models (LLMs) are and why next-token prediction at scale can capture syntax, semantics, and engineering patterns (Sect. 2.1). Next, we describe the *transformer* using a dataflow perspective (Sect. 2.2), where tokenization and embeddings define the input interface, multihead self-attention mixes information under a causal mask, and feed-forward blocks with residual connections and normalization refine and stabilize each token. We then survey different model families and deployment options (Sect. 2.3), including general-purpose, code-centric, and multimodal models, along with scaling and context-window trade-offs, and we compare open versus hosted models in terms of capability, control, and compliance. After that, we turn to data and training (Sect. 2.4), introducing domain-adaptive pretraining (DAPT) and task-adaptive pretraining (TAPT) for continued pretraining, emphasizing a retrieval-first approach before fine-tuning, and defining a clean hardware corpus consisting of hardware description languages (HDLs), assertions, testbenches, logs, and specifications, all validated through linting, compilation, and simulation. Next, we describe the workflow (Sect. 2.5), using prompts that resemble interface contracts, a header-then-body generation style, small smoke tests, and retrieval-augmented generation (RAG) with proper citations, supported by a lightweight loop involving a designer, planner, programmer, reviewer, and evaluator with a human in the loop. We then define evaluation criteria (Sect. 2.6), including compile and simulation pass rates, meaningful coverage, and use of formal methods, combined into a stable score with full traceability. Finally, we examine risks (Sect. 2.8), considering both functional and security aspects, and

I. Tashdid · K. Tasnia · S. Rahman (✉)
University of Central Florida, Orlando, FL, USA
e-mail: mohammad.rahman@ucf.edu

F. Farahmandi et al. (eds.), *LLM for SoC Design and Security*,
https://doi.org/10.1007/978-3-032-34116-7_2

discuss governance practices that ensure an LLM-assisted workflow remains fast, accountable, and suitable for silicon security.

2.1 What Are Large Language Models?

Large language models, or LLMs, are a class of artificial intelligence systems that have captured the spotlight in recent years [2, 42]. At their core, they are very large neural networks trained on enormous collections of text and code [20, 21]. Their training objective is deceptively simple: Predict the next token in a sequence, given all the tokens that came before [2, 20]. By repeating this prediction game billions of times over billions of examples, LLMs gradually internalize the structure of language, programming syntax, and even patterns of reasoning [12, 39]. Imagine reading a technical document and trying to guess the next word in every sentence. At first, you might get only obvious cases right. But after reading thousands of documents, your guesses become sharper; you not only know grammar, but also the typical flow of ideas in engineering texts. LLMs do something similar at scale, except that instead of you, it is a Transformer model running across thousands of graphics processing units (GPUs) and training on a mix of books, web pages, and code repositories [2, 12, 17]. LLMs do not actually read text or code the way humans do. Instead of taking in full words or entire lines, they break everything down into smaller building blocks called tokens [13, 26]. In English, a token might be a complete word, such as chip, or even just a fragment like trans-. In hardware description languages, tokens look a little different; a keyword such as `always`, a `4'b1010`, or even part of a long identifier. Over time, the model learns which tokens are likely to appear together, much like how hardware engineers quickly recognize that `posedge clk` usually follows a sensitivity list [4, 22]. Representative examples of widely used commercial LLMs are shown in Fig. 2.1.

Moreover, Transformers introduced the key idea of self-attention, a mechanism that allows the model to scan across an entire sequence and highlight which tokens matter most for predicting the next one [37]. In practice, this means that when reading a line of SystemVerilog code, the model can keep track of the module declaration at the beginning and still use that information to close a block many tokens later [6, 21]. Unlike older architectures that struggled with long

Fig. 2.1 Commercially available large language models

dependencies [10], self-attention gives Transformers the ability to connect distant parts of text or code with surprising accuracy [37]. To make training stable at large scale, each attention block is wrapped with residual connections and normalization layers [1, 9], ensuring that signals do not vanish or explode as the network deepens. The end result is an architecture that can be stacked hundreds of layers high, trained across thousands of GPUs [12, 35], and still generalize well across very different tasks [39]. This combination of long-range context, stability, and scalability is what makes the Transformer such a natural fit for LLMs [37, 42]. We will return to the details of tokens, embeddings, attention heads, and residual pathways in later sections, but for now it is enough to recognize that the Transformer is what makes modern LLMs tick.

However, unlike traditional electronic design automation (EDA) tools that are designed for a narrow task (say, synthesis or static timing analysis), an LLM is a general-purpose generator [42]. It can answer natural language questions, summarize specifications, generate snippets of register-transfer level (RTL) codes [4, 22], or adapt to new domains through fine-tuning or retrieval [6, 18, 21]. Think of it as a versatile copilot. Not perfect, but often useful when guided carefully [11]. So, at a high level, an LLM is nothing more than a next-token predictor scaled to extraordinary size [12, 17]. However, when combined with the right prompts, retrieval, and evaluation, it can behave like a knowledgeable assistant [18, 39]. This blend of simplicity at the core and versatility in use is why LLMs deserve a place in SoC design discussions.

2.2 Transformer Architecture

With the high-level idea of LLMs in place, we can now look under the hood at the Transformer itself. Rather than diving into dense math, we will use concepts familiar to anyone who has designed or debugged a datapath.

A Transformer can be imagined as a deep pipeline: Tokens come in as signals, flow through specialized stages, and emerge as probabilities for the next token, as illustrated in Fig. 2.2. Each stage refines the representation, adds context, and stabilizes the flow, similar to how registers, combinational blocks, and bypass paths keep information moving smoothly in hardware. This hardware analogy also shows why Transformers succeeded where older networks struggled. A recurrent neural network is like a fragile single-cycle datapath [25]: fine for short paths but easily broken by long dependencies. Transformers, on the contrary, add the equivalent of registers, feedback, and shortcuts so they can scale to hundreds of layers without collapsing. In practice, this means they can read an entire SystemVerilog file, remember the declaration at the top, and still use that knowledge much later when closing a block. The pipeline itself has several key stages. It begins with tokenization and embeddings, which turn language or code into machine-readable units and map them into dense vectors. Next comes multihead self-attention, the mechanism that lets tokens "communicate" across long sequences and capture

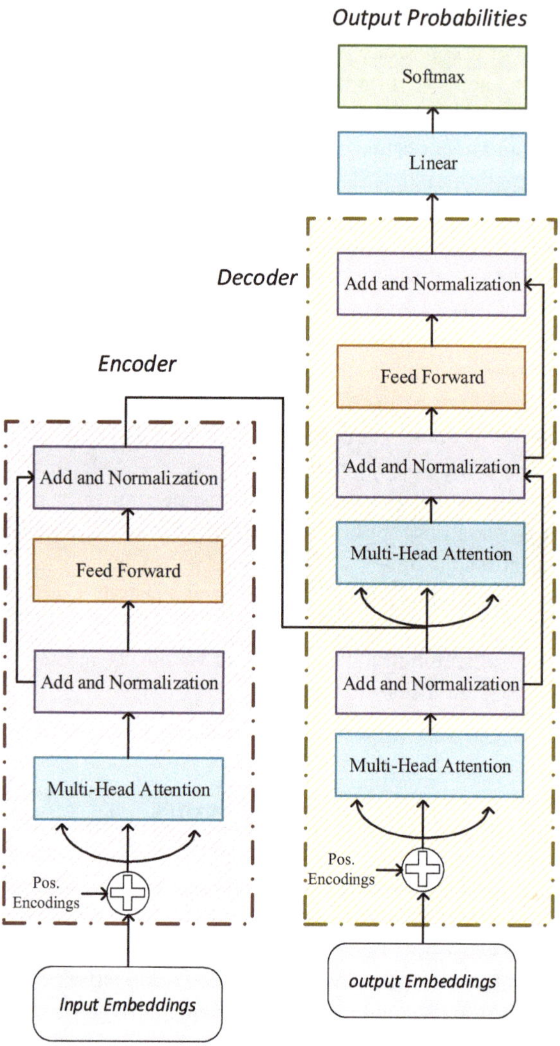

Fig. 2.2 Encoder-decoder overview. The encoder reads input tokens and builds a context-aware representation that captures meaning and long-range relationships. The decoder generates output one token at a time, consulting its own past outputs (left to right with masking). Many modern LLMs use only the decoder stack; encoder-decoder models are common for translation and other seq2seq tasks

relationships, whether between words in a sentence or signals in a module. Then comes feed-forward layers, residual connections, and normalization steps that refine and stabilize the representations, the architectural tricks that let Transformers go deep without destabilizing. Finally, the entire system is trained by playing the next-token prediction game, with fine-tuning and retrieval shaping it into a versatile assistant.

In the following subsections, we unpack these components one by one, showing how they operate and why they matter for LLMs in practice.

Listing 2.1 Tokenization for sample RTL

```
always @(posedge clk) begin
  sum <= a + b;
end
```

Listing 2.2 Tokens and their integer IDs (repeated tokens share the same ID)

```
['always','@','(','posedge','clk',')','begin','sum','<=','a','+','b',';',
 'if','(','clk',')','sum','<=','a','+','b',';','end']
[101, 57, 12, 998, 451, 13, 300, 712, 90, 22, 44, 23, 8,
 520, 12, 451, 13, 712, 90, 22, 44, 23, 8, 301]
```

2.2.1 *Tokenization and Embeddings*

Before any Transformer can process text or HDL, the raw input has to be broken down into manageable pieces. This step is called tokenization. Listings 2.1 and 2.2 show a concrete example: A short RTL snippet is split into tokens, and each token is assigned an integer identifier drawn from the model's vocabulary. Note that repeated tokens such as `clk` and `sum` receive the same ID regardless of position. Humans read in words or full lines of code; a model sees only these integer-indexed tokens. Once split, the model still needs a way to represent them numerically. This is the role of the *embedding layer*. Think of it as a dictionary that maps each token to a dense vector in a high-dimensional space. Tokens with similar usage patterns end up close together, giving the model a geometric sense of meaning. To capture order, these vectors are combined with positional encodings. Without them, a sequence like `posedge clk` could look identical to `clk posedge`. Absolute encodings use fixed sinusoidal patterns, while rotary encodings add phase shifts, but the goal is the same: to provide the model with a memory of sequence and direction. Together, tokenization and embeddings form the input interface of the Transformer pipeline. They translate human-readable text or HDL into a numerical signal on which the later stages can operate effectively.

2.2.2 *Multihead Self-Attention*

With tokenization and embeddings complete, the model now holds a sequence of vectors $X \in \mathbb{R}^{T \times d}$ one vector per token. This sequence, not the raw text, is fed in next to self-attention layer. The multiple attention heads run in parallel *inside* the attention block over the same embedded sequence, as illustrated in Fig. 2.3.

This self-attention is the core mechanism that gives Transformers their power. Instead of treating tokens one by one, the model lets every token look at the others in

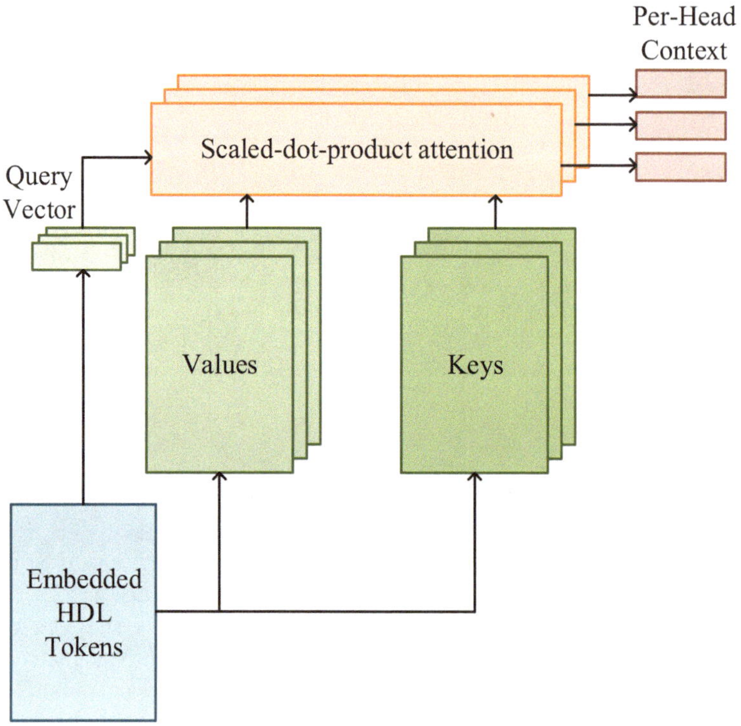

Fig. 2.3 Multihead self-attention for HDL tokens: Embeddings are projected to a query, keys, and values; parallel heads perform scaled-dot-product attention (with causal mask) to produce per-head context

the sequence and decide which ones are most relevant. You can imagine each token asking a question: "Who in this sentence or line of code should I pay attention to right now?" This process does not happen once, but in parallel across many attention heads. Each head specializes in capturing different relationships. One might learn how port declarations connect to their later use, another might track reset sequences, and another might separate control signals from data paths. For hardware description languages, that means one attention head could ensure that a signal defined early in a module is remembered when it is used much later, while another could make sure that an `if` lines up properly with an `else`.

Moreover, to keep this process realistic, the Transformer is trained with a rule called causal masking. It forces the model to look only at the tokens that have already appeared, never at the ones that come next. In other words, when the model is predicting the next symbol, it is limited to the past context, just like a digital circuit that computes the next state from the current state and inputs, but cannot peek into the future clock cycle. This constraint is what makes Transformers suitable for generation tasks while still respecting the natural flow of time and logic.

2.2.3 Feed-Forward Blocks, Residuals, and Normalization

Once attention has done its job of mixing information across tokens, the model still needs to refine each token's representation locally. This is where feed-forward blocks come in, as illustrated in Fig. 2.4. You can think of them as little arithmetic logic units (ALUs) wired into every stage of the pipeline. Each block is a small neural network, two linear layers with a nonlinearity in between, that takes the context-enriched vector for a token and sharpens it, emphasizing useful features and suppressing noise. Even though the same block is applied independently to every token, it is crucial for giving the model nonlinear capacity, much like how combinational logic extends what a datapath can do beyond just moving signals around.

Of course, when you stack hundreds of these layers, stability becomes a problem. This is why residual connections are included: They act like bypass wires in a datapath, carrying the original signal forward even if the transformation in the main path distorts it. With residuals, the model never "forgets" where it came from. Next, normalization layers step in as the balancing circuits, ensuring values do not grow out of range or collapse toward zero. In hardware analogy, they are the bias and scale correction that keep analog circuits stable, but are applied here in a digital neural context.

The trio of feedforward blocks, residuals, and normalization form the scaffolding that lets Transformers scale. Together, they allow a model to be hundreds of layers

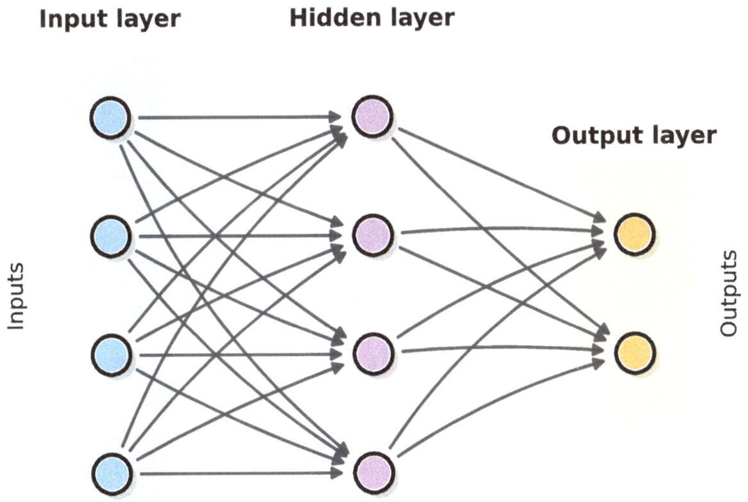

Fig. 2.4 A per-token feed-forward block refines the representation, a residual path restores the original signal, and normalization stabilizes the scale, so deep Transformer stacks remain trainable

deep without diverging during training. For hardware engineers, you can imagine a long HDL module being passed through these stages again and again; without stability mechanisms, the signal would degrade. With them, line-by-line coherence is preserved, so that the behavior of the code remains interpretable even after many transformations. This combination of local refinement, bypass stability, and global balance is what makes Transformers not just powerful, but trainable at scale.

2.3 Model Families and Design Trade-Offs

Not all large language models are created equal. Once you understand how Transformers work under the hood, the next question becomes practical: *Which model should you actually use*? The answer depends on three levers you can control: The job you want the model to do, the data you can provide, and the governance constraints you must respect. In hardware terms, it is like choosing between a soft intellectual property (IP) core, a hardened macro, or a custom RTL block; each offers different trade-offs in flexibility, performance, and trust. We will, in the following subsection, dive into the three most popular model families and talk about their strengths and trade-offs.

2.3.1 General-Purpose Models

General-purpose models are the "Swiss Army knives" of the LLM world. They are trained broadly, often on massive corpora that mix natural language, programming languages, and a wide variety of other data sources. Their strength lies in versatility. They can follow instructions, connect dots across different contexts, and handle an impressive range of tasks. For SoC engineers, this might mean quickly rephrasing a dense paragraph of a specification into plain English, generating scaffolds for a SystemVerilog module, or helping diagnose why a synthesis run produced a puzzling error message. The trade-off, however, is that these models are jacks of all trades but masters of none. They may stumble on HDL-specific quirks or produce more syntax errors compared to specialized tools. However, when speed and breadth matter, say in the early phases of design exploration or brainstorming, general-purpose models can be a reliable companion. Their outputs should always be reviewed carefully before integration into a production flow.

2.3.2 Code-Centric Models

Code-centric models take the general-purpose foundation and sharpen it for programming. They are pretrained on large volumes of source code, and in some cases further tuned on HDL and EDA logs. This specialization makes them considerably better at generating compilable outputs, respecting the finer points of syntax, and producing structured code that more closely resembles what a human engineer would write. For SoC workflows, they shine in tasks such as generating clean RTL modules, refactoring legacy Verilog, or writing assertions with fewer errors. The trade-off is that these models are narrower in scope: They may not summarize design reviews or produce free-form prose as fluidly as general-purpose models, but when precise, executable HDL is required, they often reduce cleanup effort significantly. Over time, we may see code-centric models evolve into EDA copilots that understand HDL conventions as well as seasoned verification engineers.

2.3.3 Multimodal Models

Multimodal models extend beyond text and code to handle diagrams, tables, and even waveforms, as illustrated in Fig. 2.5. This capability opens new possibilities: A block diagram of a memory subsystem could be used to request a matching RTL skeleton, or a waveform trace could be submitted to obtain a hypothesis for why reset logic misfired. For engineers, this is equivalent to having an assistant that can reason over the same visual artifacts they already work with, rather than relying solely on textual descriptions. The trade-off is complexity. These models require carefully formatted inputs, substantially larger compute resources, and are harder to debug when something goes wrong. They also raise workflow questions: How should diagram inputs be standardized, and how can a model's waveform interpretation be validated against ground truth? Nevertheless, they point toward a future where chip teams work with AI systems that can reason across multiple representations of the same design, including source code, block diagrams, logs, and waveforms, without treating them as separate silos.

2.3.4 Open vs. Hosted Models

Model type is only half the story. Deployment choices, whether to use open models that you host yourself or rely on hosted services from third parties, can matter just as much. Open models give you full control. You can run them on-premises, retrain them on proprietary corpora, and guarantee reproducibility. For IP-sensitive industries like semiconductors, this often aligns better with governance requirements. Hosted models, in contrast, offer raw capability and polished tooling. You get access to cutting-edge models with minimal setup, but often at the expense of transparency and control. The trade-off is familiar to hardware engineers: It is like deciding whether to rely on vendor-provided scripts or build your own in-house automation. Vendors can provide powerful tools quickly, but doing so requires trusting their black boxes. In practice, most teams end up blending the

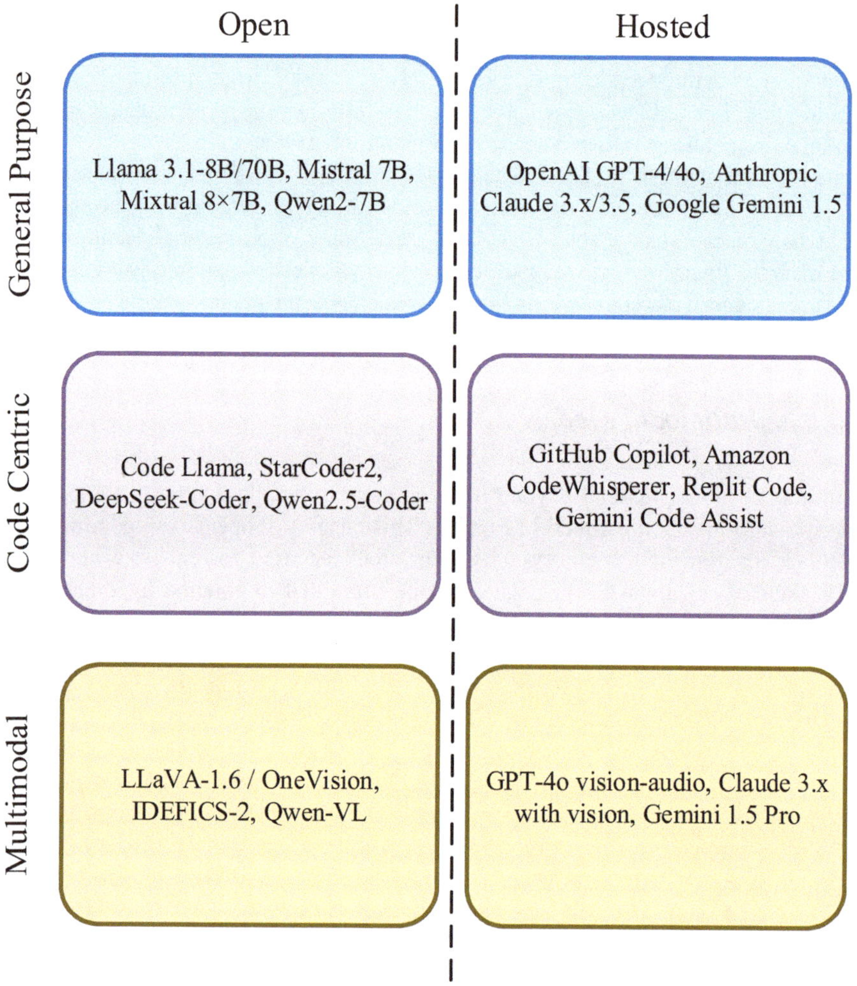

Fig. 2.5 Open vs. hosted models across general purpose, code centric, and multimodal

two approaches. A hosted model might power early exploration, such as drafting design notes, generating quick test scaffolds, or running what-if experiments, while a smaller open model, fine-tuned on internal HDL and specifications, takes over for production use where correctness and auditability are critical. Ultimately, the right choice is not about model size or benchmark scores, but about fit. Does the model reduce engineering effort while respecting your security, licensing, and governance constraints? If the answer is yes, then it has earned its place in the flow.

2.3.5 Scaling and Context Windows

Larger models are not automatically more capable, but scaling changes what they can do. One way to scale is size: More layers and parameters generally improve output quality, at the cost of greater training and serving infrastructure. Another way is context length—how many tokens the model can attend to at once. A short context covers a single `always` block; a long context spans an entire SoC specification. Longer contexts are valuable for design reviews or diagnosing lengthy logs, but memory and latency grow rapidly as the window extends. To keep large contexts practical, serving systems rely on several techniques:

- During generation, the model attends to everything it has already seen. A key-value (KV) cache stores the intermediate representations of past tokens so they do not need to be recomputed at every step, reducing the cost of next-token prediction significantly.
- Batching groups multiple requests together to maximize hardware utilization. Overall throughput increases, though individual request latency may rise slightly.
- Parallelism strategies such as tensor, pipeline, and sequence parallelism distribute computation across multiple devices, analogous to fitting a large design across multiple field-programmable gate arrays (FPGAs) to meet timing constraints.

In EDA contexts, performance ultimately comes from balancing resources effectively, not simply increasing the number of parameters.

2.4 Data and Training for Hardware Contexts

With the knowledge of different LLM families, we now shift focus to another important aspect. If there is one lever that dominates outcomes, it is data quality. Most of this work is not glamorous, curating clean corpora, building solid retrieval pipelines, and setting up ruthless evaluation loops, but it determines whether a model is a toy demo or a reliable assistant.

2.4.1 Pretraining and Continued Pretraining Strategies (DAPT/TAPT)

When you want an LLM to "think like a hardware engineer," you do not start from scratch. You bend a strong base model toward your domain. The two primary strategies for this are domain-adaptive pretraining (DAPT) and task-adaptive pretraining (TAPT). DAPT teaches the model the language of your world (HDL, specs, timing reports, tool logs). TAPT then tunes that knowledge toward the shape of the jobs you expect (explain a waveform, refactor a module, triage a lint failure). Together, they turn a general assistant into an RTL-savvy copilot.

At its core, pretraining is a large-scale next-token prediction task: Given a string of tokens, predict the one that follows. Scale makes the difference. A base model that has seen trillions of tokens develops a broad sense of syntax, semantics, and reasoning patterns. Hardware teams rarely build these large models from scratch, as the computational cost is prohibitive. Instead, existing models are adapted through continued pretraining, nudging them with domain-specific data until they shift from being generic assistants to tools that are well-suited to chip design workflows. The question then becomes: What kind of additional data should be used to guide this adaptation?

2.4.1.1 Domain-Adaptive Pretraining (DAPT)

DAPT is about teaching the model a new dialect, in this case, the language of hardware. The model continues training on a carefully curated blend of HDL, EDA logs, tool errors, and related engineering artifacts. The goal is to instill hardware-specific priors so the model no longer conflates constructs such as `posedge clk` with ordinary programming syntax. Quality checks are essential: Files should be deduplicated, licensing must be respected, and light sanity filters such as lint, compile, and simulation passes should be applied before any data reaches a GPU. This is analogous to qualifying IP blocks before integrating them into a larger SoC. Done correctly, DAPT builds a foundation where the model understands not just code structure, but the broader quirks of hardware development.

2.4.1.2 Task-Adaptive Pretraining (TAPT)

Where DAPT broadens knowledge, TAPT sharpens it. If the intended use involves specification review, errata analysis, or requirement parsing, then that is precisely the data the model should be exposed to. TAPT is typically brief, providing just enough additional training to align the model's behavior with the target task. It is less about introducing new capabilities and more about tuning context. Much like preloading a cache with the exact workload expected, TAPT primes the model so that instruction tuning and downstream fine-tuning proceed more naturally and effectively.

2.4.1.3 Curriculum and Mixing

The order in which data is presented during training matters. A staged approach often works best: Begin with general programming corpora, move into HDL, then focus on assertions, testbenches, and finally specifications. Each layer builds on the last, mirroring how engineers progress from writing small RTL blocks to handling complex SoCs. Mixing can also help; combining prose with code shows the model how specifications and HDL interact in real-world workflows. Even something as simple as starting with short modules before moving to long hierarchical ones can improve training stability. The point is not to micromanage every sample, but to guide the model's training trajectory with the same care one would apply to floorplanning macros [31].

2.4.2 *Adaptation: Retrieval First, Then Fine-Tune When It Pays*

When bringing LLMs into hardware workflows, it is tempting to jump straight to fine-tuning. In practice, however, the fastest and most reliable path starts with retrieval. Retrieval-augmented generation (RAG) lets the model pull in relevant specifications, HDL snippets, or design notes at query time, instead of relying on details that may or may not have been memorized during training. The advantages are immediate: Iteration is quick, governance is easier, and knowledge stays fresh without retraining the model weights. For many tasks, such as answering synthesis error messages or retrieving the correct SystemVerilog macro from a library, RAG alone may be sufficient.

Fine-tuning becomes worth the cost only when a stable, high-value use case emerges. Imagine your team consistently drafting assertions from specs written in a particular house style. At that point, it pays to teach the model that style directly. The good news is you do not need to retrain the entire model, which would be expensive and memory-hungry. Instead, lightweight methods such as low-rank adapters (often called LoRA) or their more memory-efficient cousin QLoRA allow you to add small "side modules" to the model while keeping the original weights frozen. In hardware terms, it is like adding a tiny coprocessor or instruction extension on top of an existing processor: The base chip does not change, but you gain faster performance for the workloads you care about. This makes fine-tuning feasible on modest compute, keeps sensitive weights local, and lets teams customize models without breaking the bank.

If the bottleneck is not raw accuracy but alignment, say, you want documentation written in a consistent tone or code suggestions that follow stricter safety rules, then preference optimization becomes important. Two common approaches are direct preference optimization (DPO) and reinforcement learning with human feedback (RLHF). Instead of just predicting the next token, the model is shown examples of "better" versus "worse" outputs and learns to prefer the better ones. You can think of this like design sign-off: The synthesis tool may produce a circuit that works, but engineers still flag whether it meets timing, follows power constraints, or violates security policies. In the same way, DPO and RLHF let humans inject style and safety constraints into the model's behavior. The catch is that this requires careful review, much like how no amount of simulation replaces the final human check before a chip goes to tape-out.

2.4.3 Building a Clean Hardware Corpus

If there is one thing that can quietly sink an otherwise good training run, it is messy data. A clean corpus is not optional. It is the foundation. Start by enforcing consistent formatting (for example, with `clang-format` for C/C++ and `verible` for SystemVerilog). Strip anything sensitive: no PII (personally identifiable information), no unreleased IP (intellectual property), no confidential logs. Keep your evaluation set walled off from training data to avoid contamination; otherwise, the numbers look great but mean little, like slipping test vectors into synthesis. Finally, version your datasets and record hashes so you can reproduce results later; if it cannot be reproduced, it cannot be trusted.

What actually belongs in a hardware-flavored dataset? HDL sources such as Verilog, VHDL, and SystemVerilog give you structure and syntax. SVA captures design intent. Testbenches show verification practice. EDA logs reveal tool behavior and common error patterns. Specifications and design reviews encode how engineers argue about correctness. Each source adds a different slice of context that the model can learn from.

But cleaning is the real work. Deduplicate aggressively to remove near-identical files. Filter out unsafe or incompatible licenses. Split data into chunks that preserve meaning, keep modules, assertion blocks, or review sections intact, so retrieval and training windows stay coherent. Over-splitting loses context; under-splitting turns every query into a haystack search. Normalize the basics (indentation, line length, naming conventions) so the model focuses on substance instead of style quirks. A well-prepared corpus will not make headlines, but it changes outcomes. It is how a general model becomes fluent in HDL, recognizes EDA messages without confusion, and follows along in design reviews.

2.4.4 HDL-Specific Augmentation and Quality Gates

Once a model has seen enough generic HDL, the real gains come from targeted augmentation. Instead of plain modules, the training data should include richer pairs that mirror real engineering practice. Bug-fix histories teach the model what incorrect versus corrected code looks like. Assertions paired with their corresponding logic blocks help it internalize design intent, not just syntax. Modular scaffolds, common templates for FIFOs, ALUs, or state machines, act like design macros that the model can later adapt. Even paraphrased specifications (e.g., turning "reset shall be synchronous" into multiple equivalent phrasings) prepare it to handle the variety of documentation styles that engineers encounter in practice.

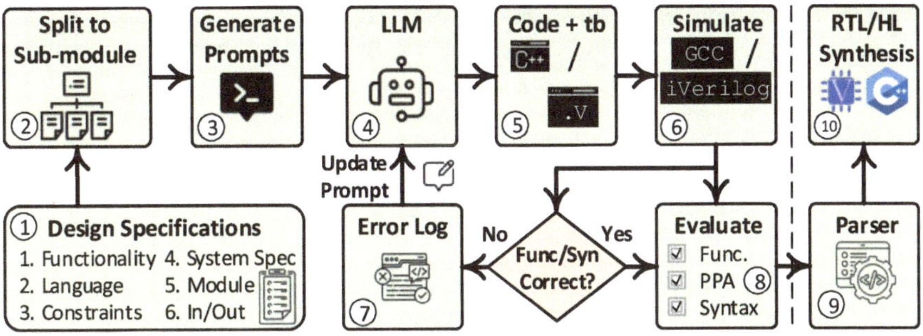

Fig. 2.6 LLM-assisted hardware design loop [33]. From design specs (1), the flow splits into submodules (2), generates prompts (3), queries the LLM (4) to produce code and testbenches (5), simulates (6) and logs errors (7). Failing checks feed back to update the prompt; passing results move to evaluation (8), parsing (9), and RTL/HL synthesis (10)

Augmentation without discipline can, however, backfire. That is where quality gates come in. Every training example should clear a basic lint, compile, or simulation pass before it reaches the GPU. These inexpensive checks catch a significant amount of problematic data and prevent the model from learning bad habits, much like design rule checking (DRC), and layout versus schematic (LVS) filters catch errors before tape-out. Lightweight formal sanity properties can add an extra layer, ensuring, for instance, that resets are reachable or that counters never underflow. These safeguards do not guarantee perfection, but they raise the baseline reliability of the data pipeline. This is also where leakage control begins. Evaluation tasks must never appear in training data. Including them would be analogous to seeding regression tests with expected answers: Pass rates appear inflated but collapse when the model encounters genuinely new inputs. The model should earn its accuracy through generalization, not by memorizing evaluation examples. Taken together, augmentation and quality gates transform the dataset into more than a collection of HDL files. The model learns not only how correct code is structured, but also how engineers diagnose, fix, and assert designs in practice. The result, illustrated in Fig. 2.6, is an assistant that reflects real engineering workflows rather than acting as a generic code generator.

2.5 Prompting, Orchestration, and Multiagent Patterns

The earlier sections established the foundational components; this section describes how they are combined into a coherent workflow. One-off prompts may suffice for demonstrations, but production SoC work requires runs that are repeatable, reviewable, and easy to debug. The guiding principle is the same one applied to build systems or regression farms: Keep inputs explicit, document every step, and ensure that each stage leaves an auditable trail. To keep the discussion concrete, we use a single running example, a small `uart_autobaud_rx` block, and show how different prompt styles and orchestration patterns affect the outcome without changing the underlying goal.

2.5.1 Prompt Patterns That Age Well

Good prompts behave like clean interface specifications: short, unambiguous, and easy to validate. Long role-play scripts become brittle as models evolve; tight contracts paired with simple validations remain stable across model versions and across team members.

2.5.1.1 State the Task Precisely

Rather than issuing a vague request such as "write a UART," the model should be given a concise specification. Listing 2.3 keeps the task well defined and testable for both engineers and continuous integration (CI) systems:

Listing 2.3 Mini-spec keeps the task crisp and testable

```
Goal: Implement a synthesizable UART receiver with autobaud detect.

Constraints:
- Module name: uart_autobaud_rx
- Parameters: CLK_HZ (int, default 100_000_000)
- Ports:
  clk (input), rst_n (input, active-low), rx (input, idle=1),
  data[7:0] (output), data_valid (output, 1-cycle pulse),
  frame_err (output, 1-cycle pulse on bad stop)
- Coding rules: no $display, no latches, nonblocking in sequential logic
- Reset: synchronous deassert, receiver idle after reset
- Behavior: detect bit period from first valid frame (start low, 8 data
    bits, stop high)

Deliverables:
- SystemVerilog module body only; do not change the signature below.
```

2.5.1.2 Start with a Manifest, Then Generate the Body

Ask for a strict header and property scaffold first, validate it, and then request the body implementation. Two small verified steps are more reliable than one large unchecked generation, as shown in Listing 2.4.

Listing 2.4 Frozen interface and property for implementation

```
module uart_autobaud_rx #(
  parameter int unsigned CLK_HZ = 100_000_000
) (
  input  logic       clk,
  input  logic       rst_n,      // active-low
  input  logic       rx,         // async serial, idle=1
  output logic [7:0] data,
  output logic       data_valid, // 1-cycle pulse per byte
  output logic       frame_err   // 1-cycle pulse on bad stop
);

  // Properties (filled by CI as SVA; listed here to guide generation):
  // P_pulse_valid: data_valid is exactly 1 cycle per received byte.
  // P_stop_high:   frame_err never when the stop bit samples high.
  // P_sync_rx:     rx is double-synchronized before any sampling.

  // === MODEL FILLS IN BELOW; HEADER MUST NOT CHANGE ===

endmodule
```

2.5.1.3 Prefer Tests to Hidden Reasoning

Provide a tiny smoke bench and ask the model to "make this pass," similar to Listing 2.5. This swaps "trust me" for "show me," which is friendlier to continuous integration (CI) and far easier to review.

Listing 2.5 Small smoke test the model must pass

```
module tb_uart_autobaud_rx;
  localparam int CLK_HZ = 100_000_000;
  logic clk = 0, rst_n = 0, rx = 1;
  logic [7:0] data; logic data_valid, frame_err;

  always #5 clk = ~clk; // 100 MHz

  uart_autobaud_rx #(.CLK_HZ(CLK_HZ)) dut (.*);

  // drive one byte LSB-first at a given baud
  task automatic send_byte(int baud, byte b);
    int cyc = CLK_HZ/baud;
    rx = 0; repeat (cyc) @(posedge clk);                // start
    for (int i=0;i<8;i++) begin rx = b[i]; repeat(cyc) @(posedge clk);
     end
    rx = 1; repeat (cyc) @(posedge clk);                // stop
    repeat (cyc) @(posedge clk);
  endtask

  initial begin
    repeat (5) @(posedge clk); rst_n = 1;
```

```
    send_byte(115200, 8'h41); // 'A'
    send_byte(115200, 8'h5A); // 'Z'
    send_byte(9600,   8'h30); // '0'
    send_byte(9600,   8'h31); // '1'
    repeat (1000) @(posedge clk); $finish;
  end

  always @(posedge clk) if (data_valid) $display("RX 0x%02h", data);
  always @(posedge clk) if (frame_err)  $display("FRAME_ERR");
endmodule
```

2.5.1.4 Version Prompts Like Source Files

Keep prompts in version control with a short comment explaining intent. Include a tiny version tag inside the prompt and always record decoding parameters next to outputs (temperature, top-p, seed). Reproducibility turns yesterday's "mysterious" win into tomorrow's reliable baseline.

2.5.2 Retrieval-Augmented Generation (RAG) Pipelines for Specs and HDL

Retrieval-augmented generation keeps the model grounded in *your* universe rather than whatever it picked up in the wild. Think of it as handing a sharp junior engineer the exact binder tabs to open before they type a line: style guide on resets, the last-reviewed block header, the short errata note everyone forgets, and that one golden testbench. The pipeline feels like a careful recipe: index, retrieve, rerank, assemble a compact context, and then generate. Do it right and answers sound like your team on a good day.

2.5.2.1 Index by Structure, Not by Length

Break things where engineers would: specs by requirement sections and figures; HDL by module and interface boundaries; logs by scenario or seed. Keep port lists, assertions, and requirement IDs intact so meaning survives the cut. Tag each chunk with light metadata you will actually use later: block name, interface, owner, commit or change-id, license, and a freshness timestamp. When someone asks for *autobaud*, prefer chunks that share surface cues (`rx`, `clk`, `rst_n`) and synonyms ("bit-center sampling," "start-bit oversampling," "baud detection"). For a PLL bring-up note, rerank chunks that mention `lock_det`, "holdover," or the exact divider parameter the query included. Structure-first indexing makes retrieval feel like memory, not a word search.

2.5.2.2 Show Your Work End to End

Every suggestion should come with breadcrumbs. Persist the retrieved snippets alongside the output with their source, commit hash, and section anchor. Record the model version, prompt tag, temperature, top-p, and seed. When a reviewer asks "why this sampling scheme?", you can point to the exact style guide paragraph that recommends double synchronization and majority voting at mid-bit, plus the UART header you retrieved that already declares the oversample factor. Less arguing, more merging.

2.5.2.3 Program the Pipeline, Not Just the Prompt

Treat RAG like a build flow: Many small, predictable passes beat one giant leap. Normalize text, retrieve, rerank, assemble a tight context window, ask for a header and a couple of properties, validate them, and then ask for the body. Small guards at each hop do most of the safety work: Drop chunks that exceed a size budget, refuse contexts missing licenses, ban nonsynthesizable constructs in the manifest, and limit the number of unrelated snippets (noise budget). For a secure-boot ROM task, retrieve the *specific* REQ IDs for key storage and boot-fuse policy, generate a tiny policy table, validate it, and then generate code that implements *only* that table. For an SPI flash controller, retrieve the timing table chunks and a reviewed init sequence, turn them into a minimal state diagram first, check that diagram for forbidden states, and only then ask for RTL. Many cheap passes; few surprises.

2.5.2.4 Handle Ambiguity Upfront

Queries are often fuzzy. Teach the pipeline to rewrite them before retrieval. "Slow fan ramp" might rewrite to "PWM duty slew limit" or "thermal hysteresis"; "bus never starves B" becomes "round-robin fairness window." Add lightweight semantic hints from the current repo: If the working directory is `arbiter_rr/`, boost retrievals that mention grant toggling or parked master. The goal is not to be clever, just helpful.

However, bad contexts happen. Make the failure mode obvious and cheap. If retrieved snippets disagree (two conflicting reset conventions), surface a red banner and stop before generation. If the top-k are weak, say so and show the empty rack rather than hallucinating parts. When logs are mismatched (old tool version, different seed), mark them stale and refuse to cite. Loud, small failures beat quiet, expensive ones every time.

For example, a UART autobaud helper asks for: the house reset guide, the UART header, and a reviewed bench that measures start-bit timing. Retrieval finds a snippet on mid-bit sampling and a note that `rst_n` is synchronous. The pipeline assembles a one-page context: reset rule, header, timing paragraph. It then asks the model for a tiny checklist and two properties ("detect start on `rx` low after idle" and "lock baud within N bits"), validates that, and *then* generates the calibration logic. If the bench later fails, you have the three sources and the exact decoding settings to replay the run. Swap in a different block and the dance is the same: For a bus monitor, retrieve the protocol table and a known-good checker, and then ask for a property list before code; for a sensor hub, retrieve the oversampling note and the latency budget, and then generate the scheduler only after a miniature schedule table is validated.

Done this way, RAG feels less like sprinkling context and more like running a tidy lab notebook. You decide what goes into the model's field of view, you keep it small and relevant, and you leave a trail you can follow six months from now when someone asks "where did this come from?"

2.5.3 Designer, Planner, Programmer, Reviewer, Evaluator (with a Human Gate)

The workflow in Fig. 2.7 is intentionally concrete: Every role hands a small, checkable artifact to the next one, and the loop continues until the design passes tests. The running example in the figure is a `signal_generator` that produces a 5-bit triangle wave. The point is not the widget itself, but the discipline of turning a vague idea into a sequence of verifiable steps.

2.5.3.1 Designer

The process starts with a compact specification that a new teammate could read in a minute: module name, ports, clocking assumptions, reset policy, and behavior in plain language. For the triangle wave, the spec fixes a 5-bit output wave, a direction bit `dir_up`, synchronous operation on `clk`, and an active-low reset `rst_n`. The wave counts between 0 and 31, incrementing when `dir_up=1` and decrementing otherwise.

2.5.3.2 Planner

The planner converts that prose into an implementation plan that a coder can follow line by line. Typical steps are: declare registers (`reg [4:0] wave;` and `reg dir_up;`), implement reset behavior (clear `wave`, set `dir_up` to start by incrementing), then write the state update with the boundary rules (increment, while $0 \leq$ `wave` < 31; when `wave==31` flip `dir_up` and prepare to decrement; symmetric behavior at zero). The plan explicitly calls out what happens at the edges so there is nothing to guess later.

DESIGNER (HUMAN IN LOOP)

Implement a Triangle Wave signal generator module that generates a waveform by incrementing and decrementing a 5-bit signal named "wave". The waveform cycles between 0 and 31, which is incremented or decremented by 1.

Module name:
signal_generator
Input ports:
clk: Clock signal used for synchronous operation.
rst_n: Active-low reset signal. Defined as 0 for reset and 1 for reset signal inactive.
Output ports:
wave: 5-bit output waveform signal representing the generated waveform.

act as a planner to execute this problem

LLM AS REVIEWER

Review Checklist Based on the Planner's Implementation Stages:

Step 1: Declare internal registers: wave [4:0], dir_up
☑ wave and dir_up correctly declared
Step 2: Reset behavior: wave <= 0, dir_up <= 1 when !rst_n
☑ Clear reset condition implemented
Step 3a: Increment logic: if dir_up == 1, increment wave until 31, then flip
☑ Handled properly with conditional toggle
Step 3b: Decrement logic: if dir_up == 0, decrement wave until 0, then flip
☑ Correct logic written and transitions handled
Step 4: Output assignment: wave is directly used as output
☑ Since wave is declared as output reg, no extra assign needed

DESIGNER (HUMAN IN LOOP)

Your design did not pass the testbench, here are the errors-
✗ Error at index 0 @time 20ns | wave= 1, expected= 0
....
✗ Error at index 99@ time 1010ns | wave= 24, expected= 5
And this is the testbench your code was checked against-

```
module tb_signal_generator (........);
 signal_generator uut (...........);
..............
 initial begin
.................
   repeat (100) begin
    if (wave !== reference[i]) begin
     error = error + 1;
...............
 end
endmodule
```

LLM AS REVIEWER

As the Reviewer, I will now evaluate whether the Programmer correctly addressed all the errors identified during the evaluation of the signal_generator design-

☑ Confirmed wave starts from 0 by initializing wave <= 0 on reset.
☑ Direction flag (dir_up) is toggled after holding at 31 or 0, not before. This allows waveform to peak/floor for a full cycle.
☑ Logic ensures proper symmetric triangle waveform with controlled toggling direction and output.

LLM AS PLANNER

Implementation Steps
[Step 1] Declare internal registers
Declare wave as a reg [4:0] (5-bit)
Declare dir_up as a reg to control direction (up/down)
[Step 2] Implement reset behavior on rst_n == 0:
Set wave <= 0
Set dir_up <= 1 (start by incrementing)
[Step 3] Implement waveform generation logic on every posedge clk:
If dir_up == 1:
If wave == 31 → change dir_up <= 0 and decrement wave
Else → increment wave
Else (dir_up == 0):
If wave == 0 → change dir_up <= 1 and increment wave
Else → decrement wave
[Step 4] Assign wave to the output since wave is a reg, it can be directly used as the output

LLM AS PROGRAMMER

```
module signal_generator (input ....., output........ );
   // [STEP 1] Declare internal registers
 reg dir_up; // Direction control: .....
   // [STEP 2] Implement logic with reset
   always @(posedge clk or negedge rst_n) begin
     if (!rst_n) begin
      ..................
   // [STEP 3] Waveform generation logic
       if (dir_up) begin  // [STEP 3a]
         if (wave == 5'd31) begin
        ..............................
        end else begin   // [STEP 3b]
         if (wave == 5'd0) begin
       wave <= wave + 1;
        ................................
       end
   // [STEP 4] Output wave is already declared as reg [4:0]
endmodule
```

LLM AS EVALUATOR

Let's dive into analyzing the provided errors and testbench behavior to find out possible fixes for the design-

🔍 Issue Analysis from Evaluation
From the testbench and waveform:
Fix 1: The wave signal skips the initial 0 (starts from 1).
Fix 2: It does not hold at peaks (31 and 0) for an extra cycle, which is expected in a clean triangle waveform.
Fix 3: It changes direction too quickly at boundary values — we need a full cycle where wave = 31 and wave = 0.

LLM AS PROGRAMMER

```
module signal_generator (
   input ..........);
// Step 1: .........
// Step 2: always @(posedge clk or negedge rst_n) begin
   if (!rst_n) begin
     wave <= 5'd0;    // Fix 1: Start at 0
// (Previously it went to 1 immediately)
.................
// Step 3
// Fix 2: Hold at wave = 31 one extra cycle before switch to down
     if (wave == 5'd31 && dir_up) begin  .............
// Fix 3: Hold at wave = 0 one extra cycle before switching to up
     else if (wave == 5'd0 && !dir_up) begin ..............
endmodule
```

Fig. 2.7 Role-based prompting for HDL development [32]. The designer states the spec (triangle-wave generator), the planner decomposes it into steps, the programmer drafts RTL, the reviewer audits against a checklist and testbench results, and the evaluator diagnoses failures and proposes fixes; the loop iterates until the tests pass

2.5.3.3 Programmer

The programmer produces RTL that follows the plan and preserves the interface exactly. The code is sectioned to mirror the steps (`// [STEP 1] declares`, `// [STEP 2] reset`, `// [STEP 3] update`, `// [STEP 4] output`), so a reviewer can map each requirement to a line of code without reverse-engineering intent.

2.5.3.4 Reviewer

Review is checklist-driven. The reviewer confirms that the declared regs match the plan, the reset branch really clears state, the conditional logic implements "count up to 31 then switch, count down to 0 then switch," and that the output is driven as specified. The checklist lives next to the code, so the verdict is traceable: Each box corresponds to one step in the plan.

2.5.3.5 Evaluator

The evaluator runs a small, deterministic testbench and inspects the first failing moment. In the figure's example, three concrete issues are surfaced: The wave started at 1 instead of 0; the design did not hold the value for one extra cycle at the boundaries; and the direction bit toggled too early. Each item comes with a tiny counterexample snippet and a suggested fix (e.g., "initialize `wave` to 0," "hold at 31 and 0 for one cycle," "toggle `dir_up` only after the hold"). These fixes are fed straight back to the programmer, who patches the corresponding step and resubmits.

2.5.3.6 Human Gate

Nothing merges automatically. A human signs off on a single bundle: the spec, the plan, the annotated RTL, the checklist, and the evaluator's logs/counterexamples. If a rerun is needed, the artifacts and seed reproduce the failure. This gate keeps the loop lightweight but accountable, small, checkable artifacts flow forward; precise, actionable feedback flows back, until the design passes.

2.5.3.7 Stop Conditions and Backoff

Endless "one more try" loops are where schedules go to die. Decide upfront when to stop (for example, two compile failures or two test runs with no coverage gain). On stop, simplify: Shrink the prompt to the minimum contract, swap in a known-good policy (strict round-robin rather than something exotic), or try a smaller local model for a sanity pass. Archive every attempt so tomorrow's fix starts from yesterday's almost-right answer instead of from scratch.

2.5.3.8 Pass Artifacts, Not Secrets

What flows between stages are headers, notes, logs, coverage, seeds, and tiny traces—not credentials or unconstrained tool access. Tool invocations sit behind allow-lists, timeouts, and redaction. That keeps the flow fast and safe, and it makes every surprising result reproducible.

This "planner–coder–simulator–verifier–human" rhythm works just as well if you swap the fan controller for a DVFS governor, or the arbiter for a DMA with timeouts. You get fewer surprises, better tempo, and changes you can defend in a design review without opening a single waveform unless you really want to.

2.5.4 Failure Modes and Fast Fixes

When retrieval brings back junk, tighten chunking around interfaces and rerank by signal or parameter overlap. When the model drifts off spec, shorten the prompt and enforce the header-first step. When compile rates tank, turn on lint-before-generate and ban common foot-guns in the prompt. When a loop spins, enforce stop conditions, surface the best attempt with logs, and let a human make the call. The goal is not perfection on the first try—it is a process that fails loudly, recovers quickly, and leaves a paper trail you can trust.

2.6 Evaluation for Hardware

If prompting is how we ask, evaluation is how we trust. For RTL work, the scoreboard is not pretty prose: It is whether designs lint, compile, simulate against a real testbench, hit sensible coverage, and satisfy a few crisp properties. Keep it boring and replayable: fixed seeds, frozen tool versions, and artifacts you can audit later. Boring is good; boring is repeatable.

General code benchmarks make a fine smoke test. They tell you whether a model can follow instructions and keep basic syntax straight before you burn simulator cycles [4, 32]. But HDL has its own vibe, and behavior matters more than style points. That is why domain-focused suites are where comparisons get real: They grade the same pipeline you actually run—compile first, then deterministic simulation, sometimes coverage, sometimes a pinch of formal [5, 15, 16, 34]. Read the rubric on each: some count "compiled" as success, others insist that tests pass, and a few even care about coverage. Use general-code scores to catch gross regressions across model versions, and rely on HDL-flavored evaluations to compare prompts, retrieval, and fine-tunes like a grown-up lab test.

When runs go sideways (and they will), measure the *why*, not just the *what*. Repair-oriented views help separate the "fixable with a nudge" syntax flubs from the truly broken [36]. Iterative debug frameworks shorten the loop from log to fix [40]. And tools that turn synthesis errors into structured explanations make failure modes trackable at scale [19]. Fold these lenses into your evaluation so CI does not just stamp FAIL; it points to the right knob to turn.

The harness itself should feel like your day job. Start with strict lint and a synthesizability pass to flush out missing resets, accidental latches, and banned constructs before you waste simulation time. Then run short, deterministic tests with fixed decoding settings; save logs and a small waveform slice for any failure so triage is fast. Let coverage be a thermometer, not a deity: Structural coverage (line, branch, toggle, FSM) shows where execution flowed, while a couple of functional coverpoints (e.g., FIFO full/empty transitions, reset-to-first-transaction latency, fairness windows in an arbiter) check whether intent was exercised. Add a pinch of formal for high-value invariants: no dual grant, no counter underflow/overflow, no `X` after reset, and watch for vacuity so you do not "prove" a property that never actually triggered. If a single number helps you track drift, blend the signals with stable weights, for example,

$$\text{Score} = 0.35\,\text{Compile} + 0.35\,\text{SimPass} + 0.20\,\text{Coverage} + 0.10\,\text{Formal},$$

with each term normalized to [0, 1]. The constants matter less than keeping them consistent across experiments.

Hygiene makes numbers honest. Freeze temperature, top-p, max tokens, seeds, tool versions, and even the prompt revision. Keep train/val/eval splits clean and deduplicated so you do not accidentally "learn the test." Tag failures by type, lint, compile, runtime, assertion, or coverage shortfall, so fixes land in the right place (prompt, retrieval, or postchecks). Above all, be artifact-first: Every run should leave behind prompts, retrieved context, model and version, decoding settings, generated sources, tool logs, coverage, and the final score. That provenance turns "why did this pass?" into a replayable question, not a debate. Keep the cadence light, a small, representative nightly suite catches drift without burning the farm, and expand only when a new failure mode keeps showing up. Do this and the evaluation stops feeling like a magic show and starts to look like the engineering discipline you already practice, because it is.

2.7 Why LLMs for SoC Design and Security

SoC programs live in controlled chaos. Specs shift, while blocks are already in flight, tool flows have sharp edges, regressions show up after midnight, and security threads through all of it like a high-tension wire. Large language models help not by designing the chip for you but by acting as a steady copilot that turns fuzzy intent into concrete artifacts and keeps the conversation between design, verification, and security moving [8, 23, 24]. A vague line in the spec becomes a crisp, testable requirement. An RTL scaffold appears that follows your house style and compiles on the first try. A handful of assertions land next to the module they protect. Logs from failing seeds get clustered so you can jump straight to the macro at fault. For security work, the same assistant drafts an asset map that ties secrets to their interfaces, proposes misuse cases an attacker would try, and suggests properties that prove those misuses cannot happen.

None of this replaces engineering judgment. Every suggestion still rides the same rails you already trust. Retrieval keeps the model grounded in your own specs and code rather than its memory. Lint, compile, simulation, and a few sturdy invariants gate what gets to merge. Tool calls run through wrappers with allow-lists, timeouts, and redaction so a prompt cannot wander into the wrong directory or leak a key. Provenance is first class: Prompts, retrieved context, decoding settings, tool versions, and outputs are logged so you can replay or audit any result. When IP policy requires it, models run on-prem and weights never leave the building.

Used this way, an LLM buys tempo without sacrificing discipline. Design teams spend less time on boilerplate and spelunking through logs. Security reviews start from precise properties, not hand-wavy risk lists. Governance is easier because everything is reproducible. The rest of this chapter turns that idea into practice. We first peek under the hood to build shared intuition for how Transformers work (Sect. 2.2) and then choose model families that fit the job and the policies you must meet (Sect. 2.3). We assemble a clean, hardware-flavored corpus and pick sensible adaptation paths (Sect. 2.4), script prompts and orchestration with a human gate (Sect. 2.5), and define success with compile and simulation pass rates rather than pretty text (Sect. 2.6). Along the way we will keep an eye on the failure modes and the controls that keep them contained, which we summarize later in Sect. 2.8.

2.8 Risks, Safety, and Trust

Putting a model in the loop reshapes your risk surface. Two failure families matter day to day. First, functional: code that looks tidy but trips lint, hides a latch behind a missing default, or fails a corner-case waveform. Second, security: places where generated logic, prompts, or tool wiring widen the attack surface: classic CWE-style bugs, Trojan-like backdoors, data leaks through logs, or prompt injection that steers a tool wrapper off the rails [7, 11, 43]. Data hygiene is part of the threat model too: contaminated eval sets inflate scores, and poorly scrubbed corpora can leak right back out of the model, sometimes verbatim [3, 14].

On the functional side, the fixes are deliberately boring. Keep house checklists for clocks, resets, and interface idioms. Lint *before* you ever ask for "fixes." Do a quick synth dry-run to flush nonsynthesizable corners. Add a few crisp invariants (no dual grant, no counter under/overflow, no `X` after reset) so failures localize fast. Freeze seeds, tool versions, and decoding settings so a green run is something you can *replay*, not just retell.

Security deserves the same discipline. Treat every model suggestion as untrusted code: Gate merges behind policy checks, and scan diffs for suspicious control paths or dormant logic. Lock the glue around the model: Tool calls behind allow-lists and timeouts, strict argument validation, and no secrets in prompts or retrieved context. Retrieval itself is an attack surface: Guard the index, pin the exact snippets you pass to the model, and prefer vetted sources [7]. For provenance and authorization in an LLM-in-the-loop flow, borrow from multiparty authentication ideas: Require independent sign-offs and cryptographic attestations on the *artifacts* that drive automation (model version, prompt, retrieval bundle), and use hierarchical approval paths to avoid single points of trust [28, 29]. Finally, plan for rapid rollback and hotfixes of model-generated changes: Stage them behind feature flags and keep a controlled "escape hatch" so you can disable or patch outputs quickly, operationally similar to selective reconfigurability used to harden and update deployed logic [30].

Trust, finally, is a governance problem. Track licenses and provenance for any external code you ingest. Keep weights, prompts, and outputs where policy says they belong. Require approvals for model and prompt updates like you would for a toolchain bump. Never let autogenerated code land without tests. Leave an audit trail: input spec, retrieved snippets, prompt, decoding settings, tool logs, artifacts, so two questions are always answerable: what changed, and why should we believe it. In hardware security proper, keep the bigger picture in view: logic locking and SAT-resilient schemes, split manufacturing and packaging defenses, and hardware roots like PUF-backed signatures complement what your LLM pipeline can (and cannot) assure [27, 38, 41].

With the guardrails clear, the rest of the book takes that scaffold into the trenches. Chapter 3 walks block-by-block design work from IP to full SoC. Chapter 4 turns the model into a verification assistant for assertions, fuzz, and clever test ideas. Chapter 5 builds threat models and turns them into properties you can actually check. Chapter 6 focuses on vulnerability detection and analysis with an LLM-aware toolchain. Chapter 7 tackles IP protection strategies such as watermarking and locking. Chapter 8 closes with research directions that push beyond today's playbook. Used with discipline, LLMs do not replace engineering judgment. They clear the underbrush so your judgment has more room to work.

References

1. Jimmy Lei Ba, Jamie Ryan Kiros, and Geoffrey E. Hinton. Layer normalization. *arXiv preprint arXiv:1607.06450*, 2016.
2. Tom B. Brown, Benjamin Mann, Nick Ryder, Melanie Subbiah, Jared Kaplan, Prafulla Dhariwal, Arvind Neelakantan, Pranav Shyam, Girish Sastry, Amanda Askell, et al. Language models are few-shot learners. In *NeurIPS*, volume 33, pages 1877–1901, 2020.
3. Nicholas Carlini, Florian Tramer, Eric Wallace, Matthew Jagielski, Ariel Herbert-Voss, Katherine Lee, Adam Roberts, Tom Brown, Dawn Song, Ulfar Erlingsson, Alina Oprea, and Colin Raffel. Extracting training data from large language models, 2021.

4. Mark Chen, Jerry Tworek, Heewoo Jun, Qiming Yuan, Henrique Ponde de Oliveira Pinto, Jared Kaplan, Harri Edwards, Yuri Burda, Nicholas Joseph, Greg Brockman, Alex Ray, Raul Puri, Gretchen Krueger, Michael Petrov, Heidy Khlaaf, Girish Sastry, Pamela Mishkin, Brooke Chan, Scott Gray, Nick Ryder, Mikhail Pavlov, Alethea Power, Lukasz Kaiser, Mohammad Bavarian, Clemens Winter, Philippe Tillet, Felipe Petroski Such, Dave Cummings, Matthias Plappert, Fotios Chantzis, Elizabeth Barnes, Ariel Herbert-Voss, William Hebgen Guss, Alex Nichol, Alex Paino, Nikolas Tezak, Jie Tang, Igor Babuschkin, Suchir Balaji, Shantanu Jain, William Saunders, Christopher Hesse, Andrew N. Carr, Jan Leike, Josh Achiam, Vedant Misra, Evan Morikawa, Alec Radford, Matthew Knight, Miles Brundage, Mira Murati, Katie Mayer, Peter Welinder, Bob McGrew, Dario Amodei, Sam McCandlish, Ilya Sutskever, and Wojciech Zaremba. Evaluating large language models trained on code, 2021.
5. Matthew DeLorenzo, Vasudev Gohil, and Jeyavijayan Rajendran. CreativEval: Evaluating creativity of LLM-based hardware code generation, 2024.
6. Jacob Devlin, Ming-Wei Chang, Kenton Lee, and Kristina Toutanova. BERT: Pre-training of deep bidirectional transformers for language understanding. In *NAACL-HLT*, pages 4171–4186, 2019.
7. Kai Greshake, Sahar Abdelnabi, Shailesh Mishra, Christoph Endres, Thorsten Holz, and Mario Fritz. Not what you've signed up for: Compromising real-world LLM-integrated applications with indirect prompt injection. In *Proceedings of the 16th ACM Workshop on Artificial Intelligence and Security*, AISec '23, page 79–90, New York, NY, USA, 2023. Association for Computing Machinery.
8. Khan Thamid Hasan, Md Ajoad Hasan, Nashmin Alam, Md Touhidul Islam, Upoma Das, and Farimah Farahmandi. AI-assisted hardware security verification: A survey and AI accelerator case study. *arXiv preprint arXiv:2604.01572*, 2026.
9. Kaiming He, Xiangyu Zhang, Shaoqing Ren, and Jian Sun. Deep residual learning for image recognition. In *CVPR*, pages 770–778, 2016.
10. Sepp Hochreiter and Jürgen Schmidhuber. Long short-term memory. *Neural Computation*, 9(8):1735–1780, 1997.
11. Zahin Ibnat, Paul E Calzada, Dipayan Saha, Hasan Al-Shaikh, Sujan Kumar Saha, Jingbo Zhou, Farimah Farahmandi, and Mark Tehranipoor. Trusting the machine: How secure is LLM-generated RTL code? In *2025 ACM/IEEE 7th Symposium on Machine Learning for CAD (MLCAD)*, pages 1–8. IEEE, 2025.
12. Jared Kaplan, Sam McCandlish, Tom Henighan, Tom B. Brown, Benjamin Chess, Rewon Child, Scott Gray, Alec Radford, Jeffrey Wu, and Dario Amodei. Scaling laws for neural language models. *arXiv preprint arXiv:2001.08361*, 2020.
13. Taku Kudo and John Richardson. SentencePiece: A simple and language independent subword tokenizer and detokenizer for neural text processing. In *EMNLP*, pages 66–71, 2018.
14. Yucheng Li, Frank Guerin, and Chenghua Lin. LatestEval: Addressing data contamination in language model evaluation through dynamic and time-sensitive test construction, 2024.
15. Mingjie Liu, Nathaniel Pinckney, Brucek Khailany, and Haoxing Ren. Invited paper: VerilogEval: Evaluating large language models for verilog code generation. In *2023 IEEE/ACM International Conference on Computer Aided Design (ICCAD)*, pages 1–8, 2023.
16. Yao Lu, Shang Liu, Qijun Zhang, and Zhiyao Xie. RTLLM: An open-source benchmark for design RTL generation with large language model, 2023.
17. OpenAI. GPT-4 technical report. *arXiv preprint arXiv:2303.08774*, 2023.
18. Long Ouyang, Jeff Wu, Xu Jiang, Diogo Almeida, Carroll L. Wainwright, Pamela Mishkin, Chong Zhang, Sandhini Agarwal, Katarina Slama, Alex Ray, et al. Training language models to follow instructions with human feedback. In *NeurIPS*, 2022.
19. Siyu Qiu, Benjamin Tan, and Hammond Pearce. LLM-aided explanations of EDA synthesis errors. In *2024 IEEE LLM Aided Design Workshop (LAD)*, page 1–6. IEEE, June 2024.
20. Alec Radford, Jeffrey Wu, Rewon Child, David Luan, Dario Amodei, and Ilya Sutskever. Language models are unsupervised multitask learners. *OpenAI Blog*, 1(8), 2019.

21. Colin Raffel, Noam Shazeer, Adam Roberts, Katherine Lee, Sharan Narang, Michael Matena, Yanqi Zhou, Wei Li, and Peter J. Liu. Exploring the limits of transfer learning with a unified text-to-text transformer. *Journal of Machine Learning Research*, 21:1–67, 2020.
22. Baptiste Rozière, Jonas Gehring, Fabian Gloeckle, Sten Sootla, Itai Gat, Xiaoqing Ellen Tan, et al. Code LLaMA: Open foundation models for code. *arXiv preprint arXiv:2308.12950*, 2023.
23. Dipayan Saha, Hasan Al Shaikh, Shams Tarek, and Farimah Farahmandi. Special session: Threatlens: LLM-guided threat modeling and test plan generation for hardware security verification. In *2025 IEEE 43rd VLSI Test Symposium (VTS)*, pages 1–5. IEEE, 2025.
24. Dipayan Saha, Shams Tarek, Katayoon Yahyaei, Sujan Kumar Saha, Jingbo Zhou, Mark Tehranipoor, and Farimah Farahmandi. LLM for SoC security: A paradigm shift. *IEEe Access*, 12:155498–155521, 2024.
25. Hojjat Salehinejad, Sharan Sankar, Joseph Barfett, Errol Colak, and Shahrokh Valaee. Recent advances in recurrent neural networks. *arXiv preprint arXiv:1801.01078*, 2017.
26. Rico Sennrich, Barry Haddow, and Alexandra Birch. Neural machine translation of rare words with subword units. In *ACL*, pages 1715–1725, 2016.
27. Pramod Subramanyan, Sayak Ray, and Sharad Malik. Evaluating the security of logic encryption algorithms. In *2015 IEEE International Symposium on Hardware Oriented Security and Trust (HOST)*, pages 137–143, 2015.
28. Ishraq Tashdid, Tasnuva Farheen, and Sazadur Rahman. AuthenTree: A scalable MPC-based distributed trust architecture for chiplet-based heterogeneous systems, 2025.
29. Ishraq Tashdid, Tasnuva Farheen, and Sazadur Rahman. SAFE-SiP: Secure authentication framework for system-in-package using multi-party computation. In *Proceedings of the Great Lakes Symposium on VLSI 2025*, GLSVLSI '25, page 391–396, New York, NY, USA, 2025. Association for Computing Machinery.
30. Ishraq Tashdid, Dewan Saiham, Nafisa Anjum, Tasnuva Farheen, and Sazadur Rahman. Ecologic: Enabling circular, obfuscated, and adaptive logic via EFPGA-augmented SoCs, 2025.
31. Ishraq Tashdid, Valentina Terry, Jordan Merkel, Tasnuva Farheen, and Sazadur Rahman. BeyondPPA: Human-inspired reinforcement learning for post-route reliability-aware macro placement. In *2025 ACM/IEEE 7th Symposium on Machine Learning for CAD (MLCAD)*, pages 1–12, 2025.
32. Kimia Tasnia, Alexander Garcia, Tasnuva Farheen, and Sazadur Rahman. VeriOpt: PPA-aware high-quality verilog generation via multi-role LLMs, 2025.
33. Kimia Tasnia and Sazadur Rahman. *OPL4GPT: An Application Space Exploration of Optimal Programming Language for Hardware Design by LLM*, page 981–987. Association for Computing Machinery, New York, NY, USA, 2025.
34. Shailja Thakur, Baleegh Ahmad, Zhenxing Fan, Hammond Pearce, Benjamin Tan, Ramesh Karri, Brendan Dolan-Gavitt, and Siddharth Garg. Benchmarking large language models for automated verilog RTL code generation, 2022.
35. Hugo Touvron, Thibaut Lavril, Gautier Izacard, Xavier Martinet, Marie-Anne Lachaux, Timothée Lacroix, Baptiste Rozière, Naman Goyal, Eric Hambro, Faisal Azhar, et al. LLaMA: Open and efficient foundation language models. *arXiv preprint arXiv:2302.13971*, 2023.
36. Yun-Da Tsai, Mingjie Liu, and Haoxing Ren. RTLFixer: Automatically fixing RTL syntax errors with large language models, 2024.
37. Ashish Vaswani, Noam Shazeer, Niki Parmar, Jakob Uszkoreit, Llion Jones, Aidan N. Gomez, Lukasz Kaiser, and Illia Polosukhin. Attention is all you need, 2023.
38. Jason Vosatka. *Introduction to Hardware Trojans*, pages 15–51. Springer International Publishing, Cham, 2018.
39. Jason Wei, Yi Tay, Rishi Bommasani, Colin Raffel, Barret Zoph, Sebastian Borgeaud, Dani Yogatama, Maarten Bosma, Denny Zhou, Donald Metzler, et al. Emergent abilities of large language models. *Transactions on Machine Learning Research*, 2022.
40. Ke Xu, Jialin Sun, Yuchen Hu, Xinwei Fang, Weiwei Shan, Xi Wang, and Zhe Jiang. *MEIC: Re-thinking RTL Debug Automation using LLMs*. Association for Computing Machinery, New York, NY, USA, 2025.

41. Muhammad Yasin, Abhrajit Sengupta, Mohammed Thari Nabeel, Mohammed Ashraf, Jeyavijayan (JV) Rajendran, and Ozgur Sinanoglu. Provably-secure logic locking: From theory to practice. In *Proceedings of the 2017 ACM SIGSAC Conference on Computer and Communications Security*, CCS '17, page 1601–1618, New York, NY, USA, 2017. Association for Computing Machinery.
42. Wayne Xin Zhao, Kun Zhou, Junyi Li, Tianyi Tang, Xiaolei Wang, Yupeng Hou, Yingqian Min, Beichen Zhang, Junjie Zhang, Zican Dong, et al. A survey of large language models. *arXiv preprint arXiv:2303.18223*, 2023.
43. Andy Zou, Zifan Wang, Nicholas Carlini, Milad Nasr, J. Zico Kolter, and Matt Fredrikson. Universal and transferable adversarial attacks on aligned language models, 2023.

Chapter 3
Hardware Design Leveraging LLMs: From IP to SoC

Sujan Ghimire, Parsa Mirfasihi, Muhtasim Alam Chowdhury, Veeramani Pugazhenthi, Harish Kumar Dharavath, Kevin Immanuel Gubbi, Mohammadnavid Tarighat, Houman Homayoun, Rozhin Yasaei, Pratik Satam, and Soheil Salehi

This chapter explores how large language models (LLMs) can be leveraged to enhance hardware design workflows, spanning from intellectual property (IP) block development to full system-on-chip (SoC) integration. It begins by outlining the foundational aspects of hardware design, including IP-level development, SoC integration challenges, and the evolving role of security across design hierarchies. This chapter then examines how LLMs assist in IP design tasks such as RTL generation from natural language and the incorporation of security-aware design practices. At the system level, it discusses LLM-driven automation in platform design, security analysis, and design-space exploration, supported by a case study on secure AES-based SoC integration. Furthermore, this chapter introduces LLM-augmented codesign frameworks, highlighting multiagent workflows, integration with existing EDA toolchains, and emerging use cases in both academia and industry. Finally, it addresses the limitations, challenges, and risks associated with deploying LLMs in hardware design, including technical constraints, trust concerns, and human-in-the-loop considerations. Together, these perspectives illustrate the potential of LLMs to transform hardware design processes while emphasizing the need for careful and secure adoption.

S. Ghimire · P. Mirfasihi · M. A. Chowdhury · V. Pugazhenthi · H. K. Dharavath · R. Yasaei · P. Satam · S. Salehi (✉)
University of Arizona, Tucson, AZ, USA
e-mail: ssalehi@arizona.edu

K. I. Gubbi · M. Tarighat · H. Homayoun
University of California Davis, Davis, CA, USA

F. Farahmandi et al. (eds.), *LLM for SoC Design and Security*,
https://doi.org/10.1007/978-3-032-34116-7_3

3.1 Foundations of Hardware Design and SoC Integration

In modern hardware, the system-on-chip (SoC) has become the default configuration, as electronic systems have grown highly sophisticated and capable of performing a wide range of functions, including signal processing, data handling, and information management [17]. A key enabler of SoC development is the use of intellectual property (IP) blocks, which are predesigned and preverified functional units such as processors, memory controllers, interfaces, and communication modules. These IP blocks are integrated into a single chip, allowing designers to build complex systems more efficiently by reusing proven components rather than creating each function from scratch [34, 10]. A good example of SoC integration is the smartwatch, which offers functionality such as heart rate monitoring, time display, step counting, and even receiving phone calls. All of this is powered by a tightly integrated SoC that combines multiple IP blocks working together seamlessly [70]. Despite advances in IP reuse and SoC technology, the process of designing a system-on-chip remains highly structured and labor-intensive. From the early specification stages through RTL development, verification, and physical design, SoC creation requires multiple specialized teams, domain-specific expertise, and time-consuming validation cycles. As modern SoCs continue to grow in complexity with more heterogeneous cores, tighter performance requirements, and broader security expectations, the effort required for design closure has increased significantly [16, 4]. This widening productivity gap has sparked growing interest in artificial intelligence techniques, particularly large language models (LLMs), as tools to augment and support human engineering efforts [2].

3.1.1 IP-Level Design Overview

An IP core in semiconductors is a reusable unit of logic, functionality, cell, or layout design, typically designed for use as building blocks in various chip designs. Some examples of the IPs are encryption (e.g., AES cores), bus controllers, memory interfaces, or signal processing components [57]. These blocks are typically written in hardware description languages (HDLs), such as Verilog or VHDL. These IPs typically undergo several key design stages. The process begins with behavioral design, where functionality is specified using high-level constructs. The next step is followed by RTL (register-transfer level) design, which models the hardware in a synthesizable form. Next, the design is synthesized into a gate-level netlist that maps to a standard cell library, preparing it for physical implementation. Finally, validation is performed through a combination of formal verification and simulation-based testing to ensure correctness [21]. Intellectual properties are routinely developed independently, often by third-party suppliers who aim to license their products to other entities. This method presents problems with integration consistency, compatibility, and overall system security.

3.1.2 SoC Integration Challenges

The design of an SoC requires integrating numerous IP blocks, which are often created independently by various entities, into a unified platform. This integration is accompanied by multiple new obstacles and limitations that are absent in the development of standalone intellectual property. This scenario highlights the need for a connection fabric, often implemented as shared buses or networks-on-chip (NoC), alongside the assurance of protocol conformance through appropriate handshaking and timing compatibility among components [54, 6]. Further issues manifest as security wrappers and isolation techniques, like scan-chain obfuscation, firewalls, and sandboxing, designed to safeguard the SoC from dangerous or untrusted intellectual properties [17, 24, 35, 12, 50]. The incorporation of design-for-test (DfT) and design-for-debug (DfD) frameworks complicates integration, as they must be customized to ensure observability and controllability, while safeguarding security [6]. The security assurances of an independent IP may no longer be valid within an integrated SoC. A cryptographic IP that is resilient to side-channel leakage in isolation may become susceptible when subjected to common power supply networks or simultaneous switching activities in adjacent modules. A notable problem in SoC design is the integration of heterogeneous components, such as analog, digital, and mixed-signal circuits, inside a unified architecture. Each of them has unique needs for power, timing, and signal integrity, which complicates the assurance of harmonious operation [16]. Incorporating several communication protocols and interfaces merely increases complexity. Power management is more crucial when multiple components with diverse power consumption profiles are integrated into a single chip, particularly in battery-operated systems. The SoC design engineers must strike a balance between performance and energy efficiency to prevent overheating and extend device longevity [44].

Furthermore, preserving signal integrity is crucial in highly populated layouts to prevent crosstalk, electromagnetic interference, and noise-induced inaccuracies. Strategies such as shielding, optimal floor planning, and noise-aware routing are necessary to mitigate these effects [37, 29]. The verification and testing of SoCs are particularly challenging due to the system's complexity and interconnectedness. To ensure the proper functionality of all modules, both independently and within the integrated framework, comprehensive validation at various stages is essential. This involves employing simulation and formal verification methods to identify flaws early and prevent expensive postsilicon corrections [56].

3.1.3 Security Shift from IP to SoC

As IP blocks are integrated into an SoC, security considerations extend from individual component protection to system-level assurance. The integration of heterogeneous IPs, shared resources, and third-party components and complex communication fabrics introduces new interactions that can affect the overall security posture of the platform [6, 54]. However, the underlying SoC security threats, including system-level vulnerabilities, hardware Trojans, side-channel leakage mechanisms, supply-chain risks, and security verification challenges, have already been comprehensively discussed in previous chapters and are not repeated here. In this chapter, security is considered primarily from the perspective of how LLMs can assist designers in developing, integrating, and validating secure IPs and SoCs throughout the hardware design flow.

3.1.4 *The Role of Automation and Intelligence*

Given the complexity of modern SoCs, automation has become essential in managing design scale and consistency [16]. However, current automation techniques rely heavily on rule-based algorithms, constraint files, and static tools. This creates an opportunity for LLMs to augment, and in some cases, redefine how specifications are translated into secure, verifiable hardware [56, 2].

By learning from large corpora of RTL designs, EDA scripts, testbenches, and architectural diagrams, LLMs offer the promise of a design assistant capable of:

- Parsing natural language specifications into RTL scaffolding [63, 64, 71, 14].
- Detecting security vulnerabilities through code pattern analysis [46, 51]
- Generating wrappers, assertions, and DfT infrastructure intelligently [32, 49, 36]

This sets the stage for the deeper exploration in the next sections, where we investigate how LLMs can enhance both IP creation and SoC-level integration in secure, scalable design flows [56, 8].

3.2 LLM-Assisted IP Block Design

LLMs, originally developed for natural language processing, have demonstrated strong generalization to HDLs such as Verilog and VHDL [64, 63, 40]. With appropriate prompting and fine-tuning, these models can generate synthesizable RTL, detect logic vulnerabilities, and support secure design transformations [46, 8, 56]. This section explores how LLMs can accelerate, automate, and secure the IP design phase.

3.2.1 *RTL Generation from Natural Language*

Advances in LLMs have enabled synthesizable RTL code to be generated directly from natural language specifications [14, 40, 74, 19, 73]. This significantly reduces manual effort in control and datapath development [63, 25], allowing designers to specify functionality at a high abstraction level and receive complete Verilog modules [64, 52].

For example:

Design a 3-bit up/down counter with synchronous reset and enable.

An LLM can translate this into a correctly structured Verilog module, bridging human intent and formal HDL syntax.

To make this translation concrete, the brief specification above can be expanded into a fuller natural language prompt, from which the model first derives the module interface and intended behavior before emitting any code.

> ***Prompt to the LLM***
> *Design a 3-bit up/down counter in Verilog with a synchronous, active-high reset and an enable. Inputs: clk, rst, en, up_down. Output: a 3-bit count. When up_down is 1, the counter increments, otherwise it decrements; counting occurs only while en is asserted, and the value wraps around modulo 8.*

Intermediate Reasoning

1. Identify the interface: clk, rst, en, and up_down as inputs and a 3-bit count as a registered output.
2. Place the reset inside the clocked block so that it is sampled synchronously on the rising clock edge, clearing the count to 0.
3. Gate state updates with en, so the counter holds its value whenever en is deasserted.
4. Use up_down to select between increment and decrement; the 3-bit width gives a natural modulo-8 wraparound (7→0 and 0→7).
5. Emit the register updates with nonblocking assignments, the idiom for sequential logic.

Listing 3.1 Synthesizable 3-bit up/down counter generated from the prompt above

```
module up_down_counter (
    input  wire       clk,       // clock
    input  wire       rst,       // synchronous, active-high reset
    input  wire       en,        // count enable
    input  wire       up_down,   // 1 = count up, 0 = count down
    output reg  [2:0] count      // 3-bit counter output
);

always @(posedge clk) begin
    if (rst)
        count <= 3'b000;           // synchronous reset
    else if (en) begin
        if (up_down)
            count <= count + 3'd1;
```

```
            else
                count <= count - 3'd1;
        end
end

endmodule
```

The result (Listing 3.1) is a directly synthesizable module in which every element of the specification, the synchronous reset, the enable, the direction control, and the modulo-8 wraparound maps to an explicit hardware behavior.

Empirical studies illustrate the progress of this approach. Chip-Chat demonstrated the first AI-assisted tapeout of a microprocessor via conversational Verilog generation [8]. Thakur et al.'s VeriGen, fine-tuned on the largest curated Verilog dataset, improved both syntactic and functional correctness over general-purpose LLMs, while their AutoChip framework iteratively refined generated code using compiler feedback [63, 64]. Yang et al. showed GPT-4 could generate Verilog for systolic arrays and neural network accelerators, though performance tuning required manual effort [71]. Liu et al.'s RTLCoder, a lightweight quantized model, outperformed GPT-3.5 in Verilog tasks while running on commodity hardware [40]. Nadimi and Zheng's MEV-LLM assigned design tasks to specialized fine-tuned experts for a 23.9% correctness improvement [48], and Gao et al.'s AutoVCoder integrated retrieval-augmented generation (RAG) to reduce hallucinations and improve logic accuracy [25].

Collectively, these research efforts show that LLMs now support rapid prototyping and architectural exploration, making RTL development more efficient and accessible.

3.2.2 *Secure IP Design with LLMs*

Security in IP design is critical due to threats such as IP piracy, reverse engineering, and hardware Trojan insertion. LLMs such as GPT-3.5 and GPT-4 have demonstrated utility in logic generation [14, 32, 49], vulnerability detection [46, 49, 51], and secure design transformation [47, 32, 51].

Logic Obfuscation and Locking: LLMs can insert security primitives such as key-controlled multiplexers or scan-chain scrambling logic. For instance:

Add scan chain obfuscation to this design with 128-bit key-controlled scrambling.

produces wrapper logic that strengthens resistance to scan-based side-channel attacks [31, 36, 68].

A representative prompt and its generated wrapper make this concrete:

Prompt to the LLM

Add scan-chain obfuscation to this design with 128-bit key-controlled scrambling. Each scan cell should be XOR-masked with the corresponding key bit, the chain should shift the scrambled values during scan, and the serial output should expose only the masked form.

Intermediate Reasoning

1. Introduce a 128-bit key port and an internal scan-chain register of the same width.
2. XOR each chain bit with the matching key bit to produce a masked vector (a generate loop scales this to the full width).
3. On scan_en, shift the masked vector serially (concatenating the incoming scan_in); otherwise, capture functional data.
4. Drive scan_out from the most significant masked bit so the externally observable scan stream is scrambled.

Listing 3.2 Key-controlled scan-chain obfuscation wrapper

```
module secure_scan_wrapper #(
    parameter WIDTH = 128
) (
    input  wire              clk,
    input  wire              scan_en,   // scan shift enable
    input  wire              scan_in,   // serial scan input
    input  wire [WIDTH-1:0]  key,       // 128-bit obfuscation key
    input  wire [WIDTH-1:0]  data_in,   // functional data captured into chain
    output wire              scan_out   // serial scan output
);

    reg  [WIDTH-1:0] chain;             // internal scan-chain state
    wire [WIDTH-1:0] masked;

    genvar i;
    generate
        for (i = 0; i < WIDTH; i = i + 1) begin : mask_stage
            assign masked[i] = chain[i] ^ key[i];  // key-controlled scrambling
        end
    endgenerate

    always @(posedge clk) begin
        if (scan_en)
            chain <= {masked[WIDTH-2:0], scan_in};  // shift in scrambled form
        else
```

```
            chain <= data_in;                        // functional
                capture
    end

    assign scan_out = masked[WIDTH-1];               // scrambled
        serial output

endmodule
```

The wrapper in Listing 3.2 is intentionally illustrative, showing the structural idea of static key-XOR masking; production-grade schemes typically derive the mask dynamically from an LFSR or a stream cipher such as Trivium to resist removal and DOSCrack-style attacks Wu, Dizon-Paradis, et al. 2025.

Finite-State Machines (FSM) Security Enhancements LLMs can generate FSMs with decoy or deadlock states to detect stealthy logic or Trojan payloads. Using structured task decomposition and one-shot prompting, GPT-3.5 can introduce rare-condition deadlock states (e.g., `DEADLOCK_STATE`) without affecting normal functionality [62, 14]. Prompt engineering techniques such as least-to-most reasoning and self-refinement improve the reliability of these transformations [75, 43, 49].

Structural Diversity for SAT Resistance: To counter SAT-based deobfuscation and design cloning attacks, LLMs can vary state encoding schemes (e.g., Gray encoding), diversify transition paths, and enforce high Hamming distances between unprotected states [72, 60, 41, 11].

These two ideas, a deadlock trap and a high-Hamming-distance encoding, can be combined in a single-hardened access-control FSM (listing 3.3). The following prompt elicits such as design:

> ***Prompt to the LLM***
> *Add a rare-condition deadlock trap state to this FSM. Use a sparse state encoding with high Hamming distance between legal states, route every unused or illegal state code into the trap, and trap on any malformed input sequence so that recovery is possible only through reset.*

Intermediate Reasoning

1. Choose a sparse encoding for the legal states (IDLE, CHECK, GRANT_ACCESS, DEADLOCK_STATE) so that every pair differs in at least two bits.
2. Send any malformed sequence (illegal_seq) directly to DEADLOCK_STATE from the normal states.

3. Map all unused 3-bit codes (001, 010, 100, 111) to the trap via the case default, so a fault-injected illegal state cannot resume normal operation.
4. Make DEADLOCK_STATE self-looping; it can be exited only by an asynchronous reset.

Listing 3.3 Access-control FSM with a sparse encoding and a deadlock trap state

```
module secure_fsm (
    input  wire clk,
    input  wire rst_n,        // asynchronous, active-low reset
    input  wire req,          // access request
    input  wire valid_seq,    // correct unlock sequence observed
    input  wire illegal_seq,  // malformed / suspicious sequence observed
    output reg  grant         // access granted
);

    // Sparse encoding: all legal states are pairwise Hamming-distance 2.
    // Unused codes (001, 010, 100, 111) are illegal and trapped.
    localparam [2:0] IDLE           = 3'b000,
                     CHECK          = 3'b011,
                     GRANT_ACCESS   = 3'b101,
                     DEADLOCK_STATE = 3'b110;

    reg [2:0] state, next_state;

    always @(posedge clk or negedge rst_n) begin
        if (!rst_n)
            state <= IDLE;
        else
            state <= next_state;
    end

    always @(*) begin
        next_state = state;
        grant      = 1'b0;

        case (state)
            IDLE: begin
                if (illegal_seq)
                    next_state = DEADLOCK_STATE;
                else if (req)
                    next_state = CHECK;
            end

            CHECK: begin
                if (illegal_seq)
                    next_state = DEADLOCK_STATE;
                else if (valid_seq)
                    next_state = GRANT_ACCESS;
            end

```

```
            GRANT_ACCESS: begin
                grant      = 1'b1;
                next_state = IDLE;
            end

            DEADLOCK_STATE:
                next_state = DEADLOCK_STATE;   // trap: exit only
                    on reset

            default:
                next_state = DEADLOCK_STATE;   // any illegal code
                    -> trap
        endcase
    end

endmodule
```

Because the four legal codes are pairwise Hamming-distance 2 and every remaining 3-bit code falls through to the trap, a single bit-flip or an injected illegal state is driven into DEADLOCK_STATE rather than into GRANT_ACCESS, raising the bar for both SAT-based and fault-injection attacks, while leaving the normal IDLE→CHECK→GRANT path unchanged.

Vulnerability Detection: Through CWE-driven contextual prompting or custom security rules, LLMs can identify insecure structures, such as always-enabled test modes, unused flip-flops, or poorly controlled state transitions, that may act as Trojan triggers [56, 61, 22, 26].

Automated Validation: LLMs can generate targeted testbenches and SystemVerilog assertions (SVAs) from natural language security requirements, enabling automatic verification of transformations [51, 49]. For example, "password checking logic must validate all bits of a 32-bit key" can be translated into a syntactically correct SVA for integration into the verification flow.

Concretely, the requirement quoted above maps directly to a checkable property:

Prompt to the LLM
From the requirement "the password-checking logic must validate all bits of a 32-bit key," generate a SystemVerilog assertion that fires whenever a flagged comparison does not match on all 32 bits.

Intermediate Reasoning

1. Capture the trigger condition (key_valid) as the antecedent of an implication.
2. Require full 32-bit equality of entered_key and reference_key as the consequent.
3. Sample on the clock and disable the check during reset so that no spurious firing occurs at startup.

4. Attach an error action so a violation is reported by the simulator or formal tool.

Listing 3.4 SystemVerilog assertion checking a full 32-bit key match

```
module key_check_sva (
    input wire        clk,
    input wire        rst,
    input wire        key_valid,      // asserted when a key comparison occurs
    input wire [31:0] entered_key,
    input wire [31:0] reference_key
);

    // Whenever a comparison is flagged valid, ALL 32 bits must match.
    property full_32bit_match;
        @(posedge clk) disable iff (rst)
        key_valid |-> (entered_key == reference_key);
    endproperty

    assert_full_key_match: assert property (full_32bit_match)
        else $error("Key mismatch: entered_key != reference_key");

endmodule
```

The property in Listing 3.4 can be attached to the password-checking unit with a SystemVerilog bind statement and discharged either dynamically in simulation or exhaustively by a formal tool such as JasperGold, closing the loop between a security requirement stated in English and a machine-checkable guarantee.

3.3 SoC Design and Integration Using LLMs

While IP design establishes the foundational building blocks of an SoC, the integration phase is where architectural ambition meets engineering reality. At this stage, designers must navigate a dense matrix of interdependent challenges, including seamless interconnectivity, multimode timing closure, power domain isolation and retention strategies, clock domain crossing integrity, bus arbitration under varying quality of service (QoS) requirements, and the enforcement of robust security and trust anchors across heterogeneous, third-party, and potentially untrusted IPs [1, 33, 23, 7, 58]. This convergence of diverse subsystems amplifies the risk of emergent vulnerabilities, ranging from side-channel leakage to privilege escalation, that remain latent at the individual IP level but manifest when interacting at the system level. Therefore, initiating formalized and automated system-level design verification from the earliest stages of integration has shifted from being a best practice to an indispensable requirement. LLMs represent a disruptive augmentation of this process, enabling intelligent automation of interconnect generation, adaptive constraint synthesis, and context-aware vulnerability assessment. By embedding these capabilities directly into the SoC integration workflow, LLMs have the potential to accelerate secure, performance-optimized, and verification-driven platform-level design flows[32, 20].

3.3.1 Automated Platform-Level Design

LLMs can significantly enhance platform-level SoC design by automating the generation of interconnects and system glue logic. Leveraging learned knowledge of interface standards such as Advanced eXtensible Interface (AXI), Advanced High-performance Bus (AHB), Wishbone, and TileLink, LLMs can infer required modules and connections directly from natural language specifications or partial HDL descriptions. Many explorations are currently underway in this direction. For instance, Arteris FlexGen has demonstrated that AI-driven interconnect IP can automatically generate bus fabrics, crossbars, decoders, and arbiters based on IP counts and address maps, achieving substantial reductions in design iteration time, latency, and wire length compared to manual methods [5]. Similarly, in the emerging domain of LLM-aided design, LLMs have been shown to synthesize RTL modules from natural language specifications and iteratively refine generated HDL through automated feedback, demonstrating their potential to automate design components and system glue logic [64]. Benchmarking efforts such as ProtocolLLM [59] further evaluate the capability of LLMs to produce synthesizable SystemVerilog implementations for standard protocols like AXI, underscoring their potential in design scaffolding and top-level module integration. Looking ahead, such approaches could be extended to cover more complex platform integration tasks, including automated generation of power management controllers, QoS-aware arbitration policies, and security-enforcing interconnect fabrics, thereby enabling more secure and performance-optimized SoC platforms.

3.3.2 Platform-Level Security Analysis

Beyond functional integration, platform-level security verification is essential for ensuring that interactions between SoC components do not introduce systemic vulnerabilities. Unlike isolated IP-level threats, platform risks often arise from cross-component behaviors, for example, scan chains exposed via insecure test wrappers leaking sensitive control signals [18], shared power-delivery networks amplifying cross-domain side-channel leakage [76], or unprotected inter-IP communication enabling privilege escalation [54]. LLMs are emerging as powerful assistants in reasoning about such risks. Frameworks like *ThreatLens* demonstrate LLM-guided threat modeling and automated test plan generation for SoC designs, validated on platforms like NEORV32 [55], while *BugWhisperer* fine-tunes LLMs on SoC RTL vulnerability datasets to improve detection accuracy and adaptability [61]. Offensive-focused work such as *TrojanWhisper* shows that LLMs can localize hardware Trojans in obfuscated RTL without prior domain-specific training [22], and *SPICED* extends these capabilities to analog and mixed-signal circuits for syntax-level bug and Trojan identification [15]. LLM-driven frameworks like DIVAS [51] further highlight automated detection of weak isolation patterns—e.g., shared buses lacking access control, open scan chains, and missing firewall logic. These advancements illustrate that LLMs can not only infer attack surfaces but also simulate threat scenarios such as Trojan hardware insertion, SAT-based key extraction [60], and fault propagation, allowing more proactive and automated SoC security validation at the platform scale.

3.3.3 Design Optimization Trade-offs

Modern SoC design demands a careful balance between power, performance, area (PPA), and security, objectives that often conflict [17, 12]. LLMs can assist in navigating these trade-offs by suggesting alternative integration architectures, such as shared versus duplicated IPs, while reasoning about isolation boundaries and access control [8, 64]. They can estimate the impact of PPA and security of design choices such as security wrappers, isolation bridges, and secure clocking domains [56] and propose interconnect topologies that minimize logic depth while satisfying protection requirements. Recent domain-specific frameworks, such as ChipNeMo, demonstrate how the adaptation of LLM to chip design improves the precision of code synthesis and architectural reasoning [39], while VeriGen highlights the potential for using LLMs to generate optimized RTLs that align with both functional and security constraints [63]. Although current LLMs cannot directly perform static timing analysis (STA) or layout optimization, integrating them with established EDA flows and formal verification frameworks enables automated exploration of secure, high-PPA design solutions early in the SoC development cycle.

3.3.4 Case Study: LLM-Assisted Secure AES SoC Integration

Imagine an SoC incorporating an AES encryption core alongside a CPU, UART, and memory controller. In an LLM-assisted design flow, the model can automatically generate AXI4-Lite glue logic, interconnects, and address mappings conforming to SoC protocols and peripheral integration patterns. For instance, domain-specific benchmarks like SLDB evaluate LLMs for system-level SoC integration tasks, including modular SoC generation and communication patterns [3]. On the security front, frameworks such as DIVAS leverage LLMs to extract vulnerabilities from SoC specifications, generate corresponding assertions, and enforce policy-based protections across bus architectures [51]. Tools like LLM-IFT extend that reasoning into information-flow tracking, allowing hierarchical dependency analyses across modules and accurate detection of confidentiality or integrity breaches in SoCs, including AES integration paths [45]. These LLM-enabled approaches

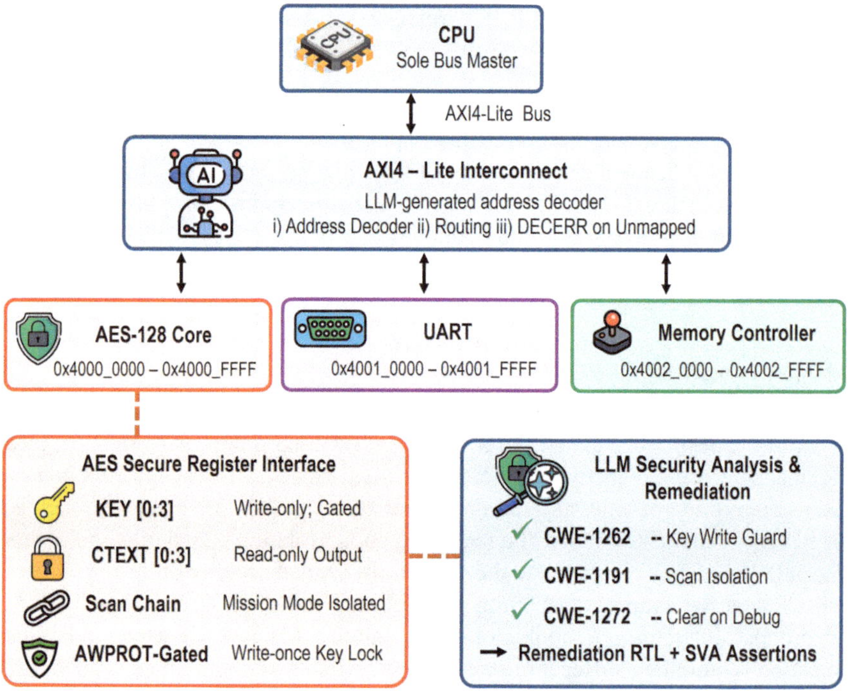

Fig. 3.1 LLM-assisted secure AES SoC integration showing an AXI4-Lite interconnect connecting a CPU, AES-128 core, UART, and memory controller with LLM-generated address decoding and glue logic. The flow also illustrates security analysis of the AES peripheral and the generation of remediation RTL and SystemVerilog assertions for identified vulnerabilities

can further propose scan-chain hardening, inter-IP access policies, or secure glue logic placement based on inferred threat models, providing a powerful, automated, and context-aware foundation for secure SoC platform construction. Figure 3.1 presents a representative case study demonstrating how LLMs can support both SoC integration and security analysis within a unified hardware design workflow.

3.3.4.1 System Composition and Glue-Logic Generation

Building on this scenario, consider the four-component SoC in more detail. The CPU is the sole bus master, while the AES-128 core, the UART, and the memory controller are memory-mapped slaves on a shared AXI4-Lite interconnect. The model is first asked to generate the address decode and glue logic that route the master's transactions to the correct slave, shown in Listing 3.5 (Table 3.1).

Table 3.1 Address map proposed for the four-component SoC

Peripheral	Base address	Size	Address range
AES-128 core	0x4000_0000	64 KB	0x4000_0000 – 0x4000_FFFF
UART	0x4001_0000	64 KB	0x4001_0000 – 0x4001_FFFF
Memory controller	0x4002_0000	64 KB	0x4002_0000 – 0x4002_FFFF
(Unmapped)	–	–	returns DECERR

Prompt 1—interconnect/address decoder

Generate AXI4-Lite glue logic for an SoC with one master and three slaves: an AES core at base 0x4000_0000, a UART at 0x4001_0000, and a memory controller at 0x4002_0000, each occupying a 64 KB region. The decoder must route read and write transactions to the correct slave based on the address, and return a DECERR response for any access that falls outside the three mapped regions.

1. Define 64 KB-aligned base addresses and a page mask that clears the low 16 bits, so an access is matched by its 64 KB page.
2. Decode the masked write address into a one-hot write-select and the masked read address into a one-hot read-select.
3. Raise a write- or read-DECERR flag whenever the masked address matches none of the three bases (the case default).
4. Keep the decode purely combinational so it adds no latency to the AXI handshake.

Listing 3.5 Combinational AXI4-Lite address decoder for the three slaves

```
module axil_decoder #(
    parameter ADDR_WIDTH = 32
) (
    input  wire [ADDR_WIDTH-1:0] awaddr,    // write address from master
    input  wire                  awvalid,
    input  wire [ADDR_WIDTH-1:0] araddr,    // read address from master
    input  wire                  arvalid,

    output reg  [2:0] wr_sel,     // one-hot write select {mem,uart,aes}
    output reg  [2:0] rd_sel,     // one-hot read  select {mem,uart,aes}
    output reg        wr_decerr,  // unmapped write -> DECERR
    output reg        rd_decerr   // unmapped read  -> DECERR
);

```

```
    // 64 KB-aligned bases; low 16 bits are the in-peripheral
       offset.
    localparam [ADDR_WIDTH-1:0] AES_BASE  = 32'h4000_0000;
    localparam [ADDR_WIDTH-1:0] UART_BASE = 32'h4001_0000;
    localparam [ADDR_WIDTH-1:0] MEM_BASE  = 32'h4002_0000;
    localparam [ADDR_WIDTH-1:0] PAGE_MASK = 32'hFFFF_0000;

    always @(*) begin
        wr_sel    = 3'b000;
        wr_decerr = 1'b0;
        if (awvalid) begin
            case (awaddr & PAGE_MASK)
                AES_BASE  : wr_sel = 3'b001;
                UART_BASE : wr_sel = 3'b010;
                MEM_BASE  : wr_sel = 3'b100;
                default   : wr_decerr = 1'b1;  // unmapped region
            endcase
        end
    end

    always @(*) begin
        rd_sel    = 3'b000;
        rd_decerr = 1'b0;
        if (arvalid) begin
            case (araddr & PAGE_MASK)
                AES_BASE  : rd_sel = 3'b001;
                UART_BASE : rd_sel = 3'b010;
                MEM_BASE  : rd_sel = 3'b100;
                default   : rd_decerr = 1'b1;  // unmapped region
            endcase
        end
    end

endmodule
```

3.3.4.2 AES Peripheral as an AXI4-Lite Slave

Next, the model wraps the AES datapath in a register-mapped AXI4-Lite slave (Listing 3.6). The register map is deliberately structured so that the key is write-only and the ciphertext is read-only, the first line of defense for the secret key.

> ***Prompt 2—AES AXI4-Lite slave***
> *Wrap a 128-bit AES core as an AXI4-Lite slave with a 32-bit data bus. Provide a control register (start, clear), a read-only status register (done, busy), a 128-bit key written as four write-only words, a 128-bit plaintext input, and a 128-bit read-only ciphertext output. Reads of the key region must return zero. Use a standard AXI4-Lite handshake.*

Table 3.2 AES peripheral register map (byte offsets within the AES region)

Offset	Register	Access	Description
0x00	CTRL	W	bit0 = start (self-clearing), bit1 = clear
0x04	STATUS	R	bit0 = done, bit1 = busy
0x10–0x1C	KEY[0:3]	W-only	128-bit key (reads return 0)
0x20–0x2C	PTEXT[0:3]	W	128-bit plaintext
0x30–0x3C	CTEXT[0:3]	R-only	128-bit ciphertext

1. Lay out the register per Table 3.2 and instantiate AES datapath (aes128_core), tying its reset to ARESETN.
2. On the write side, accept the AW and W beats together and respond on the B channel; map word writes to CTRL, the four KEY words, and the four PTEXT words; ignore WSTRB for these word-aligned registers.
3. Make CTRL bit0 a one-cycle self-clearing start pulse (the per-cycle default reassigns it to 0, and a write wins on that cycle).
4. On the read side, return STATUS and the four CTEXT words; leave KEY and PTEXT out of the read multiplexer so the default branch returns zero, and keys are never observable on the read bus.

Listing 3.6 AES-128 core wrapped as an AXI4-Lite slave with a write-only key and read-only ciphertext

```

module aes128_axil #(
    parameter ADDR_WIDTH = 8,
    parameter DATA_WIDTH = 32
) (
    input  wire                    ACLK,
    input  wire                    ARESETN,    // active-low reset

    // ---- Write address channel ----
    input  wire [ADDR_WIDTH-1:0]  AWADDR,
    input  wire                   AWVALID,
    output reg                    AWREADY,

    // ---- Write data channel ----
    input  wire [DATA_WIDTH-1:0]  WDATA,
    input  wire [DATA_WIDTH/8-1:0] WSTRB,
    input  wire                   WVALID,
    output reg                    WREADY,

    // ---- Write response channel ----
    output reg [1:0]              BRESP,
    output reg                    BVALID,
    input  wire                   BREADY,

    // ---- Read address channel ----
```

```
    input  wire [ADDR_WIDTH-1:0] ARADDR,
    input  wire                  ARVALID,
    output reg                   ARREADY,

    // ---- Read data channel ----
    output reg [DATA_WIDTH-1:0]  RDATA,
    output reg [1:0]             RRESP,
    output reg                   RVALID,
    input  wire                  RREADY
);

    // ---- Register offsets (word addresses) ----
    localparam CTRL_ADDR   = 8'h00; // [0] start (self-clearing), [1]
        clear
    localparam STATUS_ADDR = 8'h04; // [0] done, [1] busy (read-only)
    localparam KEY0_ADDR   = 8'h10; // KEY 0x10..0x1C (write-only)
    localparam PTEXT0_ADDR = 8'h20; // PTXT 0x20..0x2C
    localparam CTEXT0_ADDR = 8'h30; // CTXT 0x30..0x3C (read-only)

    // ---- Architectural registers ----
    reg  [1:0]   ctrl_reg;
    reg  [127:0] key_reg;
    reg  [127:0] ptext_reg;
    wire [127:0] ctext_reg;
    wire         core_done, core_busy;

    // ---- AES datapath (functional core abstracted away) ----
    aes128_core u_core (
        .clk        (ACLK),
        .rst_n      (ARESETN),
        .start      (ctrl_reg[0]),
        .clear      (ctrl_reg[1]),
        .key        (key_reg),
        .plaintext  (ptext_reg),
        .ciphertext (ctext_reg),
        .done       (core_done),
        .busy       (core_busy)
    );

    // ============================================================
    // Write channel
    // ============================================================
    wire write_fire = AWVALID && WVALID && ~AWREADY && ~WREADY;

    always @(posedge ACLK) begin
        if (!ARESETN) begin
            AWREADY <= 1'b0; WREADY <= 1'b0;
            BVALID  <= 1'b0; BRESP <= 2'b00;
            ctrl_reg <= 2'b00; key_reg <= 128'b0; ptext_reg <= 128'b0;
        end else begin
            AWREADY <= 1'b0;
            WREADY  <= 1'b0;
            ctrl_reg[0] <= 1'b0; // self-clearing start

```

```
            if (write_fire) begin
                AWREADY <= 1'b1;
                WREADY  <= 1'b1;
                BVALID  <= 1'b1;
                BRESP   <= 2'b00; // OKAY

                case (AWADDR)
                    CTRL_ADDR      : ctrl_reg <= WDATA[1:0];
                    KEY0_ADDR      : key_reg[31:0]   <= WDATA;
                    KEY0_ADDR+4    : key_reg[63:32]  <= WDATA;
                    KEY0_ADDR+8    : key_reg[95:64]  <= WDATA;
                    KEY0_ADDR+12   : key_reg[127:96] <= WDATA;

                    PTEXT0_ADDR      : ptext_reg[31:0]   <= WDATA;
                    PTEXT0_ADDR+4    : ptext_reg[63:32]  <= WDATA;
                    PTEXT0_ADDR+8    : ptext_reg[95:64]  <= WDATA;
                    PTEXT0_ADDR+12   : ptext_reg[127:96] <= WDATA;

                    default: ; // STATUS / CTEXT not writable
                endcase

            end else if (BVALID && BREADY) begin
                BVALID <= 1'b0;
            end
        end
    end

    // ==============================================================
    // Read channel
    // ==============================================================
    always @(posedge ACLK) begin
        if (!ARESETN) begin
            ARREADY <= 1'b0; RVALID <= 1'b0;
            RRESP   <= 2'b00; RDATA <= 32'b0;
        end else begin
            ARREADY <= 1'b0;

            if (ARVALID && ~ARREADY && ~RVALID) begin
                ARREADY <= 1'b1;
                RVALID  <= 1'b1;
                RRESP   <= 2'b00;

                case (ARADDR)
                    STATUS_ADDR    : RDATA <= {30'b0, core_busy,
                        core_done};
                    CTEXT0_ADDR    : RDATA <= ctext_reg[31:0];
                    CTEXT0_ADDR+4  : RDATA <= ctext_reg[63:32];
                    CTEXT0_ADDR+8  : RDATA <= ctext_reg[95:64];
                    CTEXT0_ADDR+12: RDATA <= ctext_reg[127:96];
                    default        : RDATA <= 32'b0; // key reads
                        return 0
                endcase

            end else if (RVALID && RREADY) begin
```

```
                    RVALID <= 1'b0;
                end
            end
        end

endmodule
```

3.3.4.3 LLM-Driven Security Analysis and Remediation

With the integration of RTL in hand, a DIVAS-style analysis pass (Paria et al. [51]) reasons over the address map and register policy, confirming the intended protections and then surfacing the platform-level weaknesses that emerge only once the AES core is wired into SoC. A representative analysis report is shown in Listing 3.7.

Listing 3.7 Representative LLM security analysis report for the AES integration

```
================================================================
LLM-Assisted SoC Security Analysis (AES-128 integration)
Sources: aes128_axil.v, axil_decoder.v     Bus: AXI4-Lite
================================================================
[INFO ] Parsing address map and register access policy ...
[INFO ] AES=0x40000000  UART=0x40010000  MEM=0x40020000

[PASS ] Read of KEY region (0x10--0x1C) returns 0x00000000
        -> key not exposed on the read data path
[PASS ] CTEXT region (0x30--0x3C) is read-only
        -> ciphertext cannot be overwritten by a master
[PASS ] Unmapped accesses return DECERR
        -> no silent address aliasing

----------------------------------------------------------------
FINDINGS
----------------------------------------------------------------
[HIGH] CWE-1262: Improper write protection of a security-
        critical register. KEY[0:3] (0x10--0x1C) is writable by
        ANY bus master; a non-secure master can overwrite the
        AES key during operation.
        -> gate key writes with a secure/access-control flag.

[HIGH] CWE-1191: Exposed debug/test access. The AES core
        shares a scan chain with the UART and scan_en is not
        forced inactive in mission mode, so key material is
        observable via scan shift after an encryption.
        -> isolate/disable AES scan in mission mode.

[MED ] CWE-1272: Sensitive state not cleared before debug or
        low-power. key_reg and ptext_reg are not zeroized on a
        mode transition.
        -> clear key/plaintext on debug/retention entry.

----------------------------------------------------------------
3 issue(s): 2 HIGH, 1 MEDIUM. Emitting remediation RTL + SVA.
================================================================
```

The findings are precisely the class of issues that are invisible at the isolated-IP level. The write-only key register (CWE-1262) is still writable by any master because the slave never checks the bus-security signal; the shared scan chain (CWE-1191) re-exposes the key during test; and sensitive state is not cleared on debug or low-power entry (CWE-1272). The model then emits remediation RTL in Listing 3.8, which gates key writes with the AXI protection signal and a post-boot key lock.

Listing 3.8 Remediation: gate key writes with a secure-access check and a write-once key lock

```

// ---- Added port (AXI protection signal) ----
input  wire [2:0] AWPROT;      // AxPROT[1] = 0 -> Secure access (TrustZone)

// ---- Added control: write-once key lock, set after secure boot ----
reg key_locked;

wire secure_axi = (AWPROT[1] == 1'b0);                   // secure master?
wire key_region = (AWADDR >= KEY0_ADDR) &&
                  (AWADDR <= KEY0_ADDR + 12);            // 0x10..0x1C

wire key_wr_ok  = key_region && secure_axi && ~key_locked;

always @(posedge ACLK) begin
  if (!ARESETN)
    key_locked <= 1'b0;
  else if (write_fire && (AWADDR == CTRL_ADDR) && WDATA[2])
    key_locked <= 1'b1;     // ctrl_reg[2] freezes the key post-boot
end

// ---- Guarded key writes (replace the unguarded KEY cases) ----
if (write_fire) begin
  // CTRL / PTEXT writes unchanged ...
  if (key_wr_ok) begin
    case (AWADDR)
      KEY0_ADDR      : key_reg[31:0]   <= WDATA;
      KEY0_ADDR+4    : key_reg[63:32]  <= WDATA;
      KEY0_ADDR+8    : key_reg[95:64]  <= WDATA;
      KEY0_ADDR+12   : key_reg[127:96] <= WDATA;
    endcase
  end
  // A non-secure or post-lock KEY write is dropped
  // (optionally answer SLVERR on the B channel instead).
end
```

Finally, the fix is paired with two formal properties so the guarantees are checkable rather than assumed: a write-protection property for the key register (Listing 3.9) and a confidentiality property for the read path (Listing 3.10):

Listing 3.9 Write-protection assertion for the AES key register (CWE-1262)

```
 property key_write_protected;
@(posedge ACLK) disable iff (!ARESETN)
(write_fire && key_region && (~secure_axi || key_locked))
  |=> (key_reg == $past(key_reg));
  endproperty
```

Listing 3.10 Confidentiality/read-path property

```

  // Formal property: the key must never appear on the read
      data bus.
  property key_confidentiality;
    @(posedge ACLK) disable iff (!ARESETN)
      (RVALID && (ARADDR >= KEY0_ADDR) && (ARADDR <= KEY0_ADDR
          + 12))
      |-> (RDATA == 32'b0);
  endproperty

  assert_key_conf: assert property (key_confidentiality)
    else $error("CWE-1262: KEY register observable on read bus"
        );
```

Taken together, the walkthrough shows the model acting across the integration flow: generating glue logic of the kind benchmarked by suites such as SLDB (Alvanaki et al. [3]), extracting weak isolation findings in the manner of DIVAS (Paria et al. [51]), and supporting the information-flow reasoning formalized by LLM-IFT (Mashnoor et al. [45]), with every machine-generated change gated behind human review and formal checks.

3.4 LLM-Augmented Codesign Frameworks

The role of LLMs in hardware design extends far beyond generating isolated RTL snippets or security wrappers. They can be embedded into holistic codesign ecosystems that interact with established EDA workflows, simulation engines, formal verifiers, synthesis flows, and security analysis tools to accelerate and enhance the design cycle [66, 42]. This section outlines the architectural components and operational paradigms of such frameworks.

3.4.1 Role of LLMs in the Design Cycle

In a conventional hardware development cycle, the process typically involves: (i) natural language specification; (ii) manual RTL entry; (iii) simulation and functional verification; (iv) synthesis and constraints definition; and (v) iterative redesign based on verification or physical feedback. By embedding LLMs into this cycle, many feedback paths can be automated or shortened. For instance, LLMs can translate high-level specifications into partial RTL [19, 73], generate self-checking testbenches for simulation [66], propose synthesis constraints (e.g., SDC clock groups, false paths), and emit formal assertions compatible with tools such as JasperGold or OneSpin [67]. This enables rapid exploration of design alternatives, early trade-off simulation, and tighter verification-design iteration cycles.

3.4.2 Multiagent LLM Workflows

Recent works focus more on multiagent LLM frameworks where each agent specializes in a distinct phase of the design flow: A *Spec Agent* decomposes functionality into modules; an *RTL Agent* generates synthesizable HDL; a *Verification Agent* generates SVAs, coverage models, and testbenches; a *Constraints Agent* produces synthesis/placement directives; and a *Security Agent* injects logic locking or Trojan detection assertions [73, 67, 51, 27]. As illustrated in Fig. 3.2, these agents can operate on a shared intermediate representation (e.g., a design graph) or communicate via structured prompts, effectively mimicking a collaborative engineering team with machine-level speed and precision.

3.4.3 Integration with Toolchains and EDA

LLMs themselves are generative reasoning engines, not hardware verifiers. However, tight coupling with existing EDA tools can form robust hybrid flows. For instance, an RTL generation agent can interface with Verilator or Synopsys VCS for immediate syntax and simulation checks; a verification agent can feed ModelSim or Synopsys VCS for functional coverage closure; a constraints agent can target Synopsys Design Compiler or PrimeTime for timing convergence; and a security agent can leverage JasperGold for formal property validation [30]. Wrapping these tools inside code-executing LLM agents enables semiautonomous loops in which the model generates code, triggers synthesis or simulation, parses reports, and refines its design under human oversight. Figure 3.3 illustrates this semiautonomous interaction between LLM-based design generation and conventional EDA toolchains.

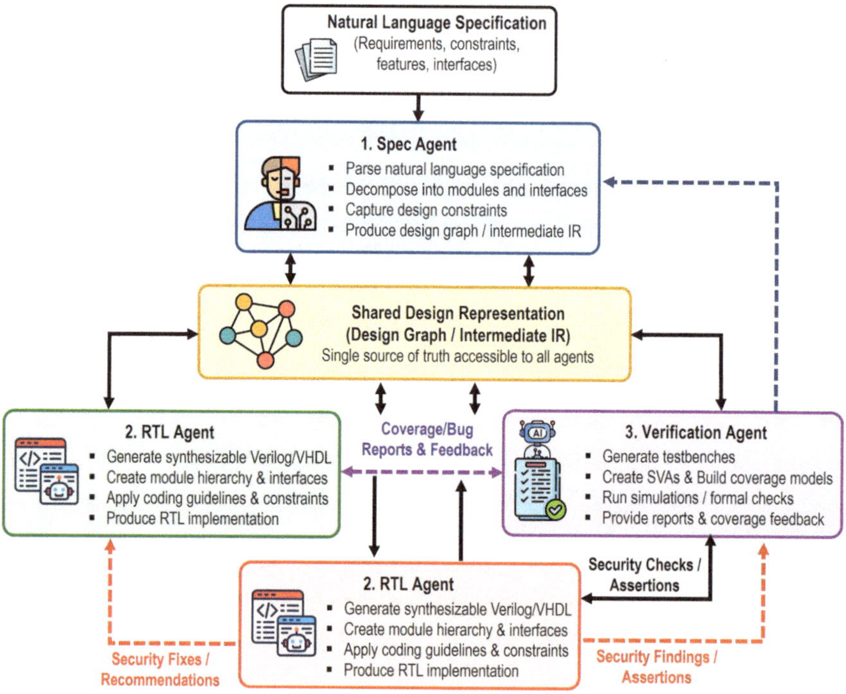

Fig. 3.2 Multiagent LLM architecture for hardware design automation, illustrating collaboration among specification, RTL generation, verification, and security agents through a shared design representation and iterative feedback loops. The framework enables coordinated design refinement, verification-driven debugging, and security-aware hardware development

Because an LLM cannot itself perform elaboration or static timing analysis, the tool report serves as the ground truth in these loops: The model's role is to parse the report, classify the problem, and apply the corresponding RTL transformation. Two worked examples make this concrete.

3.4.3.1 Worked Example: Resolving an Inferred Latch (Design Compiler)

A combinational decoder is synthesized, and design compiler reports an unintended latch, as shown in Listing 3.11:

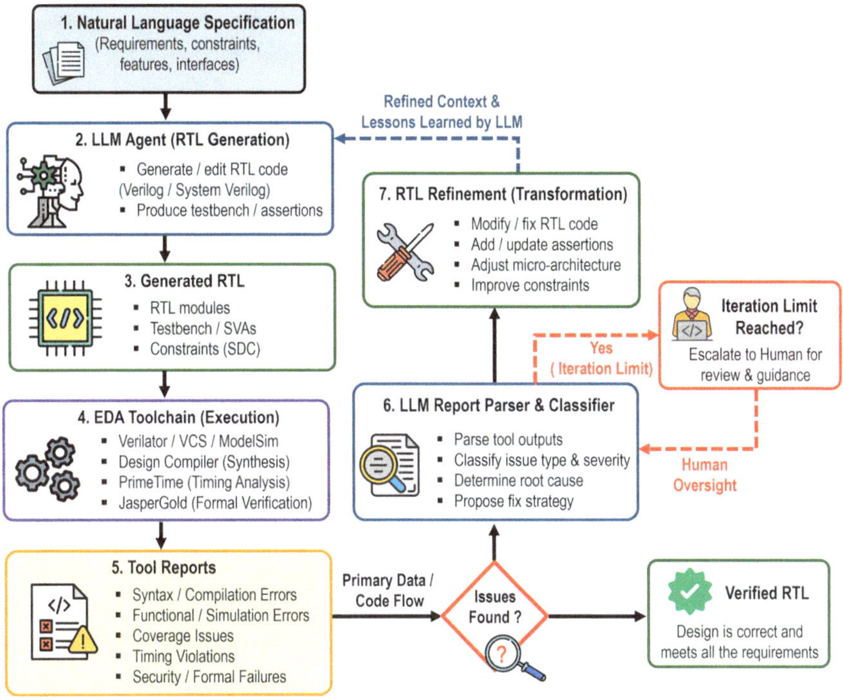

Fig. 3.3 LLM-EDA feedback loop for hardware design automation, illustrating iterative RTL generation, tool-driven validation, report analysis, and design refinement under human oversight. The workflow integrates simulation, synthesis, timing, and formal verification feedback to progressively improve design correctness and quality

Listing 3.11 Design compiler elaboration report flagging an inferred latch (message text and codes vary by tool version)

```

****************************************************
Report : elaborate
Design : alu_decode
Tool   : Design Compiler
****************************************************

Inferred memory devices in process
    in routine alu_decode line 14 in file
        './rtl/alu_decode.v'
===============================================================
|  Register Name   | Type   | Width | Bus | AR | AS | SR | SS |
===============================================================
|  decoded_op_reg  | Latch  | 4     | Y   | N  | N  | N  | N  |
===============================================================

Warning: ./rtl/alu_decode.v:14: Variable 'decoded_op' is not
assigned in all branches of the case statement; a latch
is inferred to hold its value.  (ELAB-368)
```

From the Report, the Model Extracts Three Facts

1. A storage element was inferred where none was intended, and the report names it a Latch.
2. It is on the signal `decoded_op` (register `decoded_op_reg`, width 4).
3. The cause is an incomplete case statement: `decoded_op` is not assigned in all branches, so the tool adds a latch to hold the old value.

Mapping "incomplete case → inferred latch" to the standard remedy, the model assigns a default value before the case and adds an explicit default branch so that every path is covered; Listings 3.12 and 3.13 show the code before and after this fix.

Listing 3.12 Before incomplete case (latch inferred)

```
// BEFORE: incomplete case -> a latch is inferred on decoded_op
always @(*) begin
  case (opcode)
    3'b000: decoded_op = 4'b0001;   // ADD
    3'b001: decoded_op = 4'b0010;   // SUB
    3'b010: decoded_op = 4'b0100;   // AND
    // remaining opcodes leave decoded_op unassigned -> latch
  endcase
end
```

Listing 3.13 After complete assignment, purely combinational

```
// AFTER: leading default + explicit default branch ->
    combinational
always @(*) begin
  decoded_op = 4'b0000;             // default: no value is held
  case (opcode)
    3'b000: decoded_op = 4'b0001; // ADD
    3'b001: decoded_op = 4'b0010; // SUB
    3'b010: decoded_op = 4'b0100; // AND
    default: decoded_op = 4'b0000; // all other opcodes
  endcase
end
```

3.4.3.2 Worked Example: Closing a Setup-Timing Violation (Primetime)

A multiply-accumulate unit passes functional simulation but fails timing; PrimeTime reports a negative setup slack, as shown in Listing 3.14:

Listing 3.14 PrimeTime setup-timing report with a violated path

```

****************************************************
Report : timing  -path full  -delay max
Design : mac_unit
Tool   : PrimeTime
****************************************************

Startpoint: mult_reg_a[7] (rising edge flip-flop clocked by clk)
Endpoint:   acc_reg[15]  (rising edge flip-flop clocked by clk)
Path Group: clk
Path Type : max (setup)          Clock clk: 500 MHz / 2.00 ns

Point                                          Incr       Path
---------------------------------------------------------------
clock clk (rise edge)                          0.00       0.00
mult_reg_a[7]/CK (DFF)                         0.00       0.00
mult_reg_a[7]/Q  (DFF)                         0.18       0.18
multiply-add combinational logic               2.51       2.69
acc_reg[15]/D    (DFF)                         0.00       2.69
data arrival time                                         2.69
---------------------------------------------------------------
clock clk (rise edge)                          2.00       2.00
clock uncertainty                             -0.10       1.90
acc_reg[15] library setup time                -0.12       1.78
data required time                                       1.78
---------------------------------------------------------------
data required time                                       1.78
data arrival time                                       -2.69
---------------------------------------------------------------
slack (VIOLATED)                                       -0.91
```

From the Report, the Model Extracts

1. The failing path runs from mult_reg_a[7] (startpoint) to acc_reg[15] (endpoint) in clock group clk.
2. The path type is setup (max), and the slack is negative: −0.91 ns, a true violation, not a hold issue.
3. The dominant cost is 2.51 ns of multiply-add combinational logic in a single cycle, so the root cause is excessive logic depth rather than a clocking error.

Recognizing a logic-depth problem on a multiply-add rather than a clocking error, the model pipelines the operation, registering the product in one stage and accumulating in the next; this shortens the per-stage path at the cost of one additional cycle of latency, as shown in Listings 3.15 (single cycle) and 3.16 (pipelined).

Listing 3.15 Single-cycle multiply-accumulate (long path)

```
// BEFORE: multiply and accumulate in one cycle -> long path (2.51 ns)
always @(posedge clk) begin
    if (rst)
        acc <= 16'b0;
    else
        acc <= acc + (a * b);    // multiply + add in a single cycle
end
```

Listing 3.16 Pipelined multiply-accumulate (+1 cycle latency)

```
// AFTER: pipeline the multiply -> shorter path/stage (+1 cycle latency)
reg [15:0] prod;
always @(posedge clk) begin
    if (rst) begin
        prod <= 16'b0;
        acc  <= 16'b0;
    end else begin
        prod <= a * b;          // stage 1: register the product
        acc  <= acc + prod;     // stage 2: accumulate registered product
    end
end
```

3.4.3.3 The Generalized Refinement Loop

Both examples instantiate the same tool-in-the-loop procedure:

1. Generate or edit the RTL from the specification (or from the previous iteration's feedback).
2. Invoke the EDA tool, elaboration/synthesis in Design Compiler, or timing analysis in PrimeTime.
3. Parse the report, extracting the structured facts (warning or violation type, named signals or path endpoints, and the relevant metric such as slack).
4. Classify the issue (e.g., inferred latch, setup violation, width mismatch).
5. Map the class to a known RTL transformation and apply the edit.
6. Repeat until the report is clean or the iteration budget is reached, at which point the case is escalated to a human engineer.

This loop is effective precisely because it is hybrid. As noted among the technical limitations in Sect. 3.5.1, current models do not natively reason about timing closure or critical-path optimization; pairing the model with deterministic EDA tools supplies the ground truth it lacks, so that each iteration is anchored in a real synthesis or timing result rather than in the model's own estimate, which is also why human oversight remains an essential part of the loop.

3.4.4 Use Cases in Academia and Industry

Illustrative academic prototypes include *ChatEDA*, which allows natural language queries for IP generation, constraint application, and synthesis invocation [66]; *AutoEDA*, a microservice-based multiagent framework for automating EDA flows from RTL to GDSII [42]; and *ORAssistant*, a RAG-powered conversational assistant for the OpenROAD physical design flow [30]. In industry, leading EDA vendors have begun integrating LLM-based assistants into their design suites, while startups are pursuing vertically integrated, AI-first hardware codesign stacks that aim to unify specification capture, RTL generation, verification, and security analysis in a single, LLM-coordinated workflow.

3.4.5 Community-Driven LLM Hardware Competitions

The academic and industrial communities have initiated educational competitions that blend artificial intelligence with hardware engineering to foster innovation and lower the barrier to entry in semiconductor design. The semiconductor industry currently faces a critical workforce crisis driven by the return of advanced chip manufacturing to the United States, with projections suggesting that 67,000 new manufacturing and design jobs risk being unfilled by 2030. This shortage poses a direct threat to global economic competitiveness, creating an urgent motivation to lower barriers to entry and diversify the talent pool. In direct response to this need, we established the AI Hardware Design League (AI-HDL), which invites around 600 students from community colleges, undergraduates, and graduates from universities worldwide to design hardware using LLMs and modern EDA tools. The initiative aims to build a sustainable talent pipeline for the semiconductor industry by removing traditional barriers, allowing individuals with little to no prior hardware experience to participate. Participants collaborate in teams and leverage AI tools to generate, debug, and optimize Verilog code. Our work is supported by similar community-driven efforts, such as the ICCD 2025 LLMs for Hardware Design Challenge for iterative PPA enhancements and the ICLAD-DAC GenAI Hackathon for evaluating lightweight versus massive language models. Furthermore, competitions like the CSAW AI Hardware Attack Challenge and the NDSS 2026 LAST-X Competition address the security frontier by challenging participants to use generative AI to embed or uncover complex system vulnerabilities. Collectively, these collaborative platforms represent a paradigm shift that standardizes chip design and ensures a robust, secure, and highly skilled semiconductor workforce for the coming decade.

3.4.6 Algorithmic Workflows of Representative Frameworks

The frameworks surveyed in the preceding sections share a small set of control structures. The pseudocode below distills each to its essential loop; the listings are conceptual and faithful to the published approach rather than verbatim implementations of any particular system.

Algorithm 3.1 Chip-Chat: conversational RTL generation with a human in the loop (Blocklove et al. [8])

```
Input : natural-language design specification  spec
Output: verified RTL module  M
-------------------------------------------------------------

history <- [ system_prompt , spec ]
repeat
    response <- LLM.chat(history)       // propose / revise RTL
    M          <- extract_RTL(response)
    (ok, feedback) <- engineer_review(M) // human in the loop
    if not ok then
        history <- history + [response, feedback]
until ok
integrate_into_flow(M)                  // lint, simulate, tape-out
return M
```

Chip-Chat keeps a running conversation: The engineer reviews each candidate, and the model revises against natural language feedback until the design is accepted and pushed into the conventional flow (Blocklove et al. [8]).

Algorithm 3.2 VeriGen: corpus curation followed by supervised fine-tuning (Thakur et al. [63])

```
Input : raw HDL corpus  C_raw , base model  theta_0
Output: fine-tuned Verilog model  theta
-------------------------------------------------------------

// Phase 1 - curate the corpus
C <- collect(C_raw)                   // GitHub + textbooks
C <- deduplicate(C)
```

(continued)

Algorithm 3.2 (continued)

```
C <- filter(C, keep = compiles(x)) // drop non-compiling samples

// Phase 2 - supervised fine-tuning
theta <- theta_0
for each minibatch B in C do
    L <- CrossEntropy(theta, B)
    theta <- theta - eta * grad(L)  // gradient step
end for

// Phase 3 - deployment
M <- theta.generate(spec)
return theta, M
```

VeriGen improves correctness up front by filtering the training corpus to compiling samples and fine-tuning a base model on it, so that the deployed model carries stronger Verilog priors before any prompting begins Thakur et al. [63].

Algorithm 3.3 AutoChip: closed-loop refinement on compiler and simulator feedback (Thakur et al. [64])

```
Input : spec , testbench TB , iteration budget N
Output: RTL module  M  (or human-review flag)
------------------------------------------------------------

prompt <- build_prompt(spec)
M      <- LLM.generate(prompt)
for i = 1 to N do
    (compiled, c_err) <- compile(M)
    if not compiled then
        feedback <- parse_compiler_errors(c_err)
    else
        (passed, s_err) <- simulate(M, TB)
        if passed then return M            // success
        else feedback <- parse_sim_mismatches(s_err)
    prompt <- prompt + M + feedback        // append context
    M      <- LLM.generate(prompt)         // regenerate
end for
flag_for_human_review(M)
return M
```

AutoChip is the canonical tool-in-the-loop pattern: It compiles and simulates each candidate, parses the resulting errors or output mismatches, appends them to the prompt, and regenerates, escalating to a human only if the iteration budget is exhausted (Thakur et al. [64]).

Algorithm 3.4 AutoVCoder: Retrieval-augmented generation with a code-aware second stage (Gao et al. [25])

```
Input : spec S , knowledge base KB , retrieval depth k
Output: refined RTL module  M
------------------------------------------------------------

q        <- embed(S)                    // query embedding
examples <- retrieve_top_k(KB, q, k) // RAG: similar RTL / snippets
prompt   <- compose(S, examples)
M        <- LLM.generate(prompt)
// second-stage, code-aware retrieval to repair / augment
ctx      <- retrieve_code_aware(KB, M)
M        <- domain_refine(M, ctx)    // domain fine-tuned pass
return M
```

AutoVCoder grounds generation in retrieved examples: It embeds the specification, retrieves similar RTL from a knowledge base to compose the prompt, and then performs a second, code-aware retrieval and a domain-specific refinement pass to curb hallucinations (Gao et al. [25]).

Algorithm 3.5 MEV-LLM: routing to a specialized expert model (Nadimi and Zheng [48])

```
Input : spec S , fine-tuned experts  E = { e_1 .. e_n }
Output: RTL module  M
------------------------------------------------------------

class  <- classify(S)           // FSM / arithmetic / memory / protocol
expert <- select(E, class)      // route to the matching expert model
M      <- expert.generate(S)
return M

// Routing to domain experts yields ~23.9% higher functional
// correctness than a single general-purpose model.
```

MEV-LLM classifies the incoming task and routes it to a matching fine-tuned expert; this mixture-of-experts routing has been reported to improve functional correctness by about 23.9% over a single general-purpose model (Nadimi and Zheng [48]).

3.5 Limitations, Challenges, and Risks

While LLMs have demonstrated considerable promise in specification analysis, RTL generation, verification assistance, and SoC integration planning, their effectiveness decreases as the design flow moves from behavioral descriptions toward implementation-driven stages such as synthesis, timing closure, physical design, and signoff. These later stages depend on quantitative optimization, technology-specific constraints, and detailed implementation feedback that current LLMs cannot model directly. Consequently, although LLMs can accelerate many front-end design activities, they remain dependent on conventional EDA tools and human expertise for implementation and deployment in production-grade hardware design flows [66, 28, 42].

3.5.1 Limited Understanding of Physical Design Constraints

One of the most significant limitations of current LLM-based design systems is their inability to natively reason about physical implementation constraints. While an LLM may generate functionally correct RTL, it generally lacks direct awareness of placement congestion, routing resources, clock-tree synthesis requirements, signal integrity issues, power-delivery constraints, process variations, and manufacturing limitations. Consequently, generated designs that appear correct at the RTL level may encounter substantial difficulties during synthesis, placement, routing, or signoff stages. This limitation becomes increasingly pronounced in large SoCs where physical effects strongly influence overall design quality, manufacturability, and feasibility [38, 42, 69].

3.5.2 Timing Closure and Performance Optimization Challenges

Modern hardware design requires extensive analysis of timing paths, setup and hold constraints, clock-domain crossings, and critical-path behavior. Current LLMs do not possess an intrinsic understanding of static timing analysis (STA) and cannot independently evaluate timing reports or accurately predict implementation-level timing violations. As a result, designs generated by LLMs may satisfy functional requirements while failing timing closure, requiring traditional synthesis, STA, and optimization tools to identify and resolve performance bottlenecks. This limitation becomes increasingly important in advanced technology nodes where timing margins are small and implementation constraints strongly influence design success [42, 28].

3.5.3 Power, Performance, and Area (PPA) Trade-offs

A fundamental objective of hardware design is the simultaneous optimization of power consumption, performance, and silicon area. Although LLMs can propose architectural alternatives, coding optimizations, and design modifications, they cannot reliably estimate the resulting PPA characteristics without feedback from synthesis, power analysis, and implementation tools. Consequently, LLM-generated solutions may appear attractive at the specification or RTL level but ultimately fail to meet design objectives once evaluated through conventional EDA workflows. Effective PPA optimization therefore continues to require iterative analysis using established design automation tools and implementation feedback.

3.5.4 Dependence on External EDA Feedback

Successful hardware design relies on continuous feedback from simulators, synthesizers, formal verification engines, timing analyzers, power estimators, and physical-design tools. Unlike conventional EDA algorithms that directly optimize against these metrics, LLMs generally operate without access to implementation results unless integrated into closed-loop workflows. Consequently, their effectiveness depends heavily on external tool feedback and iterative refinement. Current evidence suggests that LLMs are most reliable when used in conjunction with established EDA tools, serving as assistants for design generation, debugging, and workflow automation rather than functioning as standalone design systems [66, 28, 53, 65].

3.5.5 Challenges in Large-Scale and Heterogeneous SoC Design

Although LLMs have shown encouraging results for bounded RTL modules and localized design tasks, end-to-end generation of complex SoCs remains largely impractical. Modern SoCs contain numerous heterogeneous subsystems, including processors, accelerators, memory hierarchies, interconnects, security modules, and mixed-signal components that interact across multiple abstraction levels. Coordinating architectural intent, interface compatibility, performance objectives, verification requirements, and implementation constraints across such systems remains beyond the capabilities of current LLM-based approaches without substantial human guidance and tool-assisted validation. As SoC complexity continues to increase, scalable integration and system-level optimization remain important challenges for future LLM-assisted design frameworks [71, 13, 9].

References

1. Hasan Al-Shaikh, Arash Vafaei, Mridha Md Mashahedur Rahman, Kimia Zamiri Azar, Fahim Rahman, Farimah Farahmandi, and Mark Tehranipoor. Sharpen: SoC security verification by hardware penetration test. In *Proceedings of the 28th Asia and South Pacific Design Automation Conference*, ASPDAC '23, page 579–584, New York, NY, USA, 2023. Association for Computing Machinery.
2. Shadan Alsaqer, Sarah Alajmi, Imtiaz Ahmad, and Mohammad Alfailakawi. The potential of LLMs in hardware design. *Journal of Engineering Research*, 2024.
3. Elisavet Lydia Alvanaki, Kevin Lee, and Luca P. Carloni. SLDB: An end-to-end heterogeneous system-on-chip benchmark suite for LLM-aided design, 2025.
4. Alon Amid, David Biancolin, Abraham Gonzalez, Daniel Grubb, Sagar Karandikar, Harrison Liew, Albert Magyar, Howard Mao, Albert Ou, Nathan Pemberton, Paul Rigge, Colin Schmidt, John Wright, Jerry Zhao, Yakun Sophia Shao, Krste Asanović, and Borivoje Nikolić. Chipyard: Integrated design, simulation, and implementation framework for custom SoCs. *IEEE Micro*, 40(4):10–21, 2020.
5. Arteris. AI-enabled tool boosts chip interconnect design productivity. *EE News Europe*, 2025.
6. S. Bhunia, S. Ray, and Susmita Sur Kolay. *Fundamentals of IP and SoC security: Design, verification, and debug*. 01 2017.
7. Swarup Bhunia, Michael S. Hsiao, Mainak Banga, and Seetharam Narasimhan. Hardware Trojan attacks: Threat analysis and countermeasures. *Proceedings of the IEEE*, 102(8):1229–1247, 2014.
8. Jason Blocklove, Siddharth Garg, Ramesh Karri, and Hammond Pearce. Chip-chat: Challenges and opportunities in conversational hardware design. In *2023 ACM/IEEE 5th Workshop on Machine Learning for CAD (MLCAD)*, pages 1–6, 2023.
9. Jason Blocklove, Siddharth Garg, Ramesh Karri, and Hammond Pearce. Evaluating LLMs for hardware design and test. In *2024 IEEE LLM Aided Design Workshop (LAD)*, pages 1–6, 2024.
10. Rima Boumaza and Fateh Boutekkouk. Intellectual properties integration: The past, the present, and the future. *International Journal of Technology Diffusion*, 12:32–45, 04 2021.
11. Rajat Subhra Chakraborty and Swarup Bhunia. Harpoon: An obfuscation-based SoC design methodology for hardware protection. *IEEE Transactions on Computer-Aided Design of Integrated Circuits and Systems*, 28(10):1493–1502, 2009.
12. Veena S Chakravarthi. *A practical approach to VLSI system on chip (SoC) design*. Springer, 2020.
13. Kaiyan Chang, Zhirong Chen, Yunhao Zhou, Wenlong Zhu, Kun Wang, Haobo Xu, Cangyuan Li, Mengdi Wang, Shengwen Liang, Huawei Li, Yinhe Han, and Ying Wang. Natural language is not enough: Benchmarking multi-modal generative AI for verilog generation. In *Proceedings of the 43rd IEEE/ACM International Conference on Computer-Aided Design*, ICCAD '24, New York, NY, USA, 2025. Association for Computing Machinery.
14. Kaiyan Chang, Ying Wang, Haimeng Ren, Mengdi Wang, Shengwen Liang, Yinhe Han, Huawei Li, and Xiaowei Li. ChipGpt: How far are we from natural language hardware design. *CoRR*, abs/2305.14019, 2023.
15. Jayeeta Chaudhuri, Dhruv Thapar, Arjun Chaudhuri, Farshad Firouzi, and Krishnendu Chakrabarty. SPICED: Syntactical bug and Trojan pattern identification in A/MS circuits using LLM-enhanced detection, 08 2024.
16. Wen Chen, Sandip Ray, Jayanta Bhadra, Magdy Abadir, and Li-C Wang. Challenges and trends in modern SoC design verification. *IEEE Design and Test*, 34(5):7–22, 2017.
17. Marcian Cirstea, Khaled Benkrid, Andrei Dinu, Romeo Ghiriti, and Dorin Petreus. Digital electronic system-on-chip design: Methodologies, tools, evolution, and trends. *Micromachines*, 15(2), 2024.
18. Amitabh Das, Ünal Kocabaş, Ahmad-Reza Sadeghi, and Ingrid Verbauwhede. PUF-based secure test wrapper design for cryptographic SoC testing. In *2012 Design, Automation and Test in Europe Conference and Exhibition (DATE)*, pages 866–869, 2012.

19. Chenhui Deng, Yun-Da Tsai, Guan-Ting Liu, Zhongzhi Yu, and Haoxing Ren. ScaleRTL: Scaling LLMs with reasoning data and test-time compute for accurate RTL code generation, 2025.
20. Gelei Deng, Yi Liu, Víctor Mayoral-Vilches, Peng Liu, Yuekang Li, Yuan Xu, Tianwei Zhang, Yang Liu, Martin Pinzger, and Stefan Rass. PentestGPT: evaluating and harnessing large language models for automated penetration testing. In *Proceedings of the 33rd USENIX Conference on Security Symposium*, SEC '24, USA, 2024. USENIX Association.
21. Ridha Djemal, Mohamed Ayoub Dhouib, Samuel Dellacherie, and Rached Tourki. A novel formal verification approach for RTL hardware IP cores. *Computer Standards and Interfaces*, 27(6):637–651, 2005.
22. Md Omar Faruque, Peter Jamieson, Ahmad Patooghy, and Abdel-Hameed A. Badawy. Trojanwhisper: Evaluating pre-trained LLMs to detect and localize hardware trojans, 2024.
23. Nusrat Farzana, Fahim Rahman, Mark Tehranipoor, and Farimah Farahmandi. SoC security verification using property checking. In *2019 IEEE International Test Conference (ITC)*, pages 1–10, 2019.
24. Amlan Ganguly, M. Meraj Ahmed, Rounak Singh Narde, Abhishek Vashist, Md Shahriar Shamim, Naseef Mansoor, Tanmay Shinde, Suryanarayanan Subramaniam, Sagar Saxena, Jayanti Venkataraman, and Mark Indovina. The advances, challenges and future possibilities of millimeter-wave chip-to-chip interconnections for multi-chip systems. *Journal of Low Power Electronics and Applications*, 8(1), 2018.
25. Mingzhe Gao, Jieru Zhao, Zhe Lin, Wenchao Ding, Xiaofeng Hou, Yu Feng, Chao Li, and Minyi Guo. AutoVCoder: A Systematic Framework for Automated Verilog Code Generation using LLMs . In *2024 IEEE 42nd International Conference on Computer Design (ICCD)*, pages 162–169, Los Alamitos, CA, USA, November 2024. IEEE Computer Society.
26. Sujan Ghimire, Yu-Zheng Lin, Muntasir Mamun, Muhtasim Alam Chowdhury, Farhad Alemi, Shuyu Cai, Jinduo Guo, Mingyu Zhu, Honghui Li, Banafsheh Saber Latibari, Setareh Rafatirad, Pratik Satam, and Soheil Salehi. HWREx: AI-enabled hardware weakness and risk exploration and storytelling framework with LLM-assisted mitigation suggestion. *ACM Trans. Des. Autom. Electron. Syst.*, May 2025. Just Accepted.
27. Aaron Grattafiori, Abhimanyu Dubey, Abhinav Jauhri, Abhinav Pandey, Abhishek Kadian, Ahmad Al-Dahle, Aiesha Letman, Akhil Mathur, Alan Schelten, Alex Vaughan, Amy Yang, Angela Fan, Anirudh Goyal, Anthony Hartshorn, Aobo Yang, Archi Mitra, Archie Sravankumar, Artem Korenev, Arthur Hinsvark, Arun Rao, Aston Zhang, Aurelien Rodriguez, Austen Gregerson, Ava Spataru, Baptiste Roziere, Bethany Biron, Binh Tang, Bobbie Chern, Charlotte Caucheteux, Chaya Nayak, Chloe Bi, Chris Marra, Chris McConnell, Christian Keller, Christophe Touret, Chunyang Wu, Corinne Wong, Cristian Canton Ferrer, Cyrus Nikolaidis, Damien Allonsius, Daniel Song, Danielle Pintz, Danny Livshits, Danny Wyatt, David Esiobu, Dhruv Choudhary, Dhruv Mahajan, Diego Garcia-Olano, Diego Perino, Dieuwke Hupkes, Egor Lakomkin, Ehab AlBadawy, Elina Lobanova, Emily Dinan, Eric Michael Smith, Filip Radenovic, Francisco Guzmán, Frank Zhang, Gabriel Synnaeve, Gabrielle Lee, Georgia Lewis Anderson, Govind Thattai, Graeme Nail, Gregoire Mialon, Guan Pang, Guillem Cucurell, Hailey Nguyen, Hannah Korevaar, Hu Xu, Hugo Touvron, Iliyan Zarov, Imanol Arrieta Ibarra, Isabel Kloumann, Ishan Misra, Ivan Evtimov, Jack Zhang, Jade Copet, Jaewon Lee, Jan Geffert, Jana Vranes, Jason Park, Jay Mahadeokar, Jeet Shah, Jelmer van der Linde, Jennifer Billock, Jenny Hong, Jenya Lee, Jeremy Fu, Jianfeng Chi, Jianyu Huang, Jiawen Liu, Jie Wang, Jiecao Yu, Joanna Bitton, Joe Spisak, Jongsoo Park, Joseph Rocca, Joshua Johnstun, Joshua Saxe, Junteng Jia, Kalyan Vasuden Alwala, Karthik Prasad, Kartikeya Upasani, Kate Plawiak, Ke Li, Kenneth Heafield, Kevin Stone, Khalid El-Arini, Krithika Iyer, Kshitiz Malik, Kuenley Chiu, Kunal Bhalla, Kushal Lakhotia, Lauren Rantala-Yeary, Laurens van der Maaten, Lawrence Chen, Liang Tan, Liz Jenkins, Louis Martin, Lovish Madaan, Lubo Malo, Lukas Blecher, Lukas Landzaat, Luke de Oliveira, Madeline Muzzi, Mahesh Pasupuleti, Mannat Singh, Manohar Paluri, Marcin Kardas, Maria Tsimpoukelli, Mathew Oldham, Mathieu Rita, Maya Pavlova, Melanie Kambadur, Mike Lewis, Min Si, Mitesh Kumar Singh, Mona Hassan, Naman Goyal, Narjes Torabi, Nikolay Bashlykov, Nikolay Bogoychev, Niladri Chatterji, Ning

Zhang, Olivier Duchenne, Onur Çelebi, Patrick Alrassy, Pengchuan Zhang, Pengwei Li, Petar Vasic, Peter Weng, Prajjwal Bhargava, Pratik Dubal, Praveen Krishnan, Punit Singh Koura, Puxin Xu, Qing He, Qingxiao Dong, Ragavan Srinivasan, Raj Ganapathy, Ramon Calderer, Ricardo Silveira Cabral, Robert Stojnic, Roberta Raileanu, Rohan Maheswari, Rohit Girdhar, Rohit Patel, Romain Sauvestre, Ronnie Polidoro, Roshan Sumbaly, Ross Taylor, Ruan Silva, Rui Hou, Rui Wang, Saghar Hosseini, Sahana Chennabasappa, Sanjay Singh, Sean Bell, Seohyun Sonia Kim, Sergey Edunov, Shaoliang Nie, Sharan Narang, Sharath Raparthy, Sheng Shen, Shengye Wan, Shruti Bhosale, Shun Zhang, Simon Vandenhende, Soumya Batra, Spencer Whitman, Sten Sootla, Stephane Collot, Suchin Gururangan, Sydney Borodinsky, Tamar Herman, Tara Fowler, Tarek Sheasha, Thomas Georgiou, Thomas Scialom, Tobias Speckbacher, Todor Mihaylov, Tong Xiao, Ujjwal Karn, Vedanuj Goswami, Vibhor Gupta, Vignesh Ramanathan, Viktor Kerkez, Vincent Gonguet, Virginie Do, Vish Vogeti, Vítor Albiero, Vladan Petrovic, Weiwei Chu, Wenhan Xiong, Wenyin Fu, Whitney Meers, Xavier Martinet, Xiaodong Wang, Xiaofang Wang, Xiaoqing Ellen Tan, Xide Xia, Xinfeng Xie, Xuchao Jia, Xuewei Wang, Yaelle Goldschlag, Yashesh Gaur, Yasmine Babaei, Yi Wen, Yiwen Song, Yuchen Zhang, Yue Li, Yuning Mao, Zacharie Delpierre Coudert, Zheng Yan, Zhengxing Chen, Zoe Papakipos, Aaditya Singh, Aayushi Srivastava, Abha Jain, Adam Kelsey, Adam Shajnfeld, Adithya Gangidi, Adolfo Victoria, Ahuva Goldstand, Ajay Menon, Ajay Sharma, Alex Boesenberg, Alexei Baevski, Allie Feinstein, Amanda Kallet, Amit Sangani, Amos Teo, Anam Yunus, Andrei Lupu, Andres Alvarado, Andrew Caples, Andrew Gu, Andrew Ho, Andrew Poulton, Andrew Ryan, Ankit Ramchandani, Annie Dong, Annie Franco, Anuj Goyal, Aparajita Saraf, Arkabandhu Chowdhury, Ashley Gabriel, Ashwin Bharambe, Assaf Eisenman, Azadeh Yazdan, Beau James, Ben Maurer, Benjamin Leonhardi, Bernie Huang, Beth Loyd, Beto De Paola, Bhargavi Paranjape, Bing Liu, Bo Wu, Boyu Ni, Braden Hancock, Bram Wasti, Brandon Spence, Brani Stojkovic, Brian Gamido, Britt Montalvo, Carl Parker, Carly Burton, Catalina Mejia, Ce Liu, Changhan Wang, Changkyu Kim, Chao Zhou, Chester Hu, Ching-Hsiang Chu, Chris Cai, Chris Tindal, Christoph Feichtenhofer, Cynthia Gao, Damon Civin, Dana Beaty, Daniel Kreymer, Daniel Li, David Adkins, David Xu, Davide Testuggine, Delia David, Devi Parikh, Diana Liskovich, Didem Foss, Dingkang Wang, Duc Le, Dustin Holland, Edward Dowling, Eissa Jamil, Elaine Montgomery, Eleonora Presani, Emily Hahn, Emily Wood, Eric-Tuan Le, Erik Brinkman, Esteban Arcaute, Evan Dunbar, Evan Smothers, Fei Sun, Felix Kreuk, Feng Tian, Filippos Kokkinos, Firat Ozgenel, Francesco Caggioni, Frank Kanayet, Frank Seide, Gabriela Medina Florez, Gabriella Sch. The Llama 3 herd of models, 2024.

28. Zesong Jiang, Qing Zhang, Cheng Liu, Huawei Li, and Xiaowei Li. IICPilot: An intelligent integrated circuit backend design framework using Open EDA, 2024.
29. Leilei Jin, Jiajie Xu, Wenjie Fu, Hao Yan, and Longxing Shi. A crosstalk-aware timing prediction method in routing, 2025.
30. Aviral Kaintura, Palaniappan R, Shui Song Luar, and Indira Iyer Almeida. Orassistant: A custom rag-based conversational assistant for openroad, 2024.
31. Hadi Mardani Kamali, Kimia Zamiri Azar, Houman Homayoun, and Avesta Sasan. On designing secure and robust scan chain for protecting obfuscated logic, 2020.
32. Rahul Kande, Hammond Pearce, Benjamin Tan, Brendan Dolan-Gavitt, Shailja Thakur, Ramesh Karri, and Jeyavijayan Rajendran. (Security) Assertions by Large Language Models. *arXiv e-prints*, page arXiv:2306.14027, June 2023.
33. Elmira Karimi, Zhen Jiang, Yunsi Fei, and David Kaeli. A timing side-channel attack on a mobile GPU. pages 67–74, 10 2018.
34. Michael Keating and Pierre Bricaud. *Reuse Methodology Manual for System-on-a-Chip Designs*. Springer Publishing Company, Incorporated, 3rd edition, 2007.
35. Chris Kwok, Priya Viswanathan, and Ping Yeung. Addressing the challenges of reset verification in SoC designs. In *Design and Verification Conference (DVCon)*, 2015.

36. Banafsheh Saber Latibari, Sujan Ghimire, Muhtasim Alam Chowdhury, Najmeh Nazari, Kevin Immanuel Gubbi, Houman Homayoun, Avesta Sasan, and Soheil Salehi. Automated hardware logic obfuscation framework using GPT. In *2024 IEEE 17th Dallas Circuits and Systems Conference (DCAS)*, pages 1–5, 2024.
37. Shu-Min Li, Yih-Huai Cherng, and Yao-Wen Chang. Noise-aware buffer planning for interconnect-driven floorplanning. In *Proceedings of the ASP-DAC Asia and South Pacific Design Automation Conference, 2003.*, pages 423–426, 2003.
38. Bingyang Liu, Haoyi Zhang, Xiaohan Gao, Zichen Kong, Xiyuan Tang, Yibo Lin, Runsheng Wang, and Ru Huang. Layoutcopilot: An LLM-powered multiagent collaborative framework for interactive analog layout design. *IEEE Transactions on Computer-Aided Design of Integrated Circuits and Systems*, 44(8):3126–3139, 2025.
39. Mingjie Liu, Teodor-Dumitru Ene, Robert Kirby, Chris Cheng, Nathaniel Pinckney, Rongjian Liang, Jonah Alben, Himyanshu Anand, Sanmitra Banerjee, Ismet Bayraktaroglu, Bonita Bhaskaran, Bryan Catanzaro, Arjun Chaudhuri, Sharon Clay, Bill Dally, Laura Dang, Parikshit Deshpande, Siddhanth Dhodhi, Sameer Halepete, Eric Hill, Jiashang Hu, Sumit Jain, Ankit Jindal, Brucek Khailany, George Kokai, Kishor Kunal, Xiaowei Li, Charley Lind, Hao Liu, Stuart Oberman, Sujeet Omar, Ghasem Pasandi, Sreedhar Pratty, Jonathan Raiman, Ambar Sarkar, Zhengjiang Shao, Hanfei Sun, Pratik P Suthar, Varun Tej, Walker Turner, Kaizhe Xu, and Haoxing Ren. Chipnemo: Domain-adapted LLMs for chip design, 2024.
40. Shang Liu, Wenji Fang, Yao Lu, Qijun Zhang, Hongce Zhang, and Zhiyao Xie. RTL-Coder: Outperforming GPT-3.5 in design RTL generation with our open-source dataset and lightweight solution. In *2024 IEEE LLM Aided Design Workshop (LAD)*, pages 1–5, 06 2024.
41. Yuntao Liu, Michael Zuzak, Yang Xie, Abhishek Chakraborty, and Ankur Srivastava. Strong anti-SAT: Secure and effective logic locking. In *2020 21st International Symposium on Quality Electronic Design (ISQED)*, pages 199–205, 2020.
42. Yiyi Lu, Hoi Ian Au, Junyao Zhang, Jingyu Pan, Yiting Wang, Ang Li, Jianyi Zhang, and Yiran Chen. AutoEDA: Enabling EDA flow automation through microservice-based LLM agents, 2025.
43. Aman Madaan, Niket Tandon, Prakhar Gupta, Skyler Hallinan, Luyu Gao, Sarah Wiegreffe, Uri Alon, Nouha Dziri, Shrimai Prabhumoye, Yiming Yang, Shashank Gupta, Bodhisattwa Prasad Majumder, Katherine Hermann, Sean Welleck, Amir Yazdanbakhsh, and Peter Clark. Self-refine: Iterative refinement with self-feedback. In A. Oh, T. Naumann, A. Globerson, K. Saenko, M. Hardt, and S. Levine, editors, *Advances in Neural Information Processing Systems*, volume 36, pages 46534–46594. Curran Associates, Inc., 2023.
44. Sumit K Mandal, Anish Krishnakumar, and Umit Y Ogras. Energy-efficient networks-on-chip architectures: Design and run-time optimization. *Network-on-Chip Security and Privacy*, pages 55–75, 2021.
45. Nowfel Mashnoor, Mohammad Akyash, Hadi Kamali, and Kimia Azar. LLM-IFT: Llm-powered information flow tracking for secure hardware, 2025.
46. Xingyu Meng, Amisha Srivastava, Ayush Arunachalam, Avik Ray, Pedro Henrique Silva, Rafail Psiakis, Yiorgos Makris, and Kanad Basu. Unlocking hardware security assurance: The potential of LLMS, 2023.
47. Xingyu Meng, Amisha Srivastava, Ayush Arunachalam, Avik Ray, Pedro Henrique Silva, Rafail Psiakis, Yiorgos Makris, and Kanad Basu. NSPG: Natural language processing-based security property generator for hardware security assurance. In *Proceedings of the 61st ACM/IEEE Design Automation Conference*, DAC '24, New York, NY, USA, 2024. Association for Computing Machinery.
48. Bardia Nadimi and Hao Zheng. A multi-expert large language model architecture for verilog code generation. *2024 IEEE LLM Aided Design Workshop (LAD)*, pages 1–5, 2024.
49. Marcelo Orenes-Vera, Margaret Martonosi, and David Wentzlaff. Using LLMS to facilitate formal verification of RTL, 2023.
50. Sudipta Paria and Swarup Bhunia. DiSPEL: Distributed security policy enforcement for bus-based SoC. *arXiv preprint arXiv:2308.02792*, 2023.

51. Sudipta Paria, Aritra Dasgupta, and Swarup Bhunia. DIVAS: An LLM-based end-to-end framework for SoC security analysis and policy-based protection, 2023.
52. Zehua Pei, Hui-Ling Zhen, Mingxuan Yuan, Yu Huang, and Bei Yu. BetterV: Controlled verilog generation with discriminative guidance, 2024.
53. Ruidi Qiu, Grace Li Zhang, Rolf Drechsler, Ulf Schlichtmann, and Bing Li. AutoBench: Automatic testbench generation and evaluation using LLMS for HDL design. In *Proceedings of the 2024 ACM/IEEE International Symposium on Machine Learning for CAD*, New York, NY, USA, 2024. Association for Computing Machinery.
54. Sandip Ray, Eric Peeters, Mark M. Tehranipoor, and Swarup Bhunia. System-on-chip platform security assurance: Architecture and validation. *Proceedings of the IEEE*, 106(1):21–37, 2018.
55. Dipayan Saha, Hasan Al Shaikh, Shams Tarek, and Farimah Farahmandi. ThreatLens: LLM-guided threat modeling and test plan generation for hardware security verification, 2025.
56. Dipayan Saha, Shams Tarek, Katayoon Yahyaei, Sujan Kumar Saha, Jingbo Zhou, Mark Tehranipoor, and Farimah Farahmandi. LLM for SoC security: A paradigm shift. *IEEE Access*, 12:155498–155521, 2024.
57. Khaled Salah and Mohamed AbdelSalam. IP cores design from specifications to production: Modeling, verification, optimization, and protection. In *2013 25th International Conference on Microelectronics (ICM)*, pages 1–4, 2013.
58. Hassan Salmani, Mark Tehranipoor, Sarwono Sutikno, and Fajar Wijitrisnanto. Trust-hub Trojan benchmark for hardware Trojan detection model creation using machine learning, 2022.
59. Arnav Sheth, Ivaxi Sheth, and Mario Fritz. ProtocolLLM: RTL benchmark for systemverilog generation of communication protocols. *arXiv preprint arXiv:2506.07945*, 2025.
60. Pramod Subramanyan, Sayak Ray, and Sharad Malik. Evaluating the security of logic encryption algorithms. In *2015 IEEE International Symposium on Hardware Oriented Security and Trust (HOST)*, pages 137–143, 2015.
61. Shams Tarek, Dipayan Saha, Sujan Saha, and Farimah Farahmandi. BugWhisperer: Fine-tuning LLMs for SoC hardware vulnerability detection, 05 2025.
62. Mohammad Tehranipoor and Farinaz Koushanfar. A survey of hardware Trojan taxonomy and detection. *IEEE Design and Test of Computers*, 27(1):10–25, 2010.
63. Shailja Thakur, Baleegh Ahmad, Hammond Pearce, Benjamin Tan, Brendan Dolan-Gavitt, Ramesh Karri, and Siddharth Garg. VeriGen: A large language model for verilog code generation. *ACM Trans. Des. Autom. Electron. Syst.*, 29(3), April 2024.
64. Shailja Thakur, Jason Blocklove, Hammond Pearce, Benjamin Tan, Siddharth Garg, and Ramesh Karri. Autochip: Automating HDL generation using LLM feedback. *CoRR*, abs/2311.04887, 2023.
65. Kiran Thorat, Jiahui Zhao, Yaotian Liu, Hongwu Peng, Xi Xie, Bin Lei, Jeff Zhang, and Caiwen Ding. Advanced language model-driven verilog development: Enhancing power, performance, and area optimization in code synthesis. *arXiv preprint arXiv:2312.01022*, 2023.
66. Haoyuan Wu, Zhuolun He, Xinyun Zhang, Xufeng Yao, Su Zheng, Haisheng Zheng, and Bei Yu. ChatEDA: A large language model powered autonomous agent for EDA. *IEEE Transactions on Computer-Aided Design of Integrated Circuits and Systems*, 43(10):3184–3197, October 2024.
67. Haoyuan Wu, Haisheng Zheng, Zhuolun He, and Bei Yu. Divergent thoughts toward one goal: LLM-based multi-agent collaboration system for electronic design automation, 2025.
68. Jiaming Wu, Olivia Dizon-Paradis, Sazadur Rahman, Damon L. Woodard, and Domenic Forte. Protecting dynamically obfuscated scan chain architecture from doscrack with Trivium pseudo-random number generation. *Cryptography*, 9(1), 2025.
69. Zhe Xiao, Xu He, Haoying Wu, Bei Yu, and Yang Guo. EDA-copilot: A rag-powered intelligent assistant for EDA tools. *ACM Trans. Des. Autom. Electron. Syst.*, January 2025. Just Accepted.
70. Geng Yang, Li Xie, Matti Mäntysalo, Xiaolin Zhou, Zhibo Pang, Li Da Xu, Sharon Kao-Walter, Qiang Chen, and Li-Rong Zheng. A health-IoT platform based on the integration of intelligent packaging, unobtrusive bio-sensor, and intelligent medicine box. *IEEE Transactions on Industrial Informatics*, 10(4):2180–2191, 2014.

71. Kaiyuan Yang, Haotian Liu, Yuqin Zhao, and Tiantai Deng. A new design approach of hardware implementation through natural language entry. *IET Collaborative Intelligent Manufacturing*, 5(4):e12087, 2023.
72. Muhammad Yasin, Abhrajit Sengupta, Mohammed Thari Nabeel, Mohammed Ashraf, Jeyavijayan (JV) Rajendran, and Ozgur Sinanoglu. Provably-secure logic locking: From theory to practice. In *Proceedings of the 2017 ACM SIGSAC Conference on Computer and Communications Security*, CCS '17, page 1601–1618, New York, NY, USA, 2017. Association for Computing Machinery.
73. Zhongzhi Yu, Mingjie Liu, Michael Zimmer, Yingyan Celine Lin, Yong Liu, and Haoxing Ren. Spec2rtl-agent: Automated hardware code generation from complex specifications using LLM agent systems, 2025.
74. Yang Zhao, Di Huang, Chongxiao Li, Pengwei Jin, Muxin Song, Yinan Xu, Ziyuan Nan, Mingju Gao, Tianyun Ma, Lei Qi, Yansong Pan, Zhenxing Zhang, Rui Zhang, Xishan Zhang, Zidong Du, Qi Guo, and Xing Hu. CodeV: Empowering LLMs with HDL generation through multi-level summarization, 2025.
75. Denny Zhou, Nathanael Schärli, Le Hou, Jason Wei, Nathan Scales, Xuezhi Wang, Dale Schuurmans, Claire Cui, Olivier Bousquet, Quoc Le, and Ed Chi. Least-to-most prompting enables complex reasoning in large language models, 2023.
76. Huifeng Zhu, Xiaolong Guo, Yier Jin, and Xuan Zhang. PowerScout: A security-oriented power delivery network modeling framework for cross-domain side-channel analysis. In *2020 Asian Hardware Oriented Security and Trust Symposium (AsianHOST)*, pages 1–6, 2020.

Chapter 4
Functional Verification Using Large Language Models

Emma Andrews and Prabhat Mishra

Functional verification is widely acknowledged as a major bottleneck in the system-on-chip (SoC) design methodology. SoC designs as well as intellectual property (IP) cores must be verified to ensure their functional correctness. Functional verification today utilizes a combination of automated methods and manual intervention of verification engineers with deep insights of the design under verification. Due to the increasing complexity and reduced time-to-market constraints, there is a critical need to automate functional verification methods. Large language models (LLMs) are promising to alleviate the need for the manual intervention by assisting in a wide variety of functional verification tasks. This chapter explores effective utilization of LLMs in different functional verification scenarios, including formal verification, assertion-based validation, fuzzing, and penetration testing.

4.1 Functional Verification in Modern SoC Design

System on chip (SoC) integrates diverse intellectual property (IP) designs in a single integrated circuit. For example, a typical SoC consists of multiple IP cores, including processors, memories, controllers, converters, etc. Functional verification is a critical step in ensuring that SoC and IP designs operate correctly based on its design specifications. Existing validation methods utilize simulation-based validation [25, 26, 31, 48] as well as formal verification [13, 18, 19, 49] to verify functional behaviors as well as nonfunctional requirements [12, 20, 21, 30, 33–35]. While SoC verification utilizes automated validation methods, it still relies on manual intervention of design experts at various steps [14]. Due to increasing

E. Andrews · P. Mishra (✉)
University of Florida, Gainesville, FL, USA
e-mail: prabhat@ufl.edu

F. Farahmandi et al. (eds.), *LLM for SoC Design and Security*,
https://doi.org/10.1007/978-3-032-34116-7_4

SoC complexity coupled with reduced time-to-market constraints, there is a critical need to avoid manual intervention using fully automated verification flow. Large language models (LLMs) provide a promising avenue to replace manual effort by trained LLM agents [16, 43, 44]. For example, LLMs can generate tests and understand design specifications, acting as a valuable assistant in functional verification workflows [15, 42]. Moreover, when the LLM processes data to generate items for a verification task, it can analyze the data for more complex relationships and use that understanding in its generation for better results than even a human could do.

In this chapter, we survey the use of LLMs in diverse functional verification activities, including assertion generation [28, 46], test generation [10, 42], fuzz testing [41], penetration testing [4], model checking [50], and theorem proving [39]. The remainder of this chapter is organized as follows. Section 4.2 presents simulation-based validation using LLM-guided generation of tests and assertions. Section 4.3 examines verification of unexpected behaviors using LLM-guided fuzzing and penetration testing. Section 4.4 describes formal verification using LLM-guided model checking and theorem proving. Finally, Sect. 4.5 concludes the chapter.

4.2 Simulation-Based Validation Using LLMs

This section describes simulation-based validation using LLM-guided generation of tests and assertions. We first discuss assertion generation using LLMs. Next, we present automated test generation using LLMs.

4.2.1 Assertion Generation Using LLMs

Assertions play an important role in verification of IPs and SoCs, ensuring that a specified condition is always true at a given point in the underlying code [48]. There are two popular forms of assertions, SystemVerilog assertions (SVAs) and property specification language (PSL). SVAs are used in register-transfer level (RTL) designs to check temporal logic in the IP and SoC designs [48]. As the languages of SVAs are well-defined, the LLMs are able to draw on knowledge to craft syntactically valid SVAs. An example SVA is `assert(x == 1)`, which says that the value located in signal `x` must be equal to 1. There are many different frameworks using LLMs to effectively generate SVAs, each with their own pros and cons, as well as specific application usages. The frameworks can be divided based on their general strategy, such as prompt engineering, model fine-tuning, and retrieval-augmented generation.

4.2.1.1 Prompt Engineering

ChIRAAG [28] is a prompt-engineering-based framework with feedback for generating SVAs automatically. The LLM is initially given the relevant RTL design and description to guide the LLM to produce coherent assertions. Without design specifications, the LLM cannot produce any complex assertions that are tailor-made for the design in question; instead, it can only produce basic assertions that can be relevant at a generalized level to any design. Once the LLM produces a set of SVAs, they are inserted into the RTL design as appropriate and simulated with a set of testbenches. From the simulated results, there can be four possible scenarios:

1. *All passing tests*, meaning that all assertions are valid and thus comprise the total assertion suite
2. *Failed assertions*, implying that the generated assertions contain an issue, such as syntactically incorrect or irrelevant to the RTL design under test
3. *Incorrect output*, indicating a potential bug in the design that needs to be examined
4. *Time-out*, where the LLM was not able to produce sufficient assertions in the given time frame. The prompt is adjusted, and the process repeats.

Of these scenarios, the failed assertions and time-out result in further interactions with the LLM. In the case of failed assertions, the results from the simulations of the RTL design with the generated assertions are given to the LLM as feedback. This feedback is vital for having the LLM understand what to improve and to expand its knowledge on how the RTL design works to produce more coherent assertions. For time-out cases, the prompt given to the LLM is adjusted so that it can learn at a faster rate and not be stuck in a specific chain of thought. An overview of the entire framework is shown in Fig. 4.1.

The prompts given to the LLM as part of the automated prompt engineering framework affect the quality of the desired output. In the case of SVAs, constructing the prompts with different information can lead to the generation of SVAs that are of higher quality for specific application usages. For example, the RTL design from which the assertions are being generated for may include comments by developers. These comments are related to assertions that are not written into the design yet. [22] can automatically extract these assertion-related comments from the RTL design and construct a prompt to the LLM based on the comment information. The LLM can thus accurately generate an SVA based on feedback from the developer, as well as place it in the appropriate location within the RTL design for best effectiveness.

4.2.1.2 Model Fine-Tuning

Beyond prompt engineering, fine-tuning can be utilized to improve the model's knowledge on a specific domain. In case of assertion generation, this can take the form of knowledge of the designs and SVAs. AssertionLLM fine-tunes an LLM for the SVA generation task to craft SVAs that perform better than those from LLMs

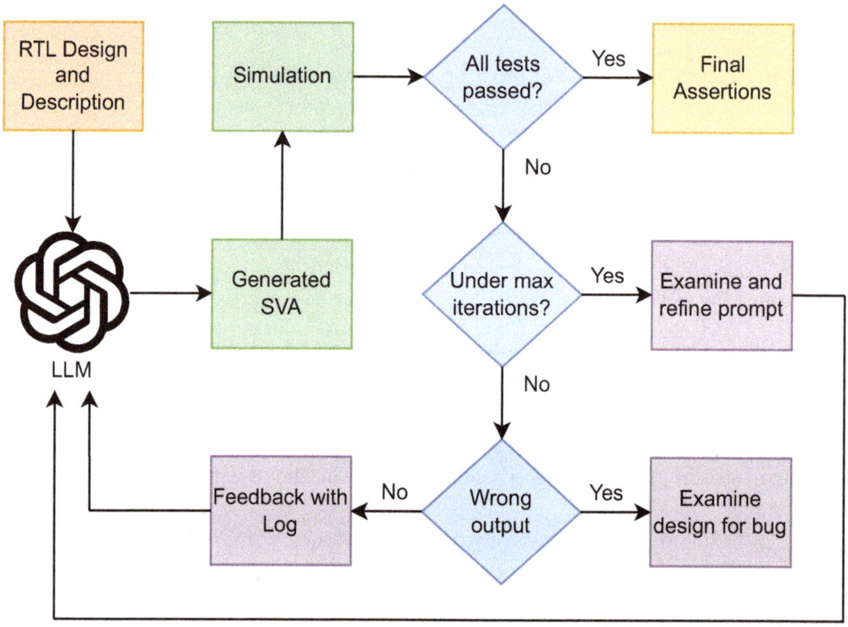

Fig. 4.1 Overview of the ChIRAAG framework [28]

without any fine-tuning [37, 38]. The training process takes a commercial LLM, such as LLaMA or GPT, and trains it specifically on assertion data. In this case, the training data consisted of pairs of Verilog designs from OpenCores [2] along with their SVAs and verified with Cadence JasperGold [1].

In addition to the fine-tuned model, the authors also present an evaluation framework, AssertionBench, for evaluating the assertions generated by an LLM [37, 38]. This assertion evaluation framework replaces having to manually check if the assertions generated by the LLM are of sufficient quality with an automated process. In addition, it provides a standardized way of determining if assertion generation techniques using LLMs are improving. AssertionBench uses the training data pairs of the Verilog designs and their associated verified SVAs as the training data. The evaluation process is shown in Fig. 4.2 that consists of four major steps:

1. Use of 1-shot and 5-shot *in-context learning (ICL)* to give the LLM detailed knowledge of examples prior to generating SVAs. The prepared Verilog designs and verified assertions are used as the training data.
2. *Generation of SVAs* using the LLM after ICL. Unseen RTL designs are given to the LLM to generate assertions.
3. *Syntax correction* using an LLM to correct syntax errors in the SVAs generated by the LLM.

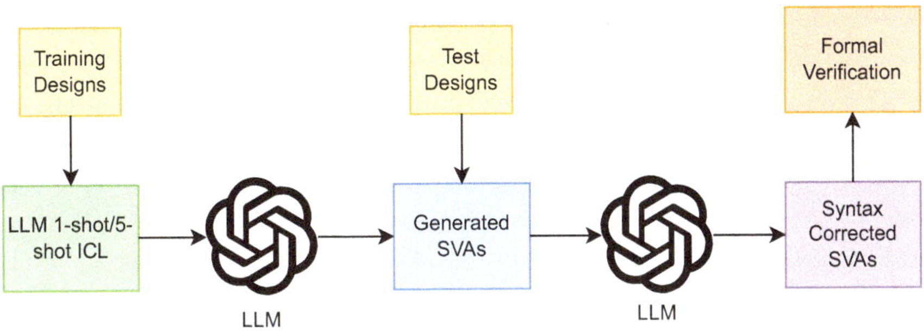

Fig. 4.2 Overview of AssertionBench to evaluate LLMs, including AssertionLLM and their generated SVAs [37, 38]

4. *Evaluation* of generated SVAs using tools such as JasperGold. If the assertions are verified using JasperGold, then the generated assertions pass the evaluation; else, they need to be regenerated as they are not sufficient.

4.2.1.3 Retrieval-Augmented Generation

The most useful information to generate assertions comes from the RTL design specification. The specification presents all the necessary properties and relationships that the RTL design must obey. Implementing SVAs in accordance with the design specification would be beneficial in verifying the design and finding potential bugs and issues. The design specifications thus lead to a retrieval-augmented generation (RAG) approach, where the LLM retrieves external information from the design specification.

AssertLLM utilizes RAG alongside other techniques to process complex design specifications and utilize their information when the LLM is generating new SVAs for an RTL design [11]. The RAG portion of the framework retrieves information that is contained in the design specification or other preprocessed information about the design specification. Multiple LLM agents are used to carry out a specific task within the overall framework for generation of the SVAs. The AssertLLM framework is shown in Fig. 4.3, and the different tasks of each LLM agent are as follows:

1. *SPEC analyzer* takes the design specification, consisting of the function descriptions, architecture definitions, and diagrams, and analyzes it. Specifically, the analysis will examine each signal in the specification for any information.
2. *Signal mapper* is the LLM that examines the design specification and the signal definitions to map the signal definitions to their intended functionality in the design specification. For example, if the master clock signal is defined as "clk," the LLM can map that this signal is the master clock throughout the RTL design.

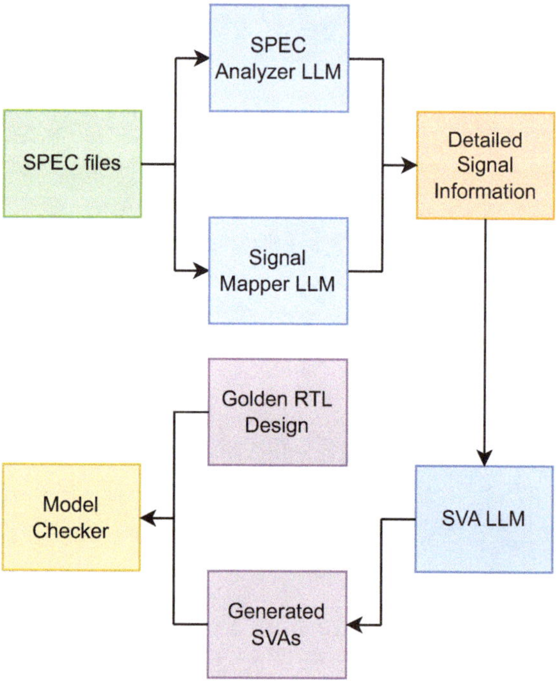

Fig. 4.3 AssertLLM framework [11]

3. *SVA generator* is the LLM responsible for generating the SVAs from the signals previously analyzed and mapped by the SPEC analyzer and the signal mapper. RAG is utilized to process this information effectively by the LLM.

Once the SVA generator creates SVAs for a given design specification, they are evaluated based on a golden RTL design for their correctness and effectiveness. This is done with a model checker and scored on different evaluation metrics, such as syntax correctness and pass/fail rate. Overall, this leads to a sufficient pipeline without losing context due to too much information at once.

While AssertLLM can extract knowledge from design specifications, it can encounter difficulty with more complex designs, not able to capture all the sophisticated relationships between data points in the design specification. AssertionForge aims to alleviate some of these challenges by taking the design specifications and extracting the information into a knowledge graph (KG) representation [7]. KGs organize information about the design specifications into a connected graph structure. The nodes are data points, while the edges between the nodes establish a relationship between the corresponding data points. RAG is used in the generation of the KG and utilized by the LLM in the construction of the SVAs to recall information in the KG.

SVA generation is an ideal task for LLMs, being able to parse out RTL designs either as code or the design specification, or both, in different types of frameworks. The application usages and setup of each RTL design allow for the development of different frameworks to generate higher quality SVAs to find potential bugs and vulnerabilities.

4.2.2 Test Generation Using LLMs

As demonstrated with assertion generation, LLMs are able to read and understand design specifications for RTL designs to carry out a specific generation task. This can be extended into test generation, where a given RTL design needs to be tested with a testbench or other similar mechanism. Crafting testbenches manually is time-consuming and often requires expert knowledge of the design under test. LLMs can speed up this process by automatically creating the testbenches, testing the necessary components of the design to find bugs or capture specific metrics, such as coverage. A majority of the work done in test generation has been with prompt engineering.

LLM-TG creates tests using an LLM for processor design under an automated prompt engineering framework [10]. Resulting tests are scored on their block and expression coverage to determine effectiveness in covering the RISC-V processor. There are two major components of the LLM-TG framework, as shown in Fig. 4.4:

1. *Prompt-based test case construction* is responsible for generating the initial test case based on the design specification. This is done by prompting the LLM with appropriate elements of the design specification.
2. *Test case optimization* is carried out by the LLM, taking the previously generated test cases and performing optimizations on them based on defined characteristics.

The prompt-based test case generation is the first step in the automated framework, creating test cases with three different types of prompt. The first prompt is *rule-based*, which provides the LLM with information about the type of task it is creating a test for and specifics on the relevant instruction. The second is instruction-set architecture (*ISA*), which informs the LLM of the specific ISA to use in the test case, such as instruction with specific syntax for correct usage. The third is *module*, which attempts to generate characteristics of a specific module in the architecture,

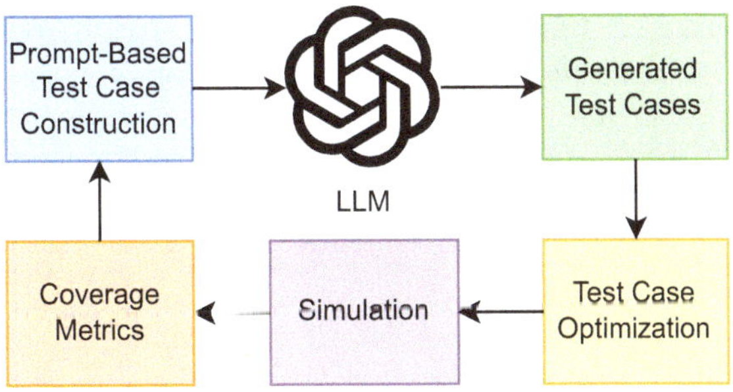

Fig. 4.4 Overview of LLM-TG [10]

such as the arithmetic logic unit (ALU). All three prompts are used to generate a test case. Once a test case is generated, it can be evaluated with the test case optimization portion of the framework. If the test case does not compile, then the LLM is prompted to fix any errors present in the test case based on the error log. Additional test cases are generated with additional instructions and modules based on coverage reports.

While LLM-TG incorporates some feedback, it does not directly inform the LLM of the coverage its test has achieved, meaning that further generated tests can potentially get stuck in a plateau and not improve any coverage metrics. VerilogReader aims to improve this by using the coverage metrics directly in the feedback given through additional prompts to the LLM [27]. The LLM is initially prompted with the RTL design under test, including its specification and additional information on what tests to generate, such as test all possible states of a specific signal within the design. The LLM generates a time sequence of all possible input values to specified signals, which is then tested through simulation. The resulting traces are analyzed according to their coverage, which is translated into a format that is easier to understand for the LLM, while still maintaining critical information. The LLM is given these coverage values in a feedback prompt, and must determine which branches remain uncovered, and attempt to generate new inputs to provide coverage.

There are many scenarios where the RTL design is implemented using a different syntax from the standard, which the LLM has limited or no knowledge. For example, a neuromorphic processor implemented in RTL design may have a specific ISA extension based on RISC-V that is defined for its specific programs. The LLM may default to only RISC-V instructions, and needs information about the new instructions, whether they are redefined from RISC-V or brand new [52]. To generate testbenches of sufficient quality for the neuromorphic processor, the LLM is taught through prompts about the new instructions and architecture of the processor. The initial prompt informs the LLM of the entire ISA of the neuromorphic processor. Once it has information on the ISA, it is then prompted to generate a testbench to target a specific functional unit within the processor architecture, such as branch prediction. As the LLM generates testbenches, feedback is given through additional prompting with error logs and coverage metrics.

Testbenches are often useful at detecting bugs in the RTL design. However, the process of actually locating the bug can also be time-consuming when done manually. Bridging the gap between LLM generation of testbenches and bug detection will provide a complete framework, with less manual intervention [8]. An overview of the testbench generation and bug detection workflow is shown in Fig. 4.5. The test generation and bug detection tasks are split up, with the LLM assisting in both. For the testbench generation, the LLM is initially prompted with information on how to write correct testbenches, such as starting lines of code and using directives to dump variables, and the RTL description of the design. The generated testbench is compiled and simulated if no errors were produced by the compiler. An additional prompt is then constructed to act as feedback to the LLM to update the testbench it previously generated. The feedback depends on the result of

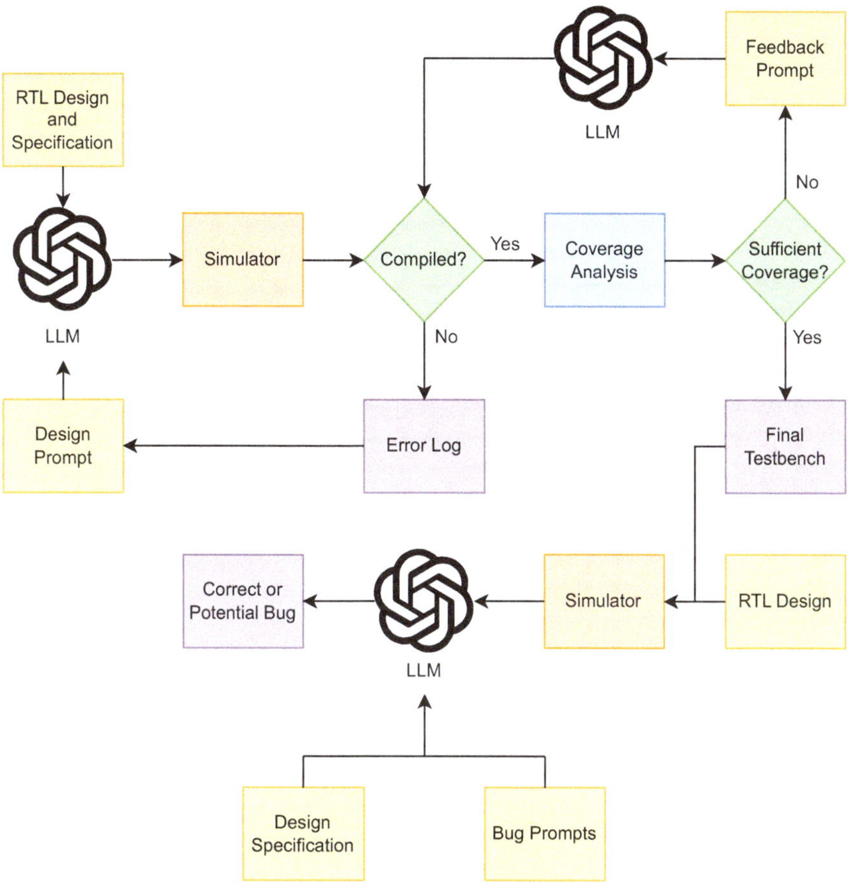

Fig. 4.5 Testbench generation and bug detection using LLMs [8]

compilation and simulation. If the testbench failed to compile, then the error log is given in the prompt. If the testbench compiled and simulated successfully, coverage metrics are obtained and given to the LLM in the feedback prompt. Once sufficient coverage has been reached, the current testbench is considered as the final version.

The bug detection task takes the testbenches and RTL designs used in the initial testbench generation task and simulates them to obtain all signals and their values in the design. This is given along with the design specification to the LLM with the goal of locating bugs, given the intended operation of the design and the real operational values when actually tested. The LLM examines this information and determines if the output traces are correct or if there is any evidence of a potential bug located within the given RTL design.

It is important that the testbenches generated by an LLM as part of an automated framework have a standardized evaluation, where appropriate. AutoEval establishes an evaluation framework for testbench generation, examining that the testbenches

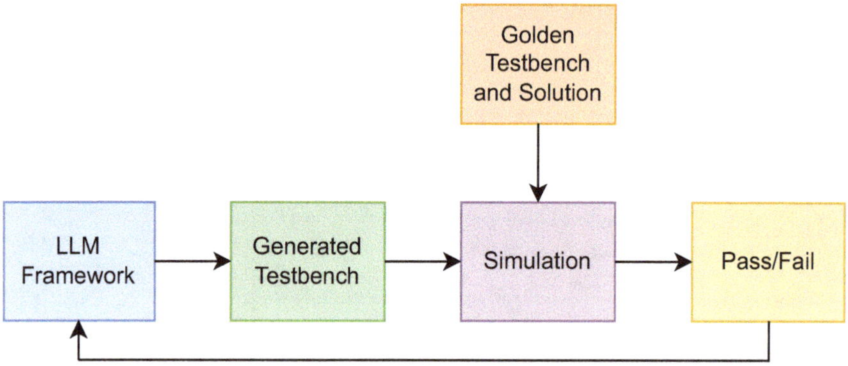

Fig. 4.6 AutoEval evaluation framework [40]

are able to detect bugs, find errors, and produce correct results [40]. This is done by scoring each LLM and the testbenches it produces on its pass rate, such as the number of testbenches that are correct syntactically and produce the expected results when compared to a golden model for a specific RTL design. This process is depicted in Fig. 4.6.

4.3 Detection of Corner Cases and Vulnerabilities Using LLMs

This section describes verification of hard-to-detect scenarios using fuzzing and penetration testing. We first discuss fuzzing based on LLMs. Next, we present penetration testing using LLMs.

4.3.1 Fuzz Testing Enhanced by LLMs

Fuzzing is a testing technique originated in software domain where many inputs are generated to find bugs, vulnerabilities, and other problems with the design [24]. These inputs can be strategically generated to have small changes from verified and correct inputs to find potential failing cases. In hardware domain, fuzzing is also useful in finding hard-to-detect bugs [23].

ChatFuzz is a reinforcement learning approach with LLMs for fuzzing RTL designs [41]. The LLM crafts machine code sequences to be used in fuzz testing of hardware, such as processors. To craft these sequences effectively, the LLM is trained as part of a reinforcement learning pipeline in several steps, as depicted in Fig. 4.7:

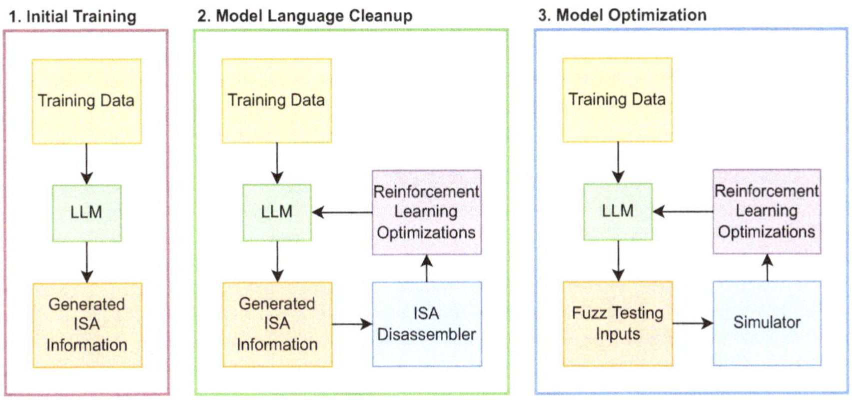

Fig. 4.7 Training process of ChatFuzz [41]

1. The *initial training* of the LLM involves fine-tuning the LLM on the ISA of the given design. At this stage, the LLM will generate information related to the ISAs, such as a series of instructions from the ISA.
2. The *model language cleanup* takes the fine-tuned LLM and processes any errors it has encountered during the generation of instructions, such as syntax errors. This is done to ensure that the LLM has a sufficient understanding of the architecture it is generating inputs for, as well as the ISA itself. The generated instructions are processed through an ISA disassembler for any errors, which informs the reinforcement learning policy of how well the LLM did on this round of instruction generation.
3. The *model optimization* has a similar reinforcement learning flow as model language cleanup except that the LLM is generating the inputs used in fuzz testing. These inputs are given to a simulator to test, with the results feeding the policy of the reinforcement learning training.

With a trained LLM, any inputs generated with the LLM can then be evaluated with its resulting output trace against the golden model trace. This comparison is done to ensure that any potential bugs with the design that have been highlighted by carrying out fuzz testing of the design are detected so that fixes can be applied.

Although advancements are still being pursued in using LLMs for hardware fuzz testing, LLMs have seen active usage in software fuzz testing. Fuzz4All automatically generates fuzzed inputs using LLMs under a prompt engineering approach [51]. Specifically, the LLM is responsible for generating inputs to test as well as the mutation engines to modify the generated input with small changes in a wide variety of programming languages. The framework is feedback-driven, meaning the results from one input generation when tested with the actual RTL design and its associated testbench are given to the LLM as a new prompt automatically. This new prompt informs the LLM on how well its previously generated inputs performed and potential avenues to explore for improving the results.

Embedded systems platforms, like BusyBox, are prime candidates for detecting bugs and vulnerabilities in designs [6]. LLMs can be used to initially generate seed sets to be used in fuzzing given a collection of targets. These targets inform the LLM of standard and nonstandard input formats, which vary based on program and specifications. The nonstandard input formats are those that are not rigorously defined or are unknown, and thus, the LLM requires additional fine-tuning for effective understanding of the input formats to generate initial seeds. Regardless of the input format type, once the LLM has generated the initial seeds, they are given to the fuzzer for testing. The automated framework will analyze the results of the tests performed with the generated initial seeds and determine if any crashes occurred, as well as potential causes. These crashes are consolidated into a new collection of targets for the LLM to generate seeds.

LLMs are effective at generating seeds for usage in fuzz testing of programs and designs in both software and hardware. While there is limited work shown in use for hardware, advancements made in software show that hardware fuzz testing is an ideal candidate for producing better seeds with LLMs.

4.3.2 Penetration Testing Strategies Improved by LLMs

Penetration testing is the process of attacking a system to detect vulnerabilities [4, 5]. While penetration testing is primarily useful in attacking software, it has achieved success in determining vulnerabilities in hardware [4]. Manual penetration testing is time-consuming and can be error prone. Automated approaches have to be structured carefully for the design under test to achieve the best results. While penetration testing for RTL designs with LLMs would have benefits, there are no published efforts as of the time of writing this chapter. Therefore, we discuss automated approaches using similar machine learning techniques to LLMs with hardware penetration testing, as well as efforts using LLMs in software penetration testing.

RePen is a gray-box hardware penetration testing technique using reinforcement learning to generate different mutation patterns to test [3]. The framework generates mutation patterns and analyzes them for vulnerabilities. Reinforcement learning is utilized to use the information obtained from testing the mutation patterns with the design to find vulnerabilities. Once the performance of the previous generation is evaluated, the framework moves onto a different state within the design, and the process repeats.

Software penetration testing contains some LLM-based approaches, which can potentially inspire future directions with hardware penetration testing with LLMs. PentestGPT performs software penetration testing using fine-tuned LLMs to find attack vectors [9], as shown in Fig. 4.8. In addition to fine-tuning the LLM, an evaluation framework to score the generation of attack vectors based on a given program specification is proposed. Most LLMs without the fine-tuning process struggled with specifications containing more complexity, as they have trouble

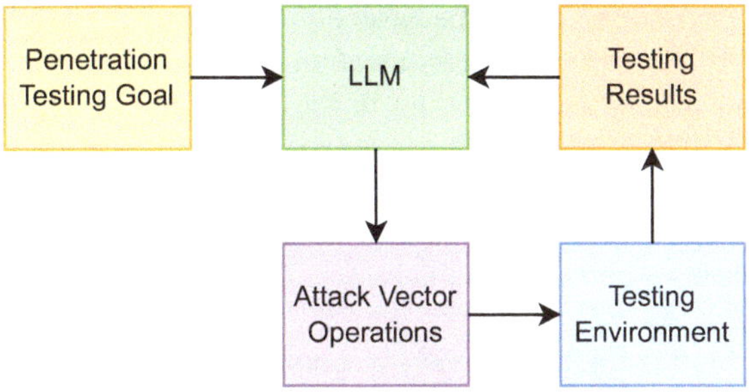

Fig. 4.8 Overview of PentestGPT [9]

maintaining a large enough context to establish relationships between complicated data points as well as remembering the context between generation tasks.

PentestAgent aims to alleviate the training burden of fine-tuning the LLM to produce sufficient attack vectors by using a retrieval-augmented generation (RAG)-based pipeline as well as multiple agents, each focused on a specific task [45]. There are four agents: a reconnaissance agent, a search agent, a planning agent, and an execution agent. These agents collaborate on their intended task to perform intelligence gathering, vulnerability analysis, and exploitation. Each agent utilizes RAG to provide context and information regarding the design and any specifics to the task. Therefore, depending on each agent's task, different documents are examined as part of retrieval-augmented generation.

4.4 Formal Verification Using LLMs

While simulation-based validation is widely used for verifying functional behaviors, it cannot provide functional correctness guarantee since the input space (the number of potential test patters) is exponential. Formal verification avoids the input space complexity and provides correctness guarantees using mathematical analysis. Note that formal verification can face state-space explosion for verifying large designs. Therefore, traditional verification frameworks use an effective combination of formal verification and simulation-based validation. This section describes formal verification using LLM-guided model checking and theorem proving. We first explore reference model generation using LLMs. Next, we present model checking using LLMs. Finally, we describe theorem proving using LLMs.

4.4.1 Model Generation Using LLMs

One aspect of formal verification is reference model generation, creating a different implementation using the same design specification and verifying if the two implementations produce equivalent outputs. The development of one implementation is a manual and expensive effort depending on the complexity of the design specification. Therefore, creation of a second one is often prohibitive. LLMs can aid in this task by aiding in the generation of reference designs, irrespective of whether LLM is assisting, while the developer is writing the first implementation or is generating the entire implementation itself.

For code generation with complex and detailed design specifications, it can often result in the LLM initially generating poor quality RTL designs. VeriAssist is a code generation LLM for Verilog RTL designs with the goal of minimizing the mistakes that the LLM generates in the corresponding code through self-verification and self-correction [17]. Thus, the LLM is prompted with the design specification it must create a Verilog RTL design for, along with a problem description to establish a testbench. These two generated items—the RTL design and the testbench—are then verified and corrected by the LLM itself, compiling and simulating the designs. If the design and/or the testbench are not syntactically correct, this is given as feedback to the LLM in a new prompt to correct. Simulation results are checked for functional correctness, and if the output is not correct, then the LLM must trace through the failed cases and verify its generated code in the next prompt iteration. If there were errors in simulation, then the LLM is asked to correct the errors in the next prompt iteration. VeriCoder is another prompt engineering approach to code generation with LLMs with a focus on the LLM generating functionally valid and correct designs [47]. It does so in a similar fashion as to VeriAssist by prompting the LLM with feedback on the simulation results from the previous iteration of generation with the goal of improving the design.

4.4.2 Model Checking Using LLMs

While simulation is useful in validation of designs, it is often necessary to symbolically check the designs against the design specification to ensure that there no bugs. For example, when we insert a systemVerilog assertion (SVA) and it passes, it does not guarantee that the design satisfies the behavior outlined in the assertion. It only indicates that the behavior was not violated during the simulation using the specific test patterns. Model checking can verify whether the design satisfies the assertion (property) without using test patterns.

Model checking can also account for SVAs that are inserted into the design, as they should be written or generated from the expected behaviors (properties) of the design. Therefore, an LLM can assist in this workflow of formal property verification by adjusting SVAs for correctness and completeness with its associated design properties [32]. This is done with a prompt engineering approach, where the LLM is prompted initially with the RTL design specification to create SVAs for certain properties for the design. The generated SVAs and the RTL design are tested with a formal property verification engine, which creates a report on the correctness and completeness of the SVAs. This report is used to manually define a new prompt for the LLM to generate new SVAs.

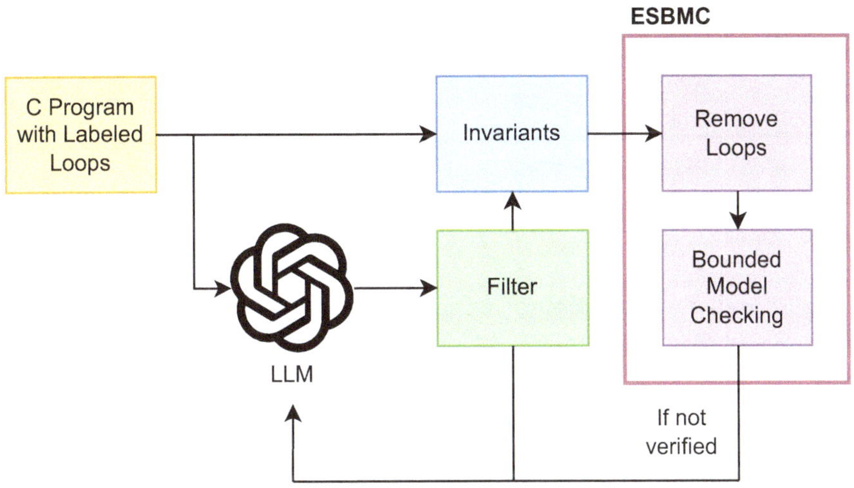

Fig. 4.9 Prompt engineering workflow of ESBMC [36]

Hardware model checking has limited exploration with LLMs, but software model checking has several use cases. These are primarily focused on bounded model checking for loops, generating invariants. We discuss two of these frameworks to showcase how they can be transferred to hardware applications. The efficient SMT-based context-bounded model checker (ESBMC) is an extension of the ESBMC verifier [29] to incorporate an LLM for generating invariants in a bounded model checking pipeline [36], as shown in Fig. 4.9. One major aspect to bounded model checking is the need to unroll any loops contained within the code, and more importantly, generate invariants appropriate to the unrolled loop. Loop unrolling is costly when using bounded model checking, and LLMs can aid in reducing this cost by generating the loop invariants necessary ahead of time. This allows for the replacement of the loops with a control-flow graph (CFG), avoiding the need to check full loops in the bounded model checker.

ESBMC specifically works with a prompt engineering approach by having the LLM iteratively generate invariants for labeled loops in C programs. The loops are labeled for the ease of prompting the LLM to target a specific loop inside the code, in the case there are multiple loops. With the invariant for the specified loop generated by the LLM, it is placed into the C program for verification, including transforming the loop and its invariant into a CFG for ESBMC to verify. The LLM generation is carried out iteratively so that the results of the model checker from its previously generated invariant are provided to the LLM. The results thus act as a feedback mechanism to inform the LLM on what the next iteration of generated invariants it needs to adjust to get better results. Note that this framework filters out any invariant that is syntactically invalid, and could be a target for future research.

LaM4Inv is another prompt engineering approach for generating invariants with LLMs as part of a loop invariant inference workflow for bounded model checking [50]. The initial prompt includes a specification of the program to verify and generate loop invariants. With the invariants generated, they are given to an SMT solver to verify they are correct according to three different conditions. These conditions are as follows:

1. *Reachability*, reaching all states possible in the loop with the invariant
2. *Inductiveness*, the new state after current execution of the loop with the loop invariant remaining within the loop invariant
3. *Provability*, verification of the specified property upon exiting the loop invariant

SMT solver creates a counterexample when a generated invariant is not satisfied. These counterexamples are used to inform the LLM on how to improve its generation of invariants. Specifically, the counterexamples are used as suggestions alongside the design specifications as strong invariants it should attempt to generate for other aspects of the program.

4.4.3 Theorem Proving Using LLMs

Theorem proving accepts the properties (theorems) of the design specification and proving that each property is satisfied by some aspect of the RTL design. Like any manual effort, manual proof generation for the properties is a time-consuming and error-prone process, often requiring expert knowledge of the design. LLMs can assist in the proof generation process, namely incremental proof generation, by defining the proof scope for the design and adding invariants [39]. These two tasks are carried out incrementally by LLMs, meaning that the scope of the proof and the corresponding invariants are small, expanding over time. Both the proof scope and the invariants are generated by the LLM through prompt engineering; however, each task has separate prompts and different prompt construction methods.

The overall workflow of incremental proof generation via prompt engineering is shown in Fig. 4.10. For the proof scope, the LLM is prompted with the design to verify and with the task of splitting the Verilog design into smaller modules. From the divided modules, the LLM focuses on one module at a time, identifying a specific module with basic functionality, such as a half-adder. The module's properties are identified by the LLM and utilized to create a model for the SMT solver. With the proof scope established in the form of the targeted module and the associate properties defined, the LLM is prompted for a third time to generate the corresponding invariants. Each invariant is checked with the SMT solver. The resulting satisfiability (or unsatisfiability) indicates whether the generated invariants within the defined scope satisfy the selected properties of the design. This process repeats incrementally until all scopes of the proof are explored by the LLM.

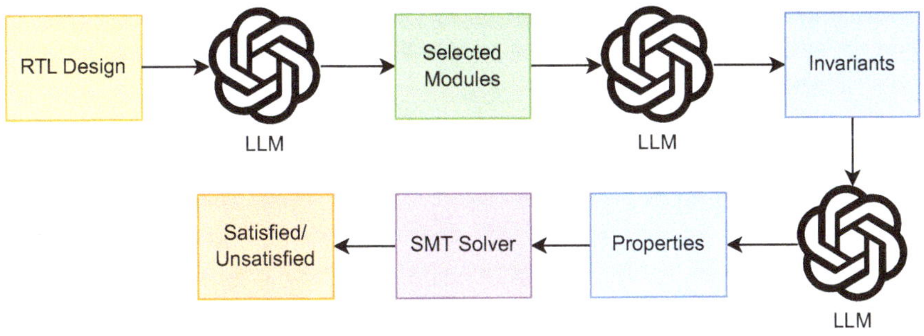

Fig. 4.10 Incremental proof generation via prompt engineering [39]

4.5 Summary

Functional verification of complex electronic systems utilizes a combination of automated methods and manual intervention of verification engineers. Due to the increasing complexity and reduced time-to-market constraints, it is necessary to automate the entire verification flow. This chapter described how large language models (LLMs) can alleviate the need for the manual intervention by assisting in a wide variety of functional verification tasks, including formal verification, assertion-based validation, fuzzing, and penetration testing. Functional verification aided by LLMs is expected to drastically reduce the overall verification effort and improve the product quality.

References

1. Jasper RTL Apps. https://www.cadence.com/en_US/home/tools/system-design-and-verification/formal-and-static-verification/jasper-verification-platform.html.
2. OpenCores. https://opencores.org/.
3. Hasan Al Shaikh, Shuvagata Saha, Kimia Zamiri Azar, Farimah Farahmandi, Mark Tehranipoor, and Fahim Rahman. Re-Pen: Reinforcement Learning-Enforced Penetration Testing for SoC Security Verification. *IEEE Transactions on Very Large Scale Integration (VLSI) Systems*, 33(3):853–866, March 2025.
4. Hasan Al-Shaikh, Arash Vafaei, Mridha Md Mashahedur Rahman, Kimia Zamiri Azar, Fahim Rahman, Farimah Farahmandi, and Mark Tehranipoor. SHarPen: SoC Security Verification by Hardware Penetration Test. In *Proceedings of the 28th Asia and South Pacific Design Automation Conference*, ASPDAC '23, pages 579–584, New York, NY, USA, January 2023. Association for Computing Machinery.
5. B. Arkin, S. Stender, and G. McGraw. Software penetration testing. *IEEE Security & Privacy*, 3(1):84–87, January 2005.
6. Asmita, Yaroslav Oliinyk, Michael Scott, Ryan Tsang, Chongzhou Fang, and Houman Homayoun. Fuzzing BusyBox: Leveraging LLM and Crash Reuse for Embedded Bug Unearthing. In *33rd USENIX Security Symposium (USENIX Security 24)*, pages 883–900, 2024.

7. Yunsheng Bai, Ghaith Bany Hamad, Syed Suhaib, and Haoxing Ren. AssertionForge: Enhancing Formal Verification Assertion Generation with Structured Representation of Specifications and RTL, May 2025.
8. Jitendra Bhandari, Johann Knechtel, Ramesh Narayanaswamy, Siddharth Garg, and Ramesh Karri. LLM-Aided Testbench Generation and Bug Detection for Finite-State Machines, June 2024.
9. Gelei Deng, Yi Liu, Víctor Mayoral-Vilches, Peng Liu, Yuekang Li, Yuan Xu, Tianwei Zhang, Yang Liu, Martin Pinzger, and Stefan Rass. PentestGPT: An LLM-empowered Automatic Penetration Testing Tool, June 2024.
10. Yifei Deng, Renzhi Chen, Chao Xiao, Zhijie Yang, Yuanfeng Luo, Jingyue Zhao, Na Li, Zhong Wan, Yongbao Ai, Huadong Dai, and Lei Wang. LLM - TG: Towards Automated Test Case Generation for Processors Using Large Language Models. In *2024 IEEE 42nd International Conference on Computer Design (ICCD)*, pages 389–396, November 2024.
11. Wenji Fang, Mengming Li, Min Li, Zhiyuan Yan, Shang Liu, Zhiyao Xie, and Hongce Zhang. AssertLLM: Generating and Evaluating Hardware Verification Assertions from Design Specifications via Multi-LLMs, November 2024.
12. Farimah Farahmandi, Yuanwen Huang, and Prabhat Mishra. *System-on-Chip Security: Validation and Verification*. Springer, 2020.
13. Farimah Farahmandi and Prabhat Mishra. Automated test generation for debugging multiple bugs in arithmetic circuits. *IEEE Transactions on Computers*, 68(2):182–197, 2018.
14. Farimah Farahmandi, M Sazadur Rahman, Sree Ranjani Rajendran, and Mark Tehranipoor. *CAD for hardware security*. Springer, 2023.
15. Khan Thamid Hasan, Md Ajoad Hasan, Nashmin Alam, Md Touhidul Islam, Upoma Das, and Farimah Farahmandi. AI-assisted hardware security verification: A survey and AI accelerator case study. *arXiv preprint arXiv:2604.01572*, 2026.
16. Md Ajoad Hasan, Dipayan Saha, Khan Thamid Hasan, Nashmin Alam, Azim Uddin, Sujan Kumar Saha, Mark Tehranipoor, and Farimah Farahmandi. Lasset: An LLM-assisted security asset identification framework for system-on-chip (SoC) verification. *arXiv preprint arXiv:2601.02624*, 2026.
17. Hanxian Huang, Zhenghan Lin, Zixuan Wang, Xin Chen, Ke Ding, and Jishen Zhao. Towards LLM-Powered Verilog RTL Assistant: Self-Verification and Self-Correction, May 2024.
18. Aruna Jayasena and Prabhat Mishra. Directed test generation for hardware validation: A survey. *ACM Computing Surveys*, 56(5):1–36, 2024.
19. Aruna Jayasena and Prabhat Mishra. Hive: Scalable hardware-firmware co-verification using scenario-based decomposition and automated hint extraction. *IEEE Transactions on Computer-Aided Design of Integrated Circuits and Systems*, 43(10):3278–3291, 2024.
20. Aruna Jayasena and Prabhat Mishra. Firmwall: Directed symbolic execution of firmware binaries for defending against unauthorized system calls. *IEEE Transactions on Information Forensics and Security (TIFS)*, 2025.
21. Aruna Jayasena, Sai Suprabhanu Nallapaneni, and Prabhat Mishra. Fuss: Coverage-directed hardware fuzzing with selective symbolic execution. *ACM Transactions on Embedded Computing Systems (TECS)*, 2025.
22. Rahul Kande, Hammond Pearce, Benjamin Tan, Brendan Dolan-Gavitt, Shailja Thakur, Ramesh Karri, and Jeyavijayan Rajendran. (Security) Assertions by Large Language Models. *IEEE Transactions on Information Forensics and Security*, 19:4374–4389, 2024.
23. Kevin Laeufer, Jack Koenig, Donggyu Kim, Jonathan Bachrach, and Koushik Sen. RFUZZ: Coverage-Directed Fuzz Testing of RTL on FPGAs. In *2018 IEEE/ACM International Conference on Computer-Aided Design (ICCAD)*, pages 1–8, November 2018.
24. Hongliang Liang, Xiaoxiao Pei, Xiaodong Jia, Wuwei Shen, and Jian Zhang. Fuzzing: State of the Art. *IEEE Transactions on Reliability*, 67(3):1199–1218, September 2018.
25. Yangdi Lyu and Prabhat Mishra. Scalable concolic testing of RTL models. *IEEE Transactions on Computers*, 70(7):979–991, 2020.

26. Yangdi Lyu, Xiaoke Qin, Mingsong Chen, and Prabhat Mishra. Directed test generation for validation of cache coherence protocols. *IEEE Transactions on Computer-Aided Design of Integrated Circuits and Systems*, 38(1):163–176, 2018.
27. Ruiyang Ma, Yuxin Yang, Ziqian Liu, Jiaxi Zhang, Min Li, Junhua Huang, and Guojie Luo. VerilogReader: LLM-Aided Hardware Test Generation. In *2024 IEEE LLM Aided Design Workshop (LAD)*, pages 1–5, June 2024.
28. Bhabesh Mali, Karthik Maddala, Vatsal Gupta, Sweeya Reddy, Chandan Karfa, and Ramesh Karri. ChIRAAG: ChatGPT Informed Rapid and Automated Assertion Generation. In *2024 IEEE Computer Society Annual Symposium on VLSI (ISVLSI)*, pages 680–683, July 2024.
29. Rafael Sá Menezes, Mohannad Aldughaim, Bruno Farias, Xianzhiyu Li, Edoardo Manino, Fedor Shmarov, Kunjian Song, Franz Brauße, Mikhail R. Gadelha, Norbert Tihanyi, Konstantin Korovin, and Lucas C. Cordeiro. ESBMC v7.4: Harnessing the Power of Intervals. In Bernd Finkbeiner and Laura Kovács, editors, *Tools and Algorithms for the Construction and Analysis of Systems*, pages 376–380, Cham, 2024. Springer Nature Switzerland.
30. Prabhat Mishra, Swarup Bhunia, and Mark Tehranipoor. *Hardware IP security and trust*. Springer, 2017.
31. Prabhat Mishra and Farimah Farahmandi. *Post-Silicon Validation and Debug*. Springer, 2019.
32. Marcelo Orenes-Vera, Margaret Martonosi, and David Wentzlaff. Using LLMs to Facilitate Formal Verification of RTL, September 2023.
33. Zhixin Pan and Prabhat Mishra. A survey on hardware vulnerability analysis using machine learning. *IEEE Access*, 10:49508–49527, 2022.
34. Zhixin Pan and Prabhat Mishra. *Explainable AI for Cybersecurity*. Springer, 2023.
35. Zhixin Pan, Jennifer Sheldon, and Prabhat Mishra. Hardware-assisted malware detection and localization using explainable machine learning. *IEEE Transactions on Computers*, 71(12):3308–3321, 2022.
36. Muhammad A. A. Pirzada, Giles Reger, Ahmed Bhayat, and Lucas C. Cordeiro. LLM-Generated Invariants for Bounded Model Checking Without Loop Unrolling. In *Proceedings of the 39th IEEE/ACM International Conference on Automated Software Engineering*, ASE '24, pages 1395–1407, New York, NY, USA, October 2024. Association for Computing Machinery.
37. Vaishnavi Pulavarthi, Deeksha Nandal, Soham Dan, and Debjit Pal. Are LLMs Ready for Practical Adoption for Assertion Generation?, February 2025.
38. Vaishnavi Pulavarthi, Deeksha Nandal, Soham Dan, and Debjit Pal. AssertionBench: A Benchmark to Evaluate Large-Language Models for Assertion Generation, February 2025.
39. Khushboo Qayyum, Muhammad Hassan, Sallar Ahmadi-Pour, Chandan Kumar Jha, and Rolf Drechsler. Late Breaking Results: LLM-assisted Automated Incremental Proof Generation for Hardware Verification. In *Proceedings of the 61st ACM/IEEE Design Automation Conference*, DAC '24, pages 1–2, New York, NY, USA, November 2024. Association for Computing Machinery.
40. Ruidi Qiu, Grace Li Zhang, Rolf Drechsler, Ulf Schlichtmann, and Bing Li. AutoBench: Automatic Testbench Generation and Evaluation Using LLMs for HDL Design. In *Proceedings of the 2024 ACM/IEEE International Symposium on Machine Learning for CAD*, MLCAD '24, pages 1–10, New York, NY, USA, September 2024. Association for Computing Machinery.
41. Mohamadreza Rostami, Marco Chilese, Shaza Zeitouni, Rahul Kande, Jeyavijayan Rajendran, and Ahmad-Reza Sadeghi. Beyond Random Inputs: A Novel ML-Based Hardware Fuzzing. In *2024 Design, Automation & Test in Europe Conference & Exhibition (DATE)*, pages 1–6, March 2024.
42. Dipayan Saha, Hasan Al Shaikh, Shams Tarek, and Farimah Farahmandi. ThreatLens: LLM-guided Threat Modeling and Test Plan Generation for Hardware Security Verification, May 2025.
43. Dipayan Saha, Shams Tarek, Hasan Al Shaikh, Khan Thamid Hasan, Pavan Sai Nalluri, Md Ajoad Hasan, Nashmin Alam, Jingbo Zhou, Sujan Kumar Saha, Mark Tehranipoor, et al. SV-LLM: An agentic approach for SoC security verification using large language models. *arXiv preprint arXiv:2506.20415*, 2025.

44. Dipayan Saha, Shams Tarek, Katayoon Yahyaei, Sujan Kumar Saha, Jingbo Zhou, Mark Tehranipoor, and Farimah Farahmandi. LLM for SoC security: A paradigm shift. *IEEe Access*, 12:155498–155521, 2024.
45. Xiangmin Shen, Lingzhi Wang, Zhenyuan Li, Yan Chen, Wencheng Zhao, Dawei Sun, Jiashui Wang, and Wei Ruan. PentestAgent: Incorporating LLM Agents to Automated Penetration Testing, May 2025.
46. Shams Tarek, Dipayan Saha, Khan Thamid Hasan, Sujan Kumar Saha, Mark Tehranipoor, and Farimah Farahmandi. Assertain: Automated security assertion generation using large language models. *arXiv preprint arXiv:2604.01583*, 2026.
47. Anjiang Wei, Huanmi Tan, Tarun Suresh, Daniel Mendoza, Thiago S. F. X. Teixeira, Ke Wang, Caroline Trippel, and Alex Aiken. VeriCoder: Enhancing LLM-Based RTL Code Generation through Functional Correctness Validation, April 2025.
48. Hasini Witharana, Yangdi Lyu, Subodha Charles, and Prabhat Mishra. A Survey on Assertion-based Hardware Verification. *ACM Comput. Surv.*, 54(11s):225:1–225:33, September 2022.
49. Hasini Witharana, Hansika Weerasena, and Prabhat Mishra. Formal verification of virtualization-based trusted execution environments. *IEEE Transactions on Computer-Aided Design of Integrated Circuits and Systems*, 43(11):4262–4273, 2024.
50. Guangyuan Wu, Weining Cao, Yuan Yao, Hengfeng Wei, Taolue Chen, and Xiaoxing Ma. LLM Meets Bounded Model Checking: Neuro-symbolic Loop Invariant Inference. In *Proceedings of the 39th IEEE/ACM International Conference on Automated Software Engineering*, ASE '24, pages 406–417, New York, NY, USA, October 2024. Association for Computing Machinery.
51. Chunqiu Steven Xia, Matteo Paltenghi, Jia Le Tian, Michael Pradel, and Lingming Zhang. Fuzz4All: Universal Fuzzing with Large Language Models. In *Proceedings of the IEEE/ACM 46th International Conference on Software Engineering*, ICSE '24, pages 1–13, New York, NY, USA, April 2024. Association for Computing Machinery.
52. Chao Xiao, Yifei Deng, Zhijie Yang, Renzhi Chen, Hong Wang, Jingyue Zhao, Huadong Dai, Lei Wang, Yuhua Tang, and Weixia Xu. LLM-based Processor Verification: A Case Study for Neuromorphic Processor. In *2024 Design, Automation & Test in Europe Conference & Exhibition (DATE)*, pages 1–6, March 2024.

Chapter 5
LLM-Based Security Verification: Vulnerability Detection and Analysis

Rui Guo, Azim Uddin, Pantha Protim Sarker, Upoma Das, Mark Tehranipoor, and Farimah Farahmandi

Modern system-on-chip (SoC) designs face growing security challenges due to increasing design complexity, extensive use of third-party intellectual property (IP), and the emergence of sophisticated hardware attacks such as hardware Trojans, fault injection, and side-channel leakage. Traditional security verification relies on electronic design automation (EDA) tools and manual analysis, which often generate large volumes of complex reports that are difficult to interpret and correlate. Recent advances in large language models (LLMs) offer new opportunities to enhance hardware security verification workflows by assisting engineers in multiple stages of the verification process. This chapter presents an overview of LLM-assisted security verification techniques, including automatic generation of SystemVerilog assertions from natural language specifications, LLM-guided hardware fuzzing for effective stimulus generation and coverage exploration, and LLM-assisted security diagnosis that interprets EDA tool outputs to identify vulnerabilities and suggest mitigation strategies. By integrating these capabilities, LLMs can act as intelligent assistants that augment existing verification tools and enable more scalable and efficient analysis of hardware security vulnerabilities.

5.1 LLM-Assisted SVA Generation

Security verification of hardware designs requires precise specification of security properties that describe expected system behavior under both normal and adversarial conditions [1, 32]. In modern verification flows, these properties are typically expressed using formal constructs such as SystemVerilog assertions (SVA).

R. Guo · A. Uddin · P. P. Sarker (✉) · U. Das · M. Tehranipoor · F. Farahmandi
University of Florida, Gainesville, FL, USA
e-mail: psarker@ufl.edu

F. Farahmandi et al. (eds.), *LLM for SoC Design and Security*,
https://doi.org/10.1007/978-3-032-34116-7_5

However, writing security assertions manually is a labor-intensive task that requires both domain expertise and a deep understanding of the design specification [13, 14]. The emergence of LLMs has provided new inspiration for automated SVA generation [3, 17, 47]. Current research in academia and industry on the problem of generating security assertions using LLMs is primarily exploring two relatively independent yet intersecting technical approaches: prompt-engineering-based approaches [56, 16] and model fine-tuning and customization-based approaches [59]. These approaches exhibit significant differences in their implementation principles, adaptability to application scenarios, and resource requirements. The core concept of the prompt-engineering-based approach is to fully exploit the raw language understanding and generation capabilities acquired by LLM during pretraining. By carefully designing and optimizing the structure, content, and context of input prompts, and employing diverse prompt engineering techniques such as chained reasoning guidance, hierarchical task decomposition, few-shot learning example embedding, and role-playing scenario setting, the model is systematically guided to complete assertion generation tasks along the intended logical path and output format. This approach does not involve any modification or adjustment of the model's internal parameter weights. While this approach offers the advantages of ease of deployment, low cost, and rapid adaptability to diverse verification scenarios, it also has significant limitations: its high reliance on the general capabilities of the underlying model. This approach can exhibit limited understanding depth and accuracy when dealing with highly specialized or domain-specific security attribute descriptions. In contrast, methods based on fine-tuning and customization focus on building high-quality annotated datasets for the field of hardware security verification. These datasets usually contain a large number of natural language security requirement descriptions reviewed by experts and their corresponding standardized assertion code pairing samples. However, the cost of this method is that it requires a lot of manpower and material resources to build high-quality training datasets, and the fine-tuning process has high requirements for computing resources, and the flexibility of model updates and iterations is relatively low [17, 3]. It is worth noting that with the continuous deepening of research, more and more researchers have begun to explore hybrid strategies that organically combine the advantages of these two methods. Table 5.1 illustrates

Table 5.1 The overview of LLM-based SVA frameworks

Frameworks/methods	Prompt engineering	Fine-tuning	Description
ChIRAAG [27]	✓	×	General specifications → SVA
LAAG-RV [26]	✓	×	Timing synchronization assertions
SANGAM [18]	✓	×	Self-correction + Multiround generation
AutoSVA2 [28]	✓	×	Annotation → SVA
SVAgent [17]	✓	×	Problem-solving chain → Security SVA
AssertLLM [12]	Partially	✓	Functional verification + waveform analysis

several existing frameworks/studies and their categorizations. In the followings sections, we thoroughly evaluate the members of each subgroup.

5.1.1 Fine-Tuning-Based Methods

Fine-tuning approaches, at the technical level, follow a systematic training process and optimization strategy. Their core concept is to start with a base language model pretrained on a large-scale general corpus. These base models typically have already learned rich language representation capabilities and basic reasoning capabilities on massive amounts of text data. Researchers then carefully construct a specialized training dataset for the specific application domain of hardware security verification [36, 35]. This dataset contains a large number of high-quality input-output paired samples, such as security property requirements described in natural language and their corresponding standardized System Verilog assertion code, along with relevant design context and domain knowledge annotations. By continuously training the base model on this domain-specific data through supervised learning, the model's internal parameter weights are gradually adjusted and optimized. This allows the model to deeply learn and internalize the specialized grammatical specifications, temporal logic expression patterns, hardware design terminology, and semantic understanding of security properties required for the assertion generation task, while maintaining its original general language capabilities [48, 47, 37]. Ultimately, this results in significant performance improvements and specialized adaptation within the target domain. This approach of enhancing model-specific task capabilities through fine-tuning on domain data has been validated and applied in multiple research projects.

5.1.1.1 AssertLLM

AssertLLM's [12] core design goal is to fully automate the conversion of complex, multimodal specification documents into high-quality System Verilog assertion code. These specifications are often highly heterogeneous and complex, containing not only long sections of natural language descriptions of functional requirements and behavioral specifications, but also often interspersed with timing waveforms, state transition diagrams, signal relationship tables, and other semistructured or unstructured technical documentation elements. These diverse information carriers together constitute a complete expression of hardware design intent.

The AssertLLM [12] workflow can be roughly divided into three phases, as shown in Fig. 5.1. In the first phase, natural language analysis, the system deploys a dedicated natural language processing module to perform in-depth semantic analysis of all text content in the specification document. This includes not only descriptive text in the form of continuous paragraphs, but also technical parameter descriptions embedded in table cells and annotations next to diagram elements.

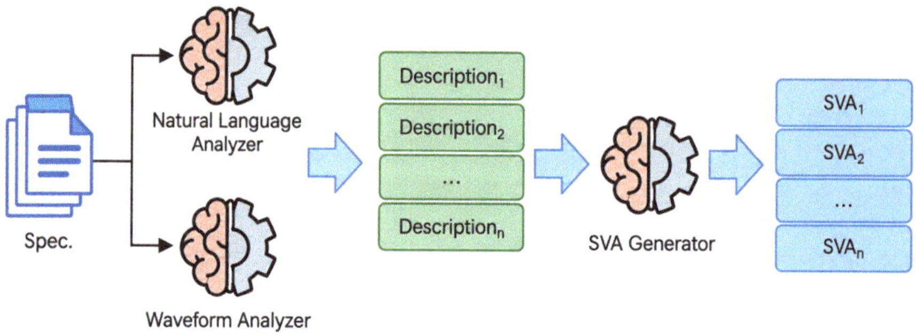

Fig. 5.1 Workflow of AssertLLM [12]

Using natural language processing techniques such as named entity recognition, relationship extraction, and semantic role labeling, this module accurately identifies and extracts key information elements related to hardware signals from massive amounts of unstructured text. It then organizes this fragmented information into structured knowledge templates. Each template corresponds to a specific hardware signal and systematically records the signal's name identifier, functional semantic description, interconnections and dependencies with other signals, timing characteristics, and other relevant technical attributes. Through this standardized template representation, the previously disorganized text information is transformed into a structured knowledge base that can be efficiently processed by computers, laying a solid data foundation for subsequent analysis. In the second phase, waveform analysis, the system introduces a specially designed visual information processing module to process the various timing waveforms and behavioral diagrams contained in the specification documents. These graphical representations often contain complex timing relationships, signal transition patterns, and state transition logic that are difficult to accurately describe in natural language. Using computer vision technologies such as image recognition, curve parsing, and pattern matching, the waveform analyzer automatically identifies signal lines, clock cycle markers, level change points, and timing constraints between signals in the waveforms. It then converts these visual behavioral descriptions into formalized timing behavior templates or pseudocode representations. The key value of this phase lies in supplementing and enriching behavioral information that may have been omitted or inadequately described in the first phase, particularly complex timing dependencies that are difficult to express clearly in natural language but are readily apparent in waveforms. This ensures that the subsequent assertion generation process receives more complete and comprehensive design behavioral specification input.

The third stage is the SVA generation stage. The system takes the structured signal description templates and timing behavior templates generated in the first two stages as input and sends them into an LLM that has been specially customized, trained, and optimized for automatic synthesis of assertion code. This core generation model does not simply use a general pretrained model but is deeply fine-tuned and trained based on the system Verilog assertion syntax specifications, hardware verification domain knowledge, and a large amount of high-quality assertion example data, so that it has a deep understanding of the grammatical structure characteristics, timing logic expression habits, and common assertion design patterns of the assertion code. During the generation process, the model can comprehensively consider the functional semantics of the signal, the timing constraint relationship, and the design context information and automatically infer and synthesize the system Verilog assertion code with standardized syntax, rigorous logic, and comprehensive coverage.

While the AssertLLM system has demonstrated impressive technical capabilities and potential for automatically generating system Verilog assertions from multi-modal specification documents, it still faces a series of severe technical challenges and inherent limitations during actual deployment and application. These challenges stem from both the objective constraints of input data quality and the inherent limitations of LLMs. Furthermore, they are closely related to the vastly varying complexity of hardware designs in real industrial scenarios.

First, from the perspective of input quality dependency, the overall performance of the AssertLLM system is highly dependent on the completeness, clarity, and structure of the specification document. However, in actual engineering practice, hardware design specification documents often suffer from numerous quality flaws. For example, due to project time pressures or personnel changes, the document content may be significantly incomplete. Key functional descriptions or timing constraints may be omitted or only implicitly mentioned. Alternatively, the specification description may be filled with vague and ambiguous language, allowing the same paragraph to be interpreted as representing multiple different design intentions. Furthermore, the document structure may be disorganized and lack a unified standard, with information scattered across different sections and lacking clear logical connections. Even more problematic, some key design behaviors may not be visually represented in waveform diagrams, but only expressed through abstract textual descriptions. These input quality issues are directly transmitted and amplified in the information extraction stage of AssertLLM, preventing the natural language analyzer and waveform analyzer from accurately and comprehensively capturing all key design information elements. This leads to problems such as incomplete information extraction, misunderstanding of signal relationships, and incorrect identification of timing constraints. Ultimately, the generated assertion code may suffer from insufficient coverage or semantic deviation.

Secondly, from the perspective of model reliability, although AssertLLM uses a specially tuned LLM as its core generation engine, the inherent problems of hallucinations and misunderstandings in LLMs remain unresolved. Hallucinations occur when a model generates seemingly reasonable but actually incorrect output based on statistical patterns without sufficient supporting evidence. In assertion generation scenarios, this phenomenon can manifest as the model fabricating nonexistent signal relationships, assuming incorrect timing constraints, or generating syntactically correct assertion code that deviates from design intent. Furthermore, the model may misinterpret certain technical terms or domain-specific expressions in specification documents, mapping them to incorrect formal semantics. These uncertainties and unreliability factors caused by the model's inherent mechanisms are difficult to completely eliminate even with multilayered verification mechanisms. In particular, existing syntax checking and formal verification tools may be unable to effectively detect subtle semantic errors or logical vulnerabilities, posing potential security risks to practical applications.

Finally, from the perspective of system generalization capability and scalability, there is still great uncertainty in the performance and robustness of AssertLLM when dealing with ultra-large-scale systems-on-chip or highly specialized industry-specific designs. Currently, the training and testing of the system are mainly concentrated in relatively standardized and medium-complexity hardware design scenarios. Modern systems-on-chip (SoCs) in real industrial environments often contain billions of transistors, hundreds of functional modules, and extremely complex hierarchical interconnection structures. Their specification documents may be thousands of pages long and involve cross-knowledge from multiple technical fields. In such extremely complex scenarios, AssertLLM's information extraction module may not be able to efficiently process massive document content due to computing resource limitations or algorithm complexity bottlenecks. Its assertion generation module may also experience performance degradation or even generation failure due to context information overload or too long relational reasoning links.

5.1.2 *Prompt-Engineering-Based Methods*

In stark contrast to the fine-tuning approach, another methodological paradigm, increasingly gaining attention among researchers, is prompt-engineering-based approaches. These two approaches exhibit fundamental differences in their technical philosophy, implementation mechanisms, and application scenarios. Prompt-engineering-based approaches adhere to the core concept of "maintaining constant model parameters while optimizing input strategies." Their fundamental characteristic is that they avoid any modification, adjustment, or retraining of the internal parameter weights of the LLM throughout the application process. Instead, they shift the focus of optimization entirely to the meticulous design and construction of input prompts. By deeply studying and understanding the model's behavioral patterns, knowledge boundaries, and reasoning mechanisms, they systematically explore the optimal prompt formulation, contextual information organization structure, task instruction expression, and example sample selection strategy, thereby maximizing and guiding the potential capabilities of the pretrained model to achieve the target task.

5.1.2.1 ChIRAAG

ChIRAAG [27] is a practical pipeline that uses GPT-4 to translate natural language design specifications into SVAs, and the framework is shown in Fig. 5.2. The core

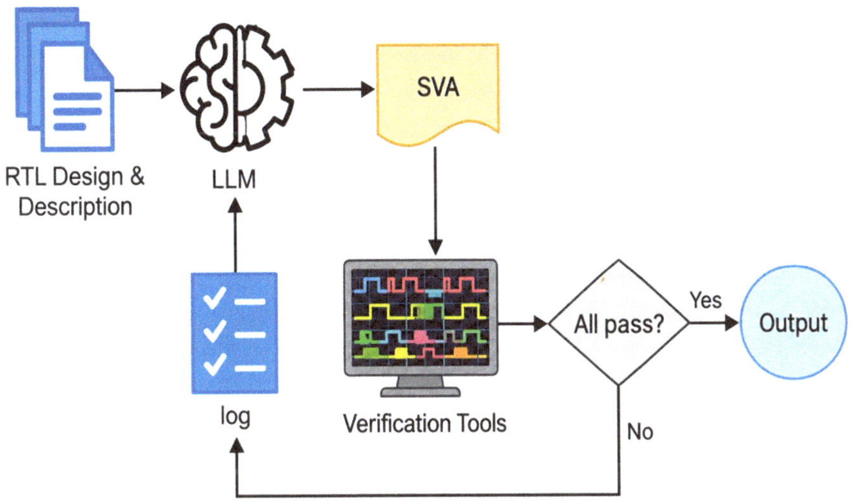

Fig. 5.2 ChIRAAG-SVA generation framework [27]

idea is to first standardize and decompose free-form specifications into a compact, machine-friendly representation and then prompt GPT-4 to synthesize candidate assertions from that structured representation. Crucially, ChIRAAG closes the loop with simulation feedback: Generated assertions are exercised against RTL using a simulator, error messages or counterexample traces are fed back to the LLM, and a small number of iterative prompt/repair cycles correct the faulty assertions.

Since it uses only off-the-shelf LLMs and no additional training, ChIRAAG demonstrates that even general-purpose language models, when guided with well-designed prompts, can produce nontrivial verification assertions with good accuracy and coverage. Moreover, its chain-of-thought prompting offers a degree of interpretability: Engineers can inspect intermediate reasoning steps to understand why a particular assertion was generated. However, the quality of SVA depends strongly on how well the specification is standardized and on prompt design; the approach still requires an engineer to review final assertions, and without careful feedback design, the LLM can produce vacuous or semantically imprecise properties.

5.1.2.2 LAAG-RV

LAAG-RV [26] targets RTL (with emphasis on processor/SoC contexts like RISC-V) and augments plain LLM prompting with engineered synchronization constructs in the generated assertions, and the workflow is shown in Fig. 5.3. The main idea of this technique is to include a one-time Verilog loop and signal-synchronization templates in the generation flow so that the SVA candidates respect handshake/timing relationships common in RTL interfaces.

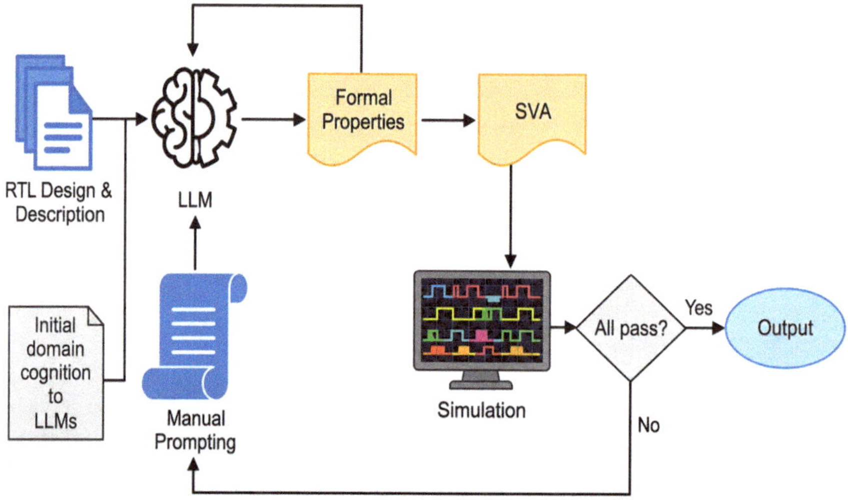

Fig. 5.3 Workflow of LAAG-RV [26]

A key strength of LAAG-RV is its domain adaptation without fine-tuning. It addresses temporal/synchronization pitfalls (handshakes, one-cycle/multicycle relations) that naive prompts often mishandle; it improves temporal correctness without full model fine-tuning by baking domain constraints into prompts/templates. Nevertheless, LAAG-RV inherits some limitations from its prompt-based nature. The approach assumes that suitable templates exist for all relevant behaviors; when specifications involve novel timing constraints not covered by predefined templates, the LLM may fail to generalize. Moreover, the system lacks iterative refinement or feedback loops, so any hallucinations or ambiguities in the prompt directly propagate to the generated assertions.

5.1.2.3 SANGAM

SANGAM [18] advances assertion synthesis by coupling LLM reasoning with a Monte Carlo tree search (MCTS)-style exploration and refinement loop, and the basic architecture is shown in Fig. 5.4. The framework is staged: (1) multimodal specification processing (Signal Mapper, SPEC Analyzer, Waveform Analyzer—implemented as LLM agents) converts spec artifacts into candidate reasoning elements; (2) MCTSr performs guided exploration per signal/property, stochastically proposing assertions, scoring them via heuristic/reward approximations (e.g., simulator feedback, coverage heuristics), and expanding/rewriting promising branches; (3) the best refined reasoning traces are composed into final SVAs. Its main advantage is that systematic exploration yields richer, higher-coverage assertion sets and integrates LLM creativity with principled search/refinement; it reduces single-pass hallucinations by scoring and pruning candidates. However,

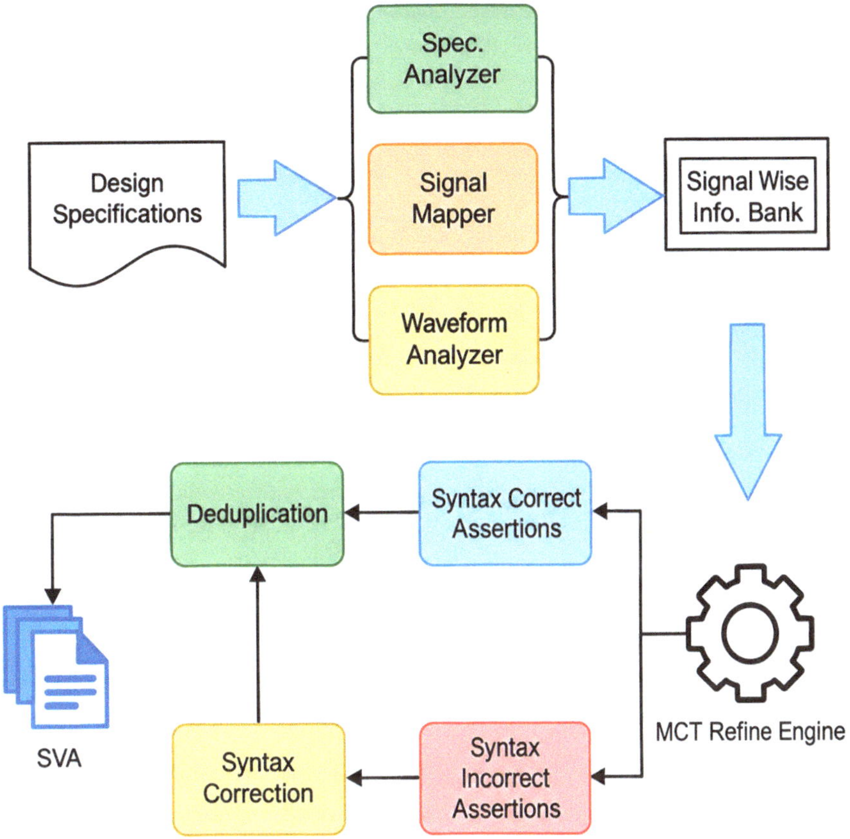

Fig. 5.4 Basic architecture of SANGAM [18]

MCTS is computationally heavier than single-shot prompting; reward estimation for LLM nodes requires approximations (which can bias search), and integrating multimodal inputs and scoring functions increases engineering complexity.

5.1.2.4 AutoSVA2

AutoSVA2 [28] extends the original AutoSVA framework to automate the generation of SVAs for hardware modules by integrating LLMs into the verification flow. Its core principle is to reduce the manual effort in writing annotations and assertions by combining rule-based templates, prompt engineering, and human-in-the-loop validation. Specifically, AutoSVA2 introduces a GPT-4-powered pipeline that first analyzes RTL modules, automatically infers interface annotations for transactions and control signals, and then generates candidate assertions in multiple batches to maximize coverage. The system uses structured prompts with timing constraints, signal dependencies, and behavioral templates to guide the LLM, while engineers review and refine the generated properties, ensuring correctness before formal verification.

The main strength of AutoSVA2 lies in its ability to significantly boost assertion coverage and diversity compared to its predecessor. By iteratively generating assertions and incorporating both internal signal properties and interface-level behaviors, AutoSVA2 can discover corner-case properties and even expose real design bugs that were missed by baseline methods. It also reduces manual annotation overhead, allowing engineers to focus on reviewing rather than writing assertions from scratch. Moreover, its modular design enables easy integration with formal verification tools, creating a semiautomated end-to-end verification flow.

However, AutoSVA2 has limitations. First, the quality of generated assertions heavily depends on prompt design and the capability of the underlying LLM; no fine-tuning or domain-specific training is performed, so complex modules or ambiguous specifications may still lead to incorrect or redundant assertions. Second, the approach struggles with scalability when dealing with very large RTL designs due to token-length constraints and potential loss of contextual information. Finally, while multiple-batch generation improves coverage, it also increases computational cost and still requires expert validation to ensure semantic correctness in safety-critical scenarios.

5.1.2.5 SVAgent

Compared to previous centralized frameworks, SVAgent [17] is the only framework that generates SVAs specifically for the hardware security. It takes a fundamentally different approach by framing SVA generation as a problem-solving chain rather than a single LLM inference task, and the workflow is shown in Fig. 5.5. Instead of

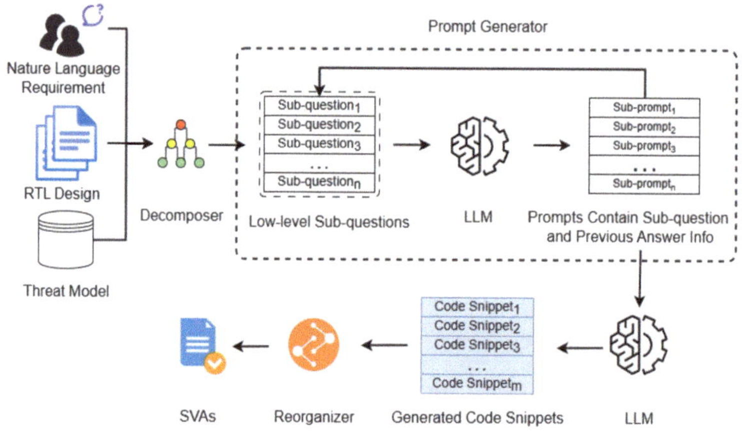

Fig. 5.5 Workflow of SVAgent [17]

asking the model to directly map specifications to assertions, SVAgent decomposes the overall problem into a sequence of smaller reasoning steps. Crucially, this decomposition is not the same as chain-of-thought prompting inside a single LLM call; rather, it structures the entire verification workflow into modular stages such as specification parsing, property decomposition, assertion synthesis, and consistency checking, with explicit information flow between stages.

The motivation behind this design is to address two persistent challenges in LLM-based verification: hallucination (fabricated or logically inconsistent assertions) and randomness-induced unreliability (different runs producing divergent results). By enforcing a deterministic problem-solving chain, SVAgent constrains each stage with well-defined inputs, outputs, and validation criteria. For example, after the property decomposition stage extracts candidate behaviors from specifications, the consistency-checking stage evaluates them for logical soundness before passing them to the synthesis stage. This structured pipeline significantly improves the trustworthiness of generated assertions compared to monolithic prompting approaches.

Key advantages of SVAgent include higher reliability, improved interpretability, and scalability to complex designs. Since each stage exposes intermediate artifacts—parsed specifications, decomposed properties, validation reports—engineers can inspect and debug the pipeline at a fine-grained level. Empirical results indicate that SVAgent produces more semantically accurate and logically consistent assertions than earlier methods, particularly in security-critical verification tasks where trustworthiness is essential. On the downside, SVAgent introduces orchestration overhead: Coordinating multiple stages and validation steps increases implementation complexity and runtime compared to single-pass methods. Additionally, while the deterministic pipeline reduces hallucinations, it may also limit the diversity of generated assertions. Finally, SVAgent currently relies on prompt engineering rather than fine-tuned models, so its performance ultimately depends on the underlying LLM's reasoning ability.

5.2 LLM-Guided Fuzzy Process

Once security properties have been defined, the next challenge is to explore the design state space to identify behaviors that violate these properties. Hardware fuzzing provides an automated method for generating input stimuli and observing system responses in order to uncover functional errors and security vulnerabilities [20, 51]. Originally developed for software security testing, fuzzing techniques have recently been adapted for hardware verification, where they can be applied through RTL simulation or FPGA-based emulation environments [11, 21, 39]. In software domain, coverage-guided fuzzers such as AFL [58] and AFLFAST [8] have shown their ability to uncover bugs in various software applications. In recent years, similar fuzzing methodologies have been adopted in the context of hardware design verification and security validation [40]. Hardware fuzzing involves applying sequences of stimuli to hardware models, either via RTL simulation or FPGA emulation, and observing outputs, assertions, or other runtime signals for detecting functional errors and security vulnerabilities. As compared to simulation-based verification, hardware fuzzing does not rely on manually crafted testbenches or predefined input sequences. Instead, it automatically generates input sequence and mutates them based on feedback to exercise different states of the design. Unlike formal verification, which requires knowledge and access to internal state of the design and suffers from state-space explosion, fuzzing can work with minimal internal knowledge and produce outputs faster. Despite all of the advantages, hardware fuzzing has still some challenges that prevent its widespread adoption in industry. Large language models (LLMs) offer an exciting new avenue to address those challenges. LLMs can process a large amount of design information like specification, documentation, source codes and make the fuzzing process more efficient. LLMs can perform automatic seed and stimulus generation and improve feedback-driven mutation and coverage analysis.

5.2.1 Hardware Fuzzing Background

Hardware fuzzing is an iterative process that automates the exploration of a hardware design's state space by applying generated input stimuli to a device under test (DUT), observing the output or coverage response, and refining the inputs based on feedback. Unlike directed testing or formal verification, fuzzing does not require detailed knowledge of the internal design or handcrafted properties.

Figure 5.6 illustrates a detailed hardware fuzzing workflow, consisting of the following components:

- **Seed Database:** A storage component that maintains an evolving pool of input test cases. It starts with a set of initial seeds and grows by incorporating inputs that lead to new coverage or bug discovery.

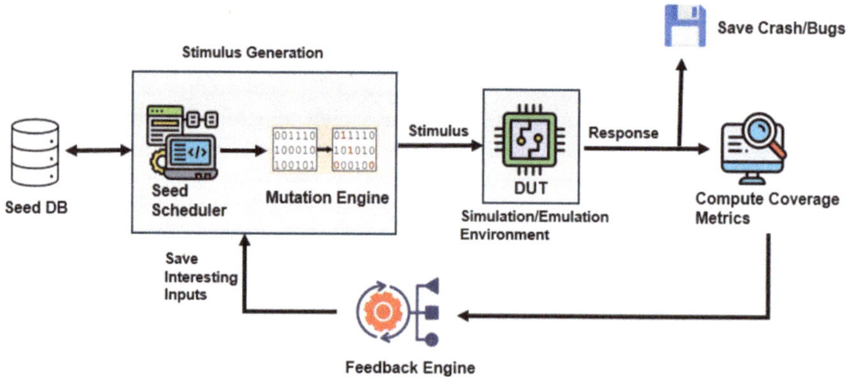

Fig. 5.6 Overview of a feedback-driven hardware fuzzing framework

- **Seed Scheduler:** Selects inputs from the seed database using heuristics such as round-robin, weighted selection, or prioritization based on past effectiveness. Its goal is to ensure diverse and impactful exploration.
- **Mutation Engine:** Applies transformations (bit flips, protocol-aware changes, instruction substitutions, etc.) to the selected inputs to generate new test stimuli. It may employ random mutations or guided strategies.
- **Device Under Test (DUT):** The hardware component being tested. It is typically instantiated within a simulation or emulation environment such as on FPGA, allowing automated stimulus application and monitoring.
- **Coverage Monitor:** Collects results of DUT execution. These may include signal toggles, code coverage, assertion failures, etc.
- **Feedback Engine:** Analyzes the collected coverage data and execution responses to evaluate the effectiveness of each test input. This evaluation is often guided by a *cost function*, which quantifies how "interesting" an input is—such as the degree of new coverage, triggering of assertions, or proximity to a target condition. Inputs meeting or exceeding a cost threshold are returned to the seed database for further mutation and exploration.
- **Bug/Crash Logger:** If a generated input causes an assertion failure, security violation, or system crash, the input and associated traces are saved for later analysis and debugging.

5.2.2 Enhancing Hardware Fuzzing with Large Language Models

Traditional hardware fuzzing techniques can be broadly categorized based on how the hardware under test is executed:

- **Fuzzing Hardware as Software:** In this approach, hardware designs are translated into software representations such as C++ models and then subjected to standard software fuzzing frameworks. This allows for the reuse of existing fuzzing tools while simulating a behavioral model of the hardware [53].
- **Fuzzing Hardware as Hardware:** Alternatively, some frameworks target the actual RTL design directly using simulation or emulation platforms. Feedback mechanisms such as signal toggles or control-flow coverage guide input generation. RFUZZ [25] and DIFUZZRTL [22] exemplify this category by instrumenting RTL designs for feedback-driven fuzzing.

Across both paradigms, input generation is typically driven by random mutations or heuristics, and feedback relies on simple coverage metrics. While effective to a degree, these strategies may fall short in uncovering deep functional bugs or corner-case vulnerabilities in complex designs. *Large language models (LLMs)* offer a promising path forward. By learning patterns from vast corpora of hardware designs, specifications, vulnerabilities, and software behavior, LLMs can generate structured, specification-consistent, and security-relevant stimuli, potentially improving exploration efficiency and vulnerability detection in complex SoC designs. For instance, *ChatFuzz* [33] employs an ML-trained model to synthesize high-quality RISC-V instruction sequences, significantly improving coverage rates compared to traditional fuzzing in a shorter time. In parallel, several fuzzing frameworks rely on manually curated cost functions to guide input selection toward "interesting" behaviors. However, crafting effective cost functions has traditionally required deep domain expertise and manual tuning. LLMs present an alternative by automatically deriving cost functions from textual specifications, signal names, or documentation.

By integrating LLMs into the fuzzing loop, both test generation and input prioritization can become more intelligent and context-aware. The following sections detail key areas where LLMs can meaningfully enhance the hardware fuzzing workflow.

5.2.2.1 Stimuli Generation

A primary application of LLMs in hardware fuzzing is intelligent stimulus generation. Instead of relying solely on random mutations or manually crafted seed inputs, an LLM can synthesize test vectors and instruction sequences that are structurally valid and semantically coherent, thereby increasing the likelihood of exercising deep state transitions or rare corner-case behaviors. By leveraging design-related artifacts such as ISA specifications, protocol documentation, microarchitectural documentation, simulation traces, and open-source RTL implementations, the model can approximate both syntactic constraints and higher-level operational intent, enabling more targeted exploration of the input space. However, because LLMs are probabilistic and may generate incorrect, nondeterministic, or partially invalid outputs, their integration into fuzzing frameworks must be coupled with strict validation mechanisms. These include grammar or protocol-level constraint checking, ISA compliance validation, static rule-based filters, differential testing against reference models, and rejection sampling to discard ill-formed or unsafe stimuli.

This concept is visually illustrated in Fig. 5.7, where the LLM consumes information from multiple sources such as the design description, specification documents, and known vulnerability patterns to generate an *initial seed database*. These seeds serve as a high-quality starting point for the fuzzer, enabling rapid convergence toward coverage-rich regions of the design space.

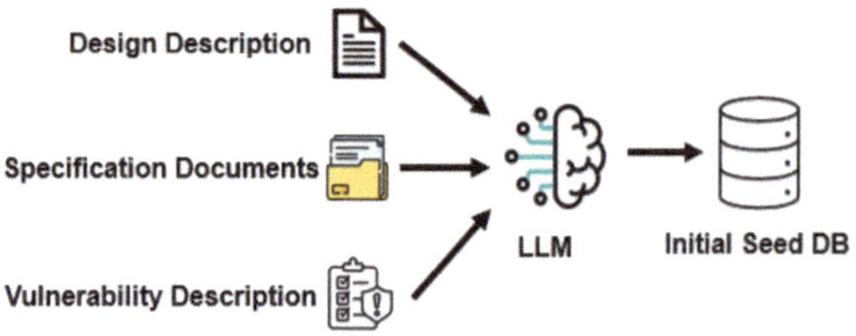

Fig. 5.7 LLM-driven initial seed generation using design descriptions, specifications, and vulnerability knowledge

An exemplary application of this idea is *ChatFuzz* [33], which employs a language-model-guided generator to produce instruction sequences for a RISC-V core. ChatFuzz achieved approximately 75% condition coverage on the Rocket core within one hour, significantly outperforming traditional-feedback-driven fuzzers. The LLM in ChatFuzz generalizes from prior instruction formats and behaviors, enabling it to craft diverse and protocol-compliant inputs that reach deeper into the state space of the DUT.

In contrast to earlier approaches that required extensive validity checks or manual encoding of grammar rules [53, 25, 23], LLM-generated stimuli are both *valid*—adhering to the interface constraints of the design—and *adaptive*—guided by learned contextual knowledge rather than random chance. This leads to faster and more thorough discovery of functional edge cases and security vulnerabilities with significantly reduced engineering effort.

5.2.2.2 Feedback Guidance

Another critical domain where LLMs can substantially enhance hardware fuzzing is *feedback-driven guidance*, especially in two key areas: (1) automatic cost function generation, and (2) recovery from fuzzing stagnation when the input corpus ceases to grow.

Cost-function-guided fuzzing augments traditional coverage-based approaches by steering test generation toward rare or semantically important states. While toggle or branch coverage helps expose shallow logic flaws, many security vulnerabilities such as privilege violations, information leaks, or deadlock conditions require more sophisticated guidance. State-of-the-art frameworks like SoCFuzzer [19] demonstrate this by manually designing cost metrics targeting unauthorized memory accesses and cryptographic key exposure. However, constructing such expert-driven metrics demands deep architectural knowledge and significant manual engineering. LLMs can alleviate this bottleneck by automatically inferring cost functions from available high-level information.

Corpus stagnation is another significant challenge in feedback-driven fuzzing. When the input seed pool ceases to produce measurable improvements in coverage metrics or cost function objectives, traditional mutation engines often fall into local minima in the search space. In these scenarios, as shown in Fig. 5.8, the fuzzing loop detects the lack of progress and triggers an LLM-assisted regeneration process. Rather than continuing ineffective mutations, the LLM analyzes recent fuzzing history and previously exercised behaviors to propose new inputs that may drive execution toward less frequently exercised logic. This could include generating instruction sequences that were not used, activate rarely used control registers, or interact with under-tested memory regions. All generated stimuli are validated before inclusion in the seed corpus through constraint checking and protocol compliance verification. In this framework, the LLM serves as a heuristic proposal mechanism that augments search diversity, while seed selection remains grounded in rigorous analysis engines.

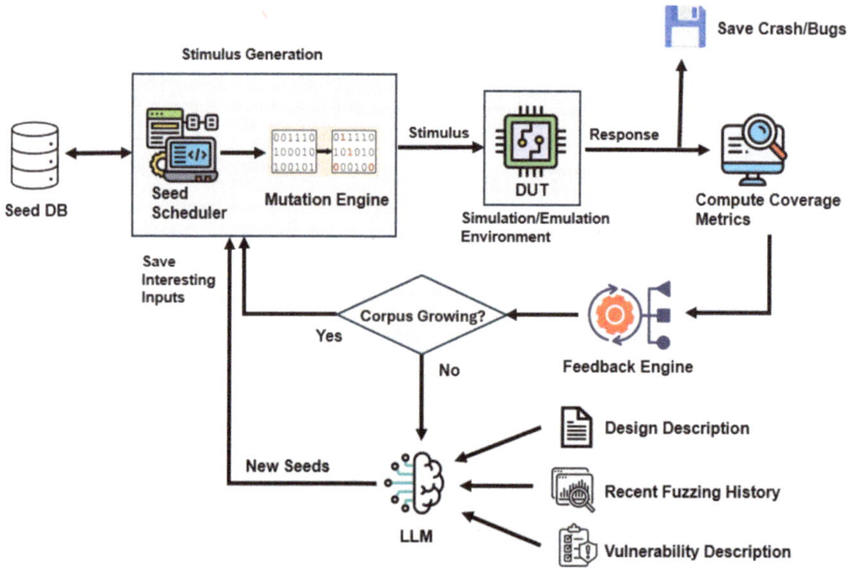

Fig. 5.8 LLM-augmented feedback loop in hardware fuzzing. When input corpus growth stalls, the LLM generates new test seeds using design documents, fuzzing history, and vulnerability knowledge to reactivate exploration

5.2.2.3 Additional Applications

Beyond test generation and cost function guidance, LLMs open up several additional applications to streamline the hardware fuzzing workflow:

- **Testbench Automation:** LLMs can automate the creation of testbenches and harness code that surround the device under test (DUT). Given an RTL module or a high-level description of a hardware block, an LLM could generate the boilerplate code to instantiate the DUT, apply stimuli, and check outputs. This capability has been hinted at by tools that use GPT-style models for code generation, and in the context of fuzzing, it means quicker setup of sophisticated test environments. In fact, ChatFuzz's use of LLM-generated instruction streams can be seen as a form of automated testbench content creation, which significantly reduced the manual effort in developing test programs [33].
- **Debugging Assistance:** Hardware fuzzing often produces a deluge of failing test cases and log data, making manual debugging labor intensive. LLMs can serve as intelligent assistants by interpreting error messages, waveform traces, and register dumps to pinpoint the likely root causes of failures. For example, an LLM could analyze an assertion failure log and suggest which signal discrepancy or state transition led to the error. This is analogous to how LLMs in software domains help developers by explaining runtime exceptions or suggesting fixes. In hardware, such an assistant could dramatically accelerate the bug triage process.

Moreover, machine learning classifiers have already been used to categorize fuzzing outcomes (e.g., distinguishing benign mismatches from true security violations in CPU fuzzing [33]), and an LLM could extend this by providing natural language explanations for each category of failure.

Overall, the infusion of LLMs into hardware fuzzing holds the potential to make verification more intelligent and automated. By generating high-quality stimuli, guiding fuzzers toward critical states, and assisting in ancillary tasks like testbench coding and debugging, LLMs address some of the biggest pain points in hardware verification. Early results from prototypes like ChatFuzz are encouraging, and ongoing research is expanding these techniques.

5.3 LLM-Assisted Security Diagnosis

The security diagnosis in modern system-on-chip (SoC) designs is a critical step to ensure that the hardware is resilient against malicious threats such as hardware Trojans [7, 57], fault injection attacks [42, 54], side-channel leakage [5, 38, 34, 6, 49], and other vulnerabilities. Traditionally, security diagnosis relies on EDA tools, which analyze the design through simulation, verification, and testing flows [45, 24, 43, 29, 30]. These tools generate extensive reports and log files that contain valuable insights about timing, power, fault coverage, and functional correctness [15]. In the context of security, these outputs can highlight anomalies or patterns that may indicate vulnerabilities at the architectural, logic, or physical levels.

However, the challenge is that these reports are massive and highly technical, making it difficult and time-consuming to spot security-relevant issues. This is where large language models (LLMs) can add real value [35]. By parsing EDA-generated reports and logs, LLMs can help engineers quickly identify potential security weaknesses, correlate results across different analysis stages, and provide a natural language summary of diagnostic findings. This section is organized as follows. We begin with a discussion of how EDA tools are currently employed in SoC security analysis, highlighting their functions, capabilities, and limitations. Next, the chapter describes a framework for LLM-assisted security diagnosis that leverages the strengths of large language models to interpret and contextualize outputs generated by EDA tools. Finally, the benefits of this approach—such as improved automation, scalability, and accessibility—are discussed, along with the challenges that may arise when integrating LLMs into hardware security verification workflows.

5.3.1 EDA Tools in SoC Security Analysis

Security analysis of SoC increasingly relies on EDA tools, which provide deep visibility into the functional and physical behavior of integrated circuits. Beyond their traditional role in design and verification, these tools can be leveraged to detect anomalies that may indicate potential security vulnerabilities. This subsection highlights three representative use cases: fault injection analysis [31, 42, 54], side-channel analysis [5, 6], and Trojan detection [50], where EDA tools play a central role.

5.3.1.1 Fault Injection Analysis

Fault injection attacks attempt to induce errors in the functional operation of a circuit by altering its environment through methods such as voltage or clock lags, or through laser or electromagnetic radiation [45]. At the presilicon stage, EDA tools can simulate these conditions to identify which parts of a design are most vulnerable [52, 31, 42, 44, 46]. Static timing analysis (STA) and fault simulation engines help engineers assess how faults propagate through the logic and whether they can bypass error-detection mechanisms. Reports generated from such analyses often include lists of critical paths, sensitive registers, or unprotected states, valuable but highly detailed information that requires careful interpretation for security diagnosis.

An example of such a tool is SOFI (security-oriented fault injection), which integrates fault modeling into the EDA design flow [55]. SOFI takes as input the fault injection specification, the security properties to be checked, and the gate-level netlist of the design. It performs three primary analyses: fault modeling, criticality analysis, and feasibility analysis. The SOFI framework and its output are shown in Figs. 5.9 and 5.11, respectively. Although the reports highlight that the program counter's instruction address registers are the most vulnerable, identifying such critical insights manually is tedious.

To address this, an LLM-based framework can be employed to automate the diagnosis process, as illustrated in Fig. 5.10. The framework first parses the extensive log files and timing reports generated by SOFI. By leveraging its natural language understanding capabilities, the LLM identifies patterns of fault propagation that lead to security property violations. For instance, it can pinpoint specific registers, such as the program counter (PC), where a single bit flip could redirect execution flow. Beyond identification, the LLM provides a security diagnosis by correlating these low-level faults with high-level security risks. Finally, it generates a human-readable summary that not only highlights the vulnerabilities but also suggests design-level mitigations, such as adding parity checks or redundancy to critical registers. This significantly reduces the manual effort required from a designer and provides a more comprehensive perspective on the design's resilience (Fig. 5.11).

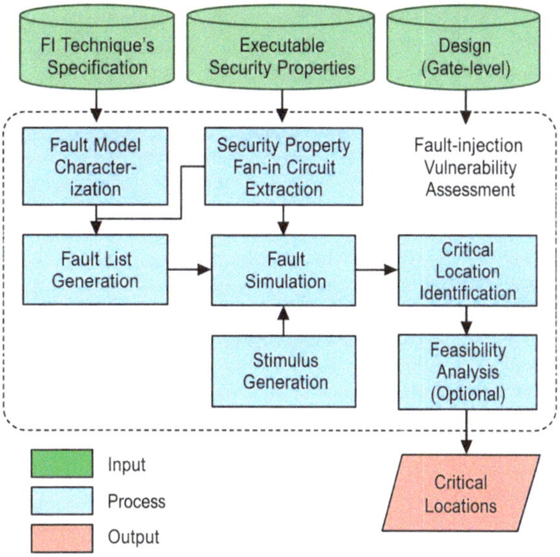

Fig. 5.9 A detailed overview of the SOFI framework

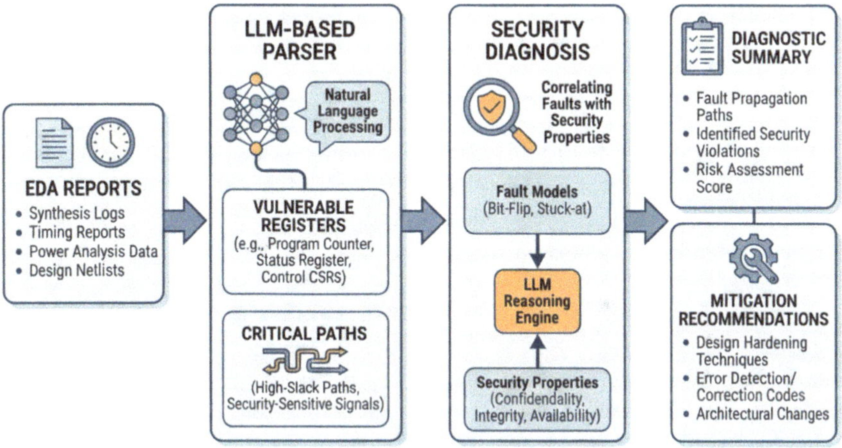

Fig. 5.10 LLM-based security diagnosis flow for fault injection analysis

5.3.1.2 Side-Channel Analysis

Side-channel attacks exploit indirect information leakage, such as variations in power consumption or electromagnetic emissions, to infer secret data such as cryptographic keys [49, 34, 6, 4]. Power analysis tools integrated into EDA workflows provide switching activity and power profiles across different operating scenarios [10]. Similarly, EM simulation frameworks can model the electromagnetic

```
< 0  0  1  0> ND ~ ([4^5]) {VARI "top.imm_generator.imm[5]"}
< 0  0  1  0> ND ~ ([5^6]) {VARI "top.imm_generator.imm[5]"}
< 0  0  1  0> ND ~ ([4^5]) {VARI "top.imm_generator.imm[6]"}
< 0  0  1  0> ND ~ ([5^6]) {VARI "top.imm_generator.imm[6]"}
< 0  0  1  0> ND ~ ([4^5]) {VARI "top.imm_generator.imm[7]"}
< 0  0  1  0> ND ~ ([5^6]) {VARI "top.imm_generator.imm[7]"}
< 0  0  1  0> ND ~ ([4^5]) {VARI "top.imm_generator.imm[8]"}
< 0  0  1  0> ND ~ ([5^6]) {VARI "top.imm_generator.imm[8]"}
< 0  0  1  0> ND ~ ([4^5]) {VARI "top.imm_generator.imm[9]"}
< 0  0  1  0> ND ~ ([5^6]) {VARI "top.imm_generator.imm[9]"}
< 1  1  1  0> DD ~ ([4^5]) {VARI "top.program_counter.instruction_address[0]"}
< 1  1  1  0> DD ~ ([5^6]) {VARI "top.program_counter.instruction_address[0]"}
< 1  1  1  0> DD ~ ([4^5]) {VARI "top.program_counter.instruction_address[1]"}
< 1  1  1  0> DD ~ ([5^6]) {VARI "top.program_counter.instruction_address[1]"}
< 1  1  1  0> DD ~ ([4^5]) {VARI "top.program_counter.instruction_address[2]"}
< 1  1  1  0> DD ~ ([5^6]) {VARI "top.program_counter.instruction_address[2]"}
< 1  1  1  0> DD ~ ([4^5]) {VARI "top.program_counter.instruction_address[3]"}
< 1  1  1  0> DD ~ ([5^6]) {VARI "top.program_counter.instruction_address[3]"}
< 1  1  1  0> DD ~ ([4^5]) {VARI "top.program_counter.instruction_address[4]"}
< 1  1  1  0> DD ~ ([5^6]) {VARI "top.program_counter.instruction_address[4]"}
< 1  1  1  0> DD ~ ([4^5]) {VARI "top.program_counter.instruction_address[5]"}
< 1  1  1  0> DD ~ ([5^6]) {VARI "top.program_counter.instruction_address[5]"}
< 1  1  1  0> DD ~ ([4^5]) {VARI "top.program_counter.instruction_address[6]"}
< 1  1  1  0> DD ~ ([5^6]) {VARI "top.program_counter.instruction_address[6]"}
< 1  1  1  0> DD ~ ([4^5]) {VARI "top.program_counter.instruction_address[7]"}
< 1  1  1  0> DD ~ ([5^6]) {VARI "top.program_counter.instruction_address[7]"}
}
#---------------------------------------------------------------------------
# Faults listed:     Prime: 376      Total: 376
#---------------------------------------------------------------------------
```

Fig. 5.11 SOFI output outlining the list of vulnerable registers

signatures of on-chip components [42]. By examining these outputs, engineers can identify potential hotspots where leakage is strongest. However, the challenge lies in correlating these detailed datasets with high-level security concerns, such as whether a particular AES implementation leaks sufficient information to compromise its keys.

An example of such a tool is Ansys RedHawk-SC security, which extends the industry-standard RedHawk-SC power integrity platform with security-driven analysis capabilities [2]. The tool takes as input the gate-level netlist, switching activity (VCD/FSDB files), and the power grid model of the design. Based on these inputs, RedHawk-SC security performs a detailed power side-channel analysis, computing time-varying power traces and generating leakage metrics such as correlation coefficients between switching activity and secret-dependent signals. The analysis results often highlight localized hotspots, such as in AES round operations where power variations are highly correlated with key-dependent transitions.

Although tools like RedHawk-SC provide deep visibility, interpreting the resulting correlation plots and large waveform datasets requires significant expertise. As shown in Fig. 5.12, an LLM can streamline this by parsing intermediate reports and leakage metrics directly. The LLM acts as an intelligent interpreter that correlates power traces with secret data, such as cryptographic keys. It can pinpoint specific S-box implementations that leak power signatures strongly aligned with key-dependent patterns. By analyzing the significance of these leakages, the LLM helps engineers understand the severity of the vulnerability. Furthermore,

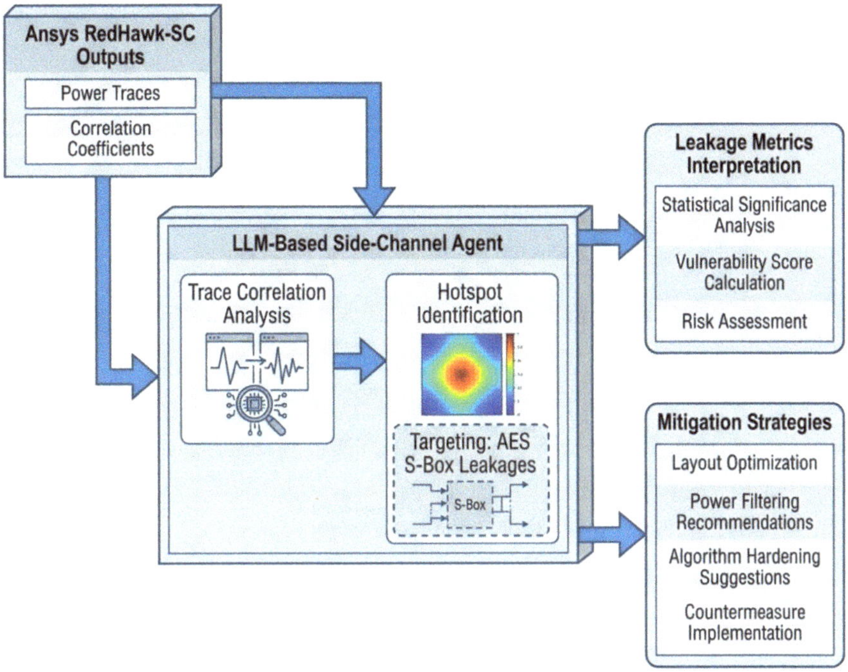

Fig. 5.12 LLM-based security diagnosis flow for side-channel analysis

it can suggest design-level mitigations tailored to the observed leakage, such as implementing side-channel-resistant logic or randomizing the execution order of sensitive operations. This automation not only accelerates the verification cycle but also ensures that subtle leakages are not overlooked during manual inspection.

5.3.1.3 Trojan Detection Analysis

Hardware Trojans inserted during design or fabrication pose a different class of threat. Unlike faults or side channels, Trojans may remain dormant until triggered under rare conditions [41, 50]. EDA-based verification tools, including equivalence checking, functional coverage analysis, and layout-versus-schematic (LVS) comparisons, are often used to detect discrepancies that could hint at unintended logic. Timing or power anomalies flagged in tool reports may also serve as indirect evidence of hidden circuits. However, these outputs are complex and rarely provide definitive answers, which require a significant manual effort to interpret.

An example of the tools used in this domain is Cadence Conformal [9]. The tool takes as input the golden RTL description, the gate-level netlist (or postlayout design), and a set of security assertions or functional specifications. With these inputs, Conformal performs a step-by-step equivalence check between

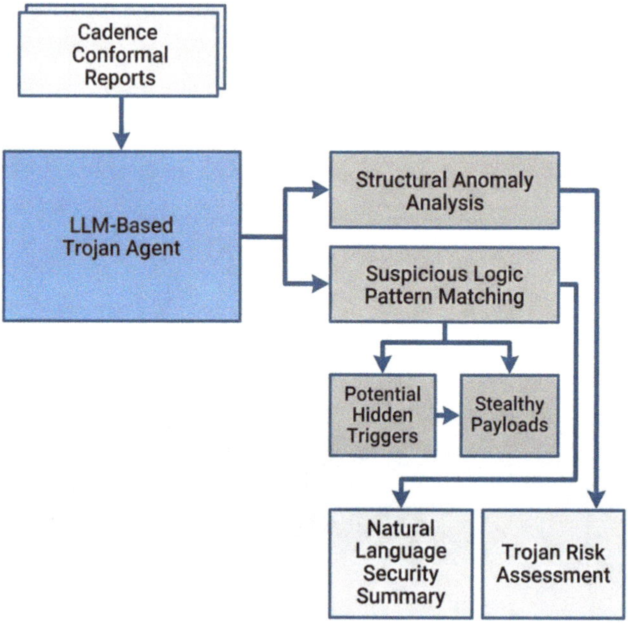

Fig. 5.13 LLM-based security diagnosis flow for Trojan detection analysis

the intended and implemented designs, ensuring that no unintended logic has been inserted. Beyond equivalence, the tool supports structural anomaly detection, such as identifying rarely activated logic, redundant states, or paths inconsistent with the design specification.

The output reports from Conformal can flag suspicious logic regions, but the sheer volume of data makes it challenging to distinguish between legitimate design optimizations and malicious Trojans. To bridge this gap, an LLM-based framework can be utilized to analyze the reports, as shown in Fig. 5.13. The LLM excels at identifying structural anomalies and suspicious logic patterns that might indicate hidden trigger conditions or stealthy payload circuits. By reading through equivalence-check results and structural logs, the LLM provides a natural language summary that highlights unusual patterns and explains the security implications of flagged anomalies. Moreover, the LLM can correlate findings across different verification stages, uncovering subtler deviations that might be overlooked during manual inspection. This automated approach provides a clearer picture of potential Trojan behavior and allows designers to focus their attention on the most critical security threats.

5.3.2 Multiagent Collaboration Architecture for SoC Security

While LLMs provide significant advantages in individual security analysis tasks, the ultimate objective is a holistic security verification of the entire SoC. To support this goal, LLM-based multiagent collaboration architectures have been explored, in which specialized agents analyze different security domains and collectively perform a comprehensive security audit. As illustrated in Fig. 5.14, such an architecture is typically centered around a manager agent that coordinates multiple specialized agents, including FaultAgent, SCAgent, and TrojanAgent.

Each specialized agent is equipped with industry-standard EDA tools—such as SoFI [55] for fault injection, Ansys RedHawk-SC for side-channel analysis, and Cadence Conformal for Trojan detection. The manager agent distributes verification tasks among these subagents based on the design's security requirements. Each agent independently performs its diagnosis, interpreting tool-specific outputs and populating a shared security evidence knowledge base. The manager agent then aggregates these findings, correlating evidence of different attack vectors to uncover complex, multistage threats. For instance, a Trojan detected by the TrojanAgent might be further analyzed by the SCAgent to determine its side-channel leakage

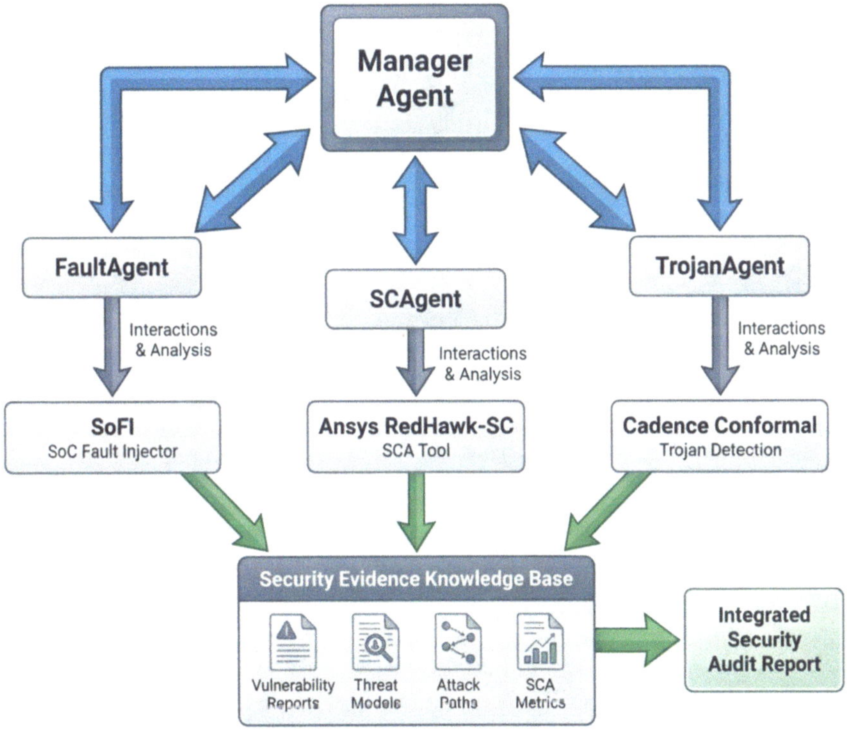

Fig. 5.14 An LLM-based multiagent collaborative framework for SoC security verification

profile. Finally, the framework produces an integrated security audit report, providing designers with a centralized and prioritized view of all potential security risks. This collaborative approach not only improves the scalability and automation of security verification but also ensures a more robust and complete analysis of the hardware design.

References

1. Sohrab Aftabjahani, Mark Tehranipoor, Farimah Farahmandi, Bulbul Ahmed, Ryan Kastner, Francesco Restuccia, Andres Meza, Kaki Ryan, Nicole Fern, Jasper Van Woudenberg, et al. Special session: CAD for hardware security-promising directions for automation of security assurance. In *2023 IEEE 41st VLSI Test Symposium (VTS)*, pages 1–10. IEEE, 2023.
2. Ansys. Redhawk-sc security, 2025. Available at: https://www.synopsys.com/implementation-and-signoff/signoff/redhawk-sc-security.html (accessed Sept 20, 2025).
3. Avinash Ayalasomayajula, Rui Guo, Jingbo Zhou, Sujan Kumar Saha, and Farimah Farahmandi. LASP: LLM assisted security property generation for SoC verification. In *Proceedings of the 2024 ACM/IEEE International Symposium on Machine Learning for CAD*, pages 1–7, 2024.
4. Md Kawser Bepary, Arunabho Basu, Sajeed Mohammad, Rakibul Hassan, Farimah Farahmandi, and Mark Tehranipoor. SPY-PMU: Side-channel profiling of your performance monitoring unit to leak remote user activity. *Cryptology ePrint Archive*, 2025.
5. Md Kawser Bepary, Tao Zhang, Farimah Farahmandi, and Mark Tehranipoor. PreSCAN: A comprehensive review of pre-silicon physical side-channel vulnerability assessment methodologies. *Chips*, 3(4):311–333, 2024.
6. Md Kawser Bepary, Tao Zhang, Jingbo Zhou, Fahim Rahman, Farimah Farahmandi, and Mark Tehranipoor. Towards efficient gate-level electromagnetic side-channel leakage modeling and vulnerability assessment. *Journal of Hardware and Systems Security*, 9(1–2):24–37, 2025.
7. Swarup Bhunia and M Tehranipoor. The hardware Trojan war. *Cham,, Switzerland: Springer*, 2018.
8. Marcel Böhme, Van-Thuan Pham, and Abhik Roychoudhury. Coverage-based greybox fuzzing as Markov chain. In *Proceedings of the 2016 ACM SIGSAC Conference on Computer and Communications Security (CCS)*, pages 1032–1043. ACM, 2016.
9. Cadence Design Systems. Conformal technologies, 2025. Available at: https://www.cadence.com/en-US/home/tools/digital-design-and-signoff/conformal-technologies.html (accessed Sept 20, 2025).
10. Upoma Das, Mohammad Rahman, Akshay Kulkarni, Mark Tehranipoor, and Farimah Farahmandi. PSCMark: Power side channel-based watermarking for SoC IPS using clock gates. *ACM Transactions on Design Automation of Electronic Systems*, 30(5):1–27, 2025.
11. Nusrat Farzana Dipu, Muhammad Monir Hossain, Kimia Zamiri Azar, Farimah Farahmandi, and Mark Tehranipoor. Formalfuzzer: Formal verification assisted fuzz testing for SoC vulnerability detection. In *2024 29th Asia and South Pacific Design Automation Conference (ASP-DAC)*, pages 355–361. IEEE, 2024.
12. W Fang, M Li, M Li, Z Yan, S Liu, H Zhang, and Z Xie. AssertLLM: Generating and evaluating hardware verification assertions from design specifications via multi-LLMs. arxiv 2024. *arXiv preprint arXiv:2402.00386*, 2024.
13. Farimah Farahmandi, M Sazadur Rahman, Sree Ranjani Rajendran, and Mark Tehranipoor. *CAD for hardware security*. Springer, 2023.
14. Farimah Farahmandi, Ankur Srivastava, Giorgio Di Natale, and Mark Tehranipoor. Introduction to the special issue on CAD for security: Pre-silicon security sign-off solutions through design cycle, 2023.

15. Domenic Forte, Swarup Bhunia, and Mark M Tehranipoor. Hardware protection through obfuscation. 2017.
16. Louie Giray. Prompt engineering with ChatGPT: a guide for academic writers. *Annals of biomedical engineering*, 51(12):2629–2633, 2023.
17. Rui Guo, Avinash Ayalasomayajula, Henian Li, Jingbo Zhou, Sujan Kumar Saha, and Farimah Farahmandi. SVAgent: AI agent for hardware security verification assertion. *arXiv preprint arXiv:2507.16203*, 2025.
18. Adarsh Gupta, Bhabesh Mali, and Chandan Karfa. SANGAM: SystemVerilog assertion generation via Monte Carlo tree self-refine. *arXiv preprint arXiv:2506.13983*, 2025.
19. Md Hossain, Amir Vafaei, Kimia Zamiri Azar, Fahim Rahman, Farimah Farahmandi, and Mark Tehranipoor. Socfuzzer: SoC vulnerability detection using cost function enabled fuzz testing. In *Design, Automation and Test in Europe Conference and Exhibition (DATE)*, pages 1–6. IEEE, 2023.
20. Muhammad Monir Hossain, Kimia Zamiri Azar, Fahim Rahman, Farimah Farahmandi, and Mark M Tehranipoor. Fuzzing for automated SoC security verification: Challenges and solution. *IEEE Design & Test*, 41(4):7–16, 2024.
21. Muhammad Monir Hossain, Nusrat Farzana Dipu, Kimia Zamiri Azar, Fahim Rahman, Farimah Farahmandi, and Mark Tehranipoor. Taintfuzzer: SoC security verification using taint inference-enabled fuzzing. In *2023 IEEE/ACM International Conference on Computer Aided Design (ICCAD)*, pages 1–9. IEEE, 2023.
22. Jaehyun Hur, Seongmin Song, Donghyun Kwon, Eunsun Baek, Joonho Kim, and Byoungyoung Lee. DifuzzRTL: Differential fuzz testing to find CPU bugs. In *2021 IEEE Symposium on Security and Privacy (SP)*. IEEE, 2021.
23. Rohit Kande, Alex Crump, Gilles Persyn, Patrick Jauernig, Ahmad-Reza Sadeghi, Akhilesh Tyagi, and Jeyavijayan Rajendran. TheHuzz: Instruction fuzzing of processors using golden-reference models for finding software-exploitable vulnerabilities. In *31st USENIX Security Symposium (USENIX Security)*, pages 3219–3236. USENIX, 2022.
24. Rasheed Kibria, Farimah Farahmandi, and Mark Tehranipoor. IFV: Information flow verification at the pre-silicon stage utilizing static-formal methodology. *Cryptology ePrint Archive*, 2026.
25. Konstantin Laeufer, John Koenig, Donggyu Kim, Jonathan Bachrach, and Koushik Sen. RFUZZ: Coverage-directed fuzz testing of RTL on FPGAs. In *2018 IEEE/ACM International Conference on Computer-Aided Design (ICCAD)*, pages 1–8. IEEE, 2018.
26. Karthik Maddala, Bhabesh Mali, and Chandan Karfa. LAAG-RV: LLM assisted assertion generation for RTL design verification. In *2024 IEEE 8th International Test Conference India (ITC India)*, pages 1–6. IEEE, 2024.
27. Bhabesh Mali, Karthik Maddala, Vatsal Gupta, Sweeya Reddy, Chandan Karfa, and Ramesh Karri. CHIRAAG: ChatGpt informed rapid and automated assertion generation. In *2024 IEEE Computer Society Annual Symposium on VLSI (ISVLSI)*, pages 680–683. IEEE, 2024.
28. Marcelo Orenes-Vera, Margaret Martonosi, and David Wentzlaff. From RTL to SVA: LLM-assisted generation of formal verification testbenches. *arXiv preprint arXiv:2309.09437*, 2023.
29. Nitin Pundir, Henian Li, Lang Lin, Norman Chang, Farimah Farahmandi, and Mark Tehranipoor. Security properties driven pre-silicon laser fault injection assessment. In *2022 IEEE International Symposium on Hardware Oriented Security and Trust (HOST)*, pages 9–12. IEEE, 2022.
30. Nitin Pundir, Jungmin Park, Farimah Farahmandi, and Mark Tehranipoor. Power side-channel leakage assessment framework at register-transfer level. *IEEE Transactions on Very Large Scale Integration (VLSI) Systems*, 30(9):1207–1218, 2022.
31. Tanvir Rahman, Shuvagata Saha, Sujan Kumar Saha, Farimah Farahmandi, and Mark Tehranipoor. EmFIA: A novel emulation-based fault injection vulnerability assessment framework at RTL level. In *2025 IFIP/IEEE 33rd International Conference on Very Large Scale Integration (VLSI-SoC)*, pages 1–5. IEEE, 2025.

32. Sree Ranjani Rajendran, Farimah Farahmandi, and Mark Tehranipoor. Cad tools pathway in hardware security. In *2024 37th International Conference on VLSI Design and 2024 23rd International Conference on Embedded Systems (VLSID)*, pages 342–347. IEEE, 2024.
33. Mohammad Rostami, Matteo Chilese, Samer Zeitouni, Rohit Kande, Jeyavijayan Rajendran, and Ahmad-Reza Sadeghi. Beyond random inputs: A novel ML-based hardware fuzzing. In *Design, Automation and Test in Europe Conference and Exhibition (DATE)*, pages 1–6. IEEE, 2024.
34. Dipayan Saha and Farimah Farahmandi. DL-SCADS: Deep learning-based post-silicon side-channel analysis using decomposed signal. In *2024 58th Asilomar Conference on Signals, Systems, and Computers*, pages 1787–1791. IEEE, 2024.
35. Dipayan Saha, Shams Tarek, Hasan Al Shaikh, Khan Thamid Hasan, Pavan Sai Nalluri, Md Ajoad Hasan, Nashmin Alam, Jingbo Zhou, Sujan Kumar Saha, Mark Tehranipoor, et al. SV-LLM: An agentic approach for SoC security verification using large language models. *arXiv preprint arXiv:2506.20415*, 2025.
36. Dipayan Saha, Shams Tarek, Katayoon Yahyaei, Sujan Kumar Saha, Jingbo Zhou, Mark Tehranipoor, and Farimah Farahmandi. LLM for SoC security: A paradigm shift. *IEEe Access*, 12:155498–155521, 2024.
37. Dipayan Saha, Katayoon Yahyaei, Sujan Kumar Saha, Mark Tehranipoor, and Farimah Farahmandi. Empowering hardware security with LLM: The development of a vulnerable hardware database. In *2024 IEEE International Symposium on Hardware Oriented Security and Trust (HOST)*, pages 233–243. IEEE, 2024.
38. Dipayan Saha, Jingbo Zhou, and Farimah Farahmandi. Physical design-aware power side-channel leakage assessment framework using deep learning. In *2025 IEEE International Symposium on Circuits and Systems (ISCAS)*, pages 1–5. IEEE, 2025.
39. Shuvagata Saha, Ahmed Alhurubi, Tanvir Rahman, Hasan Al Shaikh, Sujan Kumar Saha, Farimah Farahmandi, and Mark Tehranipoor. GEmFuzz: Uncovering system-level vulnerabilities in SoCs via emulation-based grey-box fuzzing. In *2026 31st Asia and South Pacific Design Automation Conference (ASP-DAC)*, pages 723–729. IEEE, 2026.
40. R. Saravanan and Sai Manoj Pudukotai Dinakarrao. The fuzz odyssey: A survey on hardware fuzzing frameworks for hardware design verification. In *Proceedings of the Great Lakes Symposium on VLSI*, pages 192–197, 2024.
41. Pantha Protim Sarker, Upoma Das, Nitin Varshney, Shang Shi, Akshay Kulkarni, Farimah Farahmandi, and Mark Tehranipoor. When everyday devices become weapons: A closer look at the pager and walkie-talkie attacks. *arXiv preprint arXiv:2501.17405*, 2025.
42. Pantha Protim Sarker, Tianze Kan, Jingchen Liang, Ozgur Tuncer, Bo He, Zelin Lu, Sudarshan Mallu, Lang Lin, Norman Chang, Riku Hasegawa, et al. ML-EMFI: A machine learning-driven pre-silicon electromagnetic fault injection security evaluation for robust IC design. In *2025 IEEE International Symposium on Hardware Oriented Security and Trust (HOST)*, pages 57–66. IEEE, 2025.
43. Amit Mazumder Shuvo, Nitin Pundir, Jungmin Park, Farimah Farahmandi, and Mark Tehranipoor. LDTFI: Layout-aware timing fault-injection attack assessment against differential fault analysis. In *2022 IEEE Computer Society Annual Symposium on VLSI (ISVLSI)*, pages 134–139. IEEE, 2022.
44. Amit Mazumder Shuvo, Md Latifur Rahman, Jingbo Zhou, Farimah Farahmandi, and Mark Tehranipoor. ReFID: A system-aware remote fault-injection attack detection & mitigation for secure heterogeneous system. In *2025 IEEE International Symposium on Hardware Oriented Security and Trust (HOST)*, pages 46–56. IEEE, 2025.
45. Amit Mazumder Shuvo, Tao Zhang, Farimah Farahmandi, and Mark Tehranipoor. A comprehensive survey on non-invasive fault injection attacks. *Cryptology ePrint Archive*, 2023.
46. Amit Mazumder Shuvo, Tao Zhang, Farimah Farahmandi, and Mark Tehranipoor. FLAT: Layout-aware and security property-assisted timing fault-injection attack assessment. *IEEE Transactions on Very Large Scale Integration (VLSI) Systems*, 32(6):1150–1163, 2024.

47. Shams Tarek, Dipayan Saha, Khan Thamid Hasan, Sujan Kumar Saha, Mark Tehranipoor, and Farimah Farahmandi. Assertain: Automated security assertion generation using large language models. *arXiv preprint arXiv:2604.01583*, 2026.
48. Shams Tarek, Dipayan Saha, Sujan Kumar Saha, and Farimah Farahmandi. BugWhisperer: Fine-tuning LLMs for SoC hardware vulnerability detection. In *2025 IEEE 43rd VLSI Test Symposium (VTS)*, pages 1–5. IEEE, 2025.
49. Mark Tehranipoor, N Nalla Anandakumar, and Farimah Farahmandi. EM side-channel attack on AES. In *Hardware Security Training, Hands-on!*, pages 163–181. Springer, 2023.
50. Mark Tehranipoor, N Nalla Anandakumar, and Farimah Farahmandi. Hardware Trojan insertion. In *Hardware Security Training, Hands-on!*, pages 73–92. Springer, 2023.
51. Mark Tehranipoor, Kimia Zamiri Azar, Navid Asadizanjani, Fahim Rahman, Hadi Mardani Kamali, and Farimah Farahmandi. SoC security verification using fuzz, penetration, and AI testing. In *Hardware Security: A Look into the Future*, pages 183–229. Springer, 2024.
52. Mark M Tehranipoor, Farimah Farahmandi, and Huanyu Wang. Security property-driven vulnerability assessments of ICS against fault-injection attacks, January 21 2025. US Patent 12,204,685.
53. Timothy Trippel, Kang Shin, Alex Chernyakhovsky, Garrett Kelly, Dan Rizzo, and Michael Hicks. Fuzzing hardware like software. In *31st USENIX Security Symposium (USENIX Security 22)*, pages 3237–3254. USENIX, 2022.
54. Azim Uddin, Sujan Kumar Saha, Farimah Farahmandi, and Mark Tehranipoor. Case study: Fault-injection vulnerability assessment at RTL level. In *2024 IEEE Physical Assurance and Inspection of Electronics (PAINE)*, pages 1–7. IEEE, 2024.
55. Huanyu Wang, Henian Li, Fahim Rahman, Mark M Tehranipoor, and Farimah Farahmandi. SoFI: Security property-driven vulnerability assessments of ICs against fault-injection attacks. *IEEE Transactions on Computer-Aided Design of Integrated Circuits and Systems*, 41(3):452–465, 2021.
56. Jules White, Quchen Fu, Sam Hays, Michael Sandborn, Carlos Olea, Henry Gilbert, Ashraf Elnashar, Jesse Spencer-Smith, and Douglas C Schmidt. A prompt pattern catalog to enhance prompt engineering with ChatGpt. *arXiv preprint arXiv:2302.11382*, 2023.
57. Kan Xiao, Domenic Forte, Yier Jin, Ramesh Karri, Swarup Bhunia, and Mohammad Tehranipoor. Hardware Trojans: Lessons learned after one decade of research. *ACM Transactions on Design Automation of Electronic Systems (TODAES)*, 22(1):1–23, 2016.
58. Michal Zalewski. American fuzzy lop technical details, 2026. Available at: https://lcamtuf.coredump.cx/afl/technical_details.txt (accessed Jan 4, 2026).
59. Biao Zhang, Zhongtao Liu, Colin Cherry, and Orhan Firat. When scaling meets LLM finetuning: The effect of data, model and finetuning method. *arXiv preprint arXiv:2402.17193*, 2024.

Chapter 6
Threat Modeling and Property Generation in SoC Security

Md Ajoad Hasan, Zahin Ibnat, Shams Tarek, Bulbul Ahmed, Avinash Ayalasomayajula, Upoma Das, Mark Tehranipoor, and Farimah Farahmandi

This chapter establishes an end-to-end, AI-driven pipeline for the foundational stages of hardware security verification, progressing logically from high-level system threat model analysis to implementation-specific property generation. It begins by exploring LLM-assisted threat model construction and asset identification, demonstrating how generative AI can parse complex design specifications to automatically pinpoint critical assets and formulate comprehensive threat models. Building upon this conceptual framework, the focus shifts to security-property generation from natural-language descriptions, detailing the capability of LLMs to translate abstract security objectives and threat vectors into structured, verifiable properties. Finally, the chapter addresses the critical challenge of SoC-by-SoC security property mapping, illustrating how LLMs can intelligently anchor these generalized properties to the specific microarchitectural signals, registers, and module boundaries of distinct hardware designs, seamlessly bridging the gap between high-level security intent and low-level physical implementation.

6.1 LLM-Assisted Threat Model Construction and Asset Identification

In reliable and secure hardware systems design, accurately identifying security assets early on provides the foundation for subsequent tasks, such as threat modeling, test plan generation, security verification, and countermeasure development [19, 33, 52, 54, 57, 58]. For example, bug detection tries to identify those assets that

---

M. A. Hasan (✉) · Z. Ibnat · S. Tarek · B. Ahmed · A. Ayalasomayajula · U. Das · M. Tehranipoor · F. Farahmandi
University of Florida, Gainesville, FL, USA
e-mail: md.hasan@ufl.edu

F. Farahmandi et al. (eds.), *LLM for SoC Design and Security*,
https://doi.org/10.1007/978-3-032-34116-7_6

attackers might compromise [9, 16], whereas threat modeling explores scenarios in which vulnerabilities, i.e., fault-injection or side-channel leakage, could propagate through these very assets [48, 60]. Similarly, security properties define how the system is expected to behave in protecting these assets [11, 14]. As underscored by *Security Annotation for Electronic Design Integration (SA-EDI)* standard [6], asset identification is not merely the initial but arguably the most important step toward sustainable security assurances in hardware systems.

Although the importance of security asset identification is widely acknowledged, its automation has yet to come due to the inherent need for a holistic understanding of the assets' behavior, which traditional static analysis tools are unable to provide [35, 40]. Only a few recent works have so far explored asset identification, and those that do often consider it in a limited context. For instance, the SAIF tool [28] attempts to identify assets in hardware designs by analyzing vulnerabilities and threat models, while the LASHED flow [9] leverages common weakness enumeration (CWE) vulnerabilities for the same purpose. However, both approaches identify assets only after analyzing design vulnerabilities, whereas asset identification should be the primary step in a security verification flow, as illustrated in Fig. 6.1. Similarly, the method in [47] relies heavily on RTL signal and register naming conventions, using pattern matching against known security-critical identifiers; as a result, it fails to generalize to designs that do not follow such conventions and offers no reasoning for why the detected elements should be considered assets.

LLMs can be suggested as a feasible assistant in identifying security assets in large-scale SoC designs. LLMs have already demonstrated strong potential in the hardware security verification arena, including detection and mitigation of hardware security issues, as presented in earlier works [39, 56]. Hence, integrating LLMs into the SoC security workflow represents a promising new direction. When applied to RTL designs, LLMs can analyze hardware description languages such as Verilog or VHDL with a nuanced understanding of syntax, structure, and design intent. And the natural-language specifications further facilitate this comprehension by providing contextual and functional clarity. Unlike traditional pattern-matching tools, LLMs

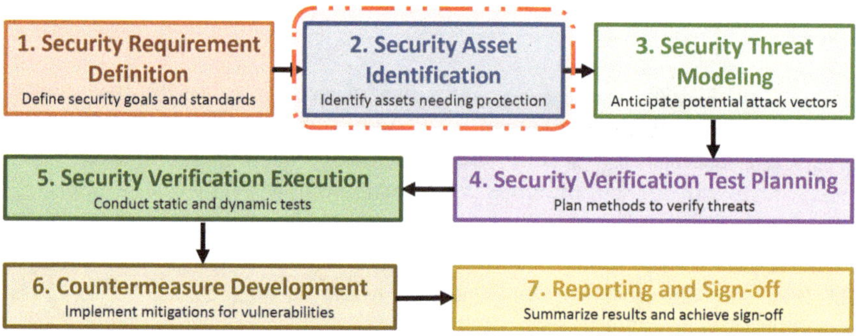

Fig. 6.1 Security verification flow in hardware design

provide contextual awareness for more accurate and scalable asset identification with significantly reduced manual effort.

Given LLM's prospects in security asset identification, two core learning paradigms of LLMs are infused within an LLM-driven asset identification framework to interpret design resources and accurately identify relevant asset information:

- **In-Context Learning**: In-context learning (ICL) allows the GPT model to respond effectively to new tasks using minimal examples, thereby significantly improving learning efficiency by reducing the need for extensive computational resources and training time. In the context of SoC security, ICL allows the model to generate context-aware responses tailored to specific security scenarios [22]. This is particularly valuable in asset identification for dynamic and complex SoC environments that demand a nuanced understanding of diverse designs and evolving threat scenarios.
- **Retrieval-Augmented Generation Learning**: Retrieval-augmented generation (RAG) learning enhances LLM capabilities by incorporating external knowledge sources into the generation process [31]. This mechanism combines a retriever, which fetches relevant and up-to-date information from external data repositories, with a generator, which integrates this information into the model's responses. In SoC security asset identification applications, RAG learning allows LLMs to access relevant security information from external specification datasheets, thereby improving the relevance and comprehensiveness of security analysis.

6.1.1 Classification of Security Asset

By definition, a security asset in SoCs is any hardware component or data element whose protection is essential to preserve the system's confidentiality, integrity, or availability (CIA). Security assets can be classified based on two criteria: (i) abstraction and (ii) dependency. According to IEEE P3164 [36], security assets can be classified as conceptual and structural depending on the abstraction level of the design:

- **Conceptual Assets**: Conceptual assets are high-level assets associated with the use case flows of the IP, such as data and system state, which involve a security objective such as confidentiality, integrity, or availability (CIA). An example of a conceptual asset is data that a user wants to identify as a secret. This data would, at least, require confidentiality as a security objective, thus making it an asset. To illustrate further, as shown in Fig. 6.2, in an AES encryption engine, the encryption key is a conceptual asset because its confidentiality must be preserved regardless of where it resides in the design.
- **Structural Assets**: Structural assets are the hardware elements that physically store or carry conceptual assets. These include registers, modules, flops/latches,

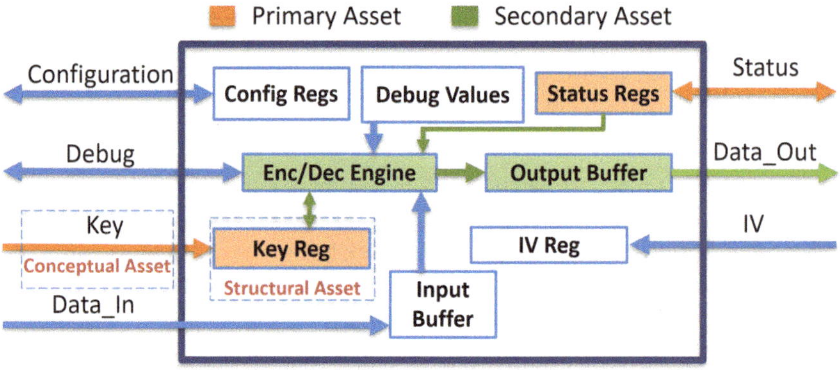

Fig. 6.2 Overview of security asset types in an AES encryption design. The key information is one of the conceptual assets, which is then mapped to the key register as a structural asset. In addition, the key register is a primary asset that can be directly attacked, whereas other registers and buffers, such as status registers and output buffers, are categorized as secondary assets, as they indirectly expose the primary asset

gates, and so on, involved in handling or storage of conceptual assets. Using the plaintext data as the conceptual asset, a structural asset would be RTL material that supports its data flow, i.e., a buffer that temporarily stores the data as it is entered into the IP or a register that contains details about this data, which should also be kept secret. In the same AES engine, the key register is a structural asset because it directly stores the encryption key value.

In addition, security assets can also be categorized into primary and secondary assets based on their dependency role in potential security breaches [28]:

- **Primary Assets**: Primary assets are the design components that are the ultimate target of an adversary or attack. These primary assets can be design components that contain hardware secrets such as hardware keys, firmware, users' passwords, and personal information. Other design components, such as PCR registers, entropy to true random number generator (TRNG), physical unclonable functions (PUFs), etc., which provide security for authentication and integrity, can also be considered primary assets. Continuing the AES example in Fig. 6.2, the encryption key itself is the primary asset since attackers seek to compromise it directly.
- **Secondary Assets**: Secondary assets are design components that interact with or propagate the secondary asset to various design regions. Secondary assets can be design components that inherit sensitive information from primary assets through functionality. These can also be design components that help propagate sensitive information from the primary asset to its target location in the design, such as system buses, peripheral ports, and internal signals/registers that carry the data of the primary asset, either fully or partially. While not directly critical, their compromise can weaken the overall security of the SoC by retrieving the

associated primary asset. In AES, elements such as the encryption/decryption engine and the output buffer qualify as secondary assets because, although they are not the final target, their compromise can indirectly expose the encryption key.

6.1.2 Security Annotation for Electronic Design Integration (SA-EDI) Standard

The SA-EDI standard provides specification guidelines for documenting the security concerns of hardware IPs as they are integrated into SoCs. The standard intends to identify known security concerns, documented in a knowledge base, associated with an asset and/or family type during IP integration. As illustrated in Fig. 6.3, security-related information—including asset definitions, security weakness databases, port- and parameter-related elements, and attack point security objectives (APSO)—is captured alongside the IP design and represented in an SA-EDI object in JSON format. The standard emphasizes that the first step is to identify conceptual assets in the design that require protection under security objectives such as confidentiality, integrity, and availability (CIA), along with their associated structural elements (e.g., ports, parameters) that could compromise these objectives. To support this process, the standard refers to the IEEE P3164 white paper [36], which introduces two methodologies for asset identification:

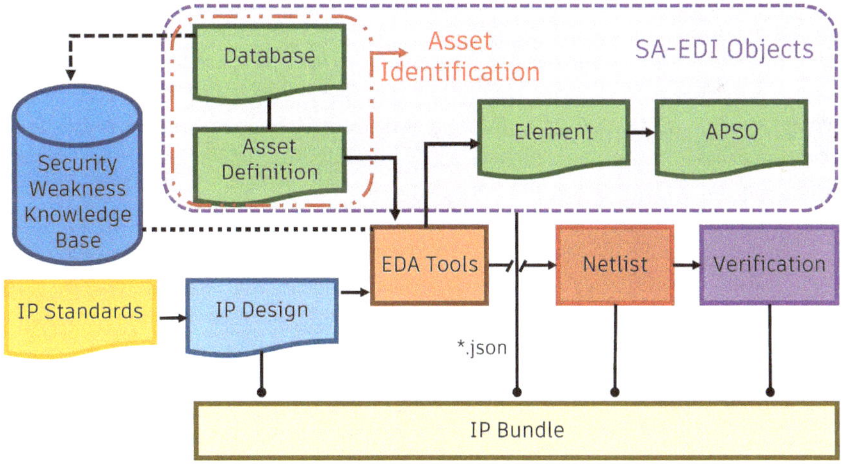

Fig. 6.3 *SA-EDI* IP bundle [6]. The standard begins with security knowledge bases and asset definitions feeding into EDA tools, along with IP designs, for generating assets that are later translated into SA-EDI objects such as element and APSO. These annotated objects, combined with design outputs, form part of the IP bundle, ensuring that security-relevant information is faithfully captured for SoC-level integration

- **Conceptual and Structural Analysis (CSA)**: The *CSA* method begins by identifying conceptual assets tied to the CIA triad and then mapping them to their corresponding RTL representations, i.e., structural assets that are passed to the next asset definition objects in SA-EDI.
- **Point of Influence Analysia (POI)**: The *POI* method starts from the IP's interfaces to find points of influence or observation where an asset could be compromised; if a security objective can be violated through such a point, its corresponding RTL component is flagged as a structural asset requiring protection and inclusion as an asset definition.

Both of the methodologies highlight the distinction between conceptual assets—the "what" that requires protection and structural assets—the "where" in the design that embodies that asset—ensuring that IP developers document security-critical assets comprehensively before integrating IPs into larger SoC designs.

6.2 LLM-Assisted Framework for Asset Identification

Identifying security assets requires a solution with several key qualities. Following the SA-EDI principle of first defining conceptual assets based on CIA objectives and then mapping them to structural design elements, it must understand both natural-language specifications and RTL to map conceptual assets to their structural counterparts; reason over design semantics, hierarchy, and intermodule dependencies to generalize across diverse SoC and IP designs; and provide verifiable, standards-aligned justifications to ensure the fidelity of identified assets. Such a task requires semantic comprehension, contextual reasoning, modular task decomposition, and explainable decision-making capabilities that traditional rule-based approaches lack. LLMs inherently offer these properties: They can jointly interpret specifications and RTL, infer security relevance from context rather than naming heuristics, be organized into specialized agents for subtasks like specification analysis, RTL analysis, asset mapping, and CWE-based rationalization, and generate human-readable rationales with self-consistency checks. Leveraging these strengths, Hasan et al. [34] designed *LAsset* as an agentic framework driven by an LLM capable of performing automated, scalable, and reliable asset identification to bridge the gap between the imperative of asset identification and the limitations of current approaches.

LAsset leverages the SA-EDI standard [6] and IEEE P3164 [36] guidelines for preliminary identification of assets using LLM, guided by structured queries framed around the confidentiality, integrity, and availability (CIA) objectives. It further validates and refines the asset list by mapping it against the CWE vulnerability database, thereby ensuring its trustworthiness. In summary, the key features of *LAsset* are as follows:

- **Novelty:** *LAsset* is the first-ever framework to utilize LLMs for the automated identification of security assets from SoC and IP designs.
- **Robustness and Versatility:** *LAsset* operates with either combined specification and RTL inputs or RTL alone, supports Verilog, SystemVerilog, and VHDL, and applies to both IP and SoC designs, ensuring broad applicability.
- **Comprehensiveness:** Beyond asset discovery, *LAsset* associates each asset with threat modeling details and security justifications, providing actionable context for downstream verification tasks.
- **Trustworthiness:** *LAsset* is supplemented by a refinement stage that validates each asset by cross-referencing the CWE database and applying self-consistency checks, yielding a high-fidelity asset list with up to 93% recall on real-world designs.

As shown in Fig. 6.4, *LAsset* operates through three sequential agents: (i) *input preprocessing*, which parses the available specification and RTL resources and aligns them under the SA-EDI conceptual-structural asset model; (ii) *asset generation*, where specialized LLMs analyze the preprocessed inputs to identify candidate conceptual assets and map them to their structural counterparts; and (iii) *asset refinement*, which validates and rationalizes each candidate through cross-referencing with the CWE database and self-consistency checks, producing a final, standards-aligned asset list. This modular agentic design enables *LAsset* to perform automated, scalable, and verifiable security asset identification across diverse SoC and IP designs.

6.2.1 Input Preprocessing

Of the two inputs to the *LAsset* framework, the RTL repository provides the internal functionalities as well as information flow tracking, whereas the corresponding technical Spec. primarily outlines the module's integration within the broader SoC architecture, providing additional design insights and interface details in natural language. Therefore, the inputs are preprocessed in terms of the security concerns of the design so that LLMs can better understand the true security context rather than hallucinating due to token limitations and constrained long-context retention. The *input preprocessing* agent performs the following three major tasks, as shown in Algorithm 6.1 (lines 3–4).

6.2.1.1 Design Modules Extraction

In this task, LLMs' reasoning is employed to (i) prune nonrelevant modules and (ii) select security-critical modules among the extracted module names from the RTL repository. For example, within the NEORV32 SoC, the *neorv32_application_image* IP module is not considered security critical

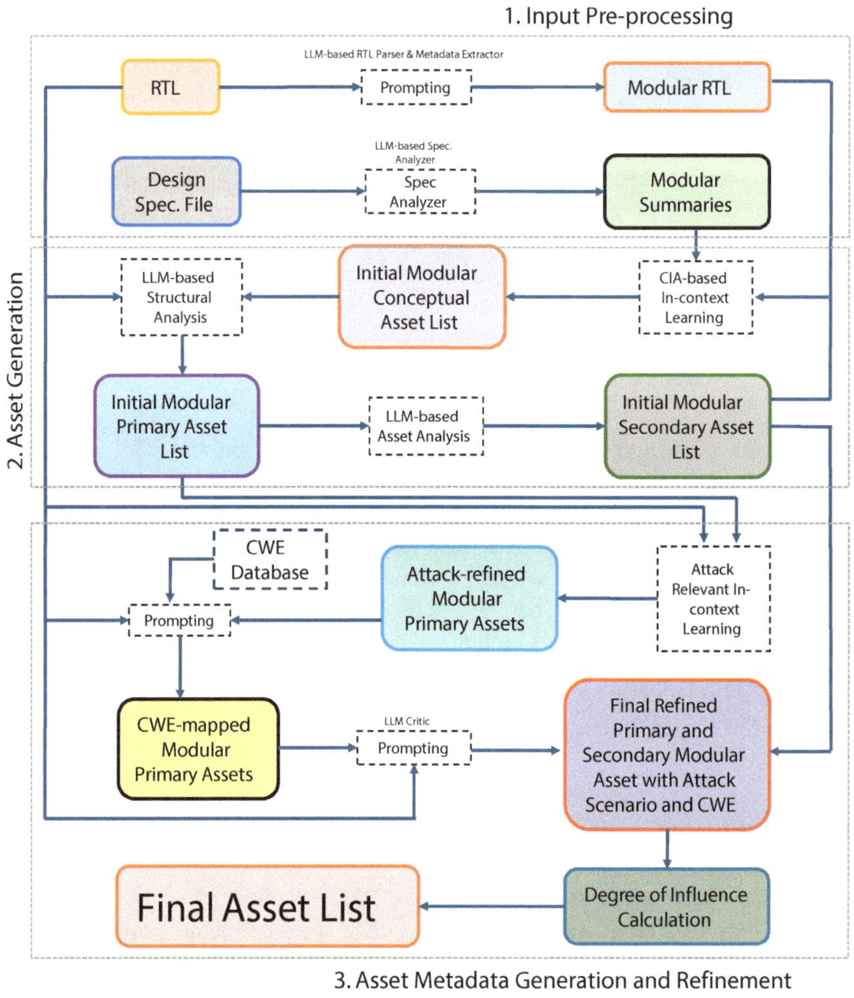

Fig. 6.4 Overview of the proposed *LAsset* framework for security asset identification. The process begins with preprocessing the Spec. and parsing the RTL to generate the modular inputs. Using confidentiality, integrity, and availability (CIA)-guided in-context learning, initial primary and secondary asset lists are extracted. These assets are then refined through attack scenario reasoning and CWE mapping, each supported by LLM-backed prompting

and was therefore excluded during this task, whereas the security-sensitive *neorv32_cpu_pmp* IP module, among others, was retained. This task is important, for it makes the flow much more resource efficient (both cost and time) by reducing the security-irrelevant design overhead in the beginning.

Algorithm 6.1 *LAsset* framework

```
Require: Design Spec S; RTL repo R = {m_1, m_2, m_3, ...}; DBASEcwe;
Require: Function Library: SpecRAG(Spec), SecAsset(Primary Asset, I/O Ports & Internal Signals,
    RTL), UnrollAsset(Asset_List, Path), ICLasset, ICLattack;
Require: Prompt Library: LLMparse, LLMasset, LLMattack, LLMcwe, LLMref.
Ensure: Group of modular asset annotations Asset_g = {Ref_Asset_m | m ∈ M}.
Ensure: List of DoI per secondary asset DOI = {DoIsec_asset_h | h ∈ H}
  M ← MOD_LISTING(R)
  Step: Modular Assets Identification
  for each m ∈ M do
  /* Input pre-processing */
      MOD_SPEC S_m ← SpecRAG(S)
      MOD_RTLPARSE {Prm_m, Sec_m} ← LLMparse(R(m))
  /* Assets generation */
      Asset_m ← LLMasset(S_m, Prm_m, R(m), ICLasset)
      Exp_Asset_m ← SecAsset(Asset_m, Sec_m, Prm_m, R(m))
  /* Assets refinement */
      Ref_Asset-1_m ← LLMattack(Exp_Asset_m, ICLattack)
      Ref_Asset-2_m ← LLMcwe(Ref_Asset-1_m, DBASEcwe)
      Ref_Asset_m ← LLMref(Ref_Asset-2_m)
  end for
  Step: DoI calculation
  H ← HIERARCH_PATH(R)
  for path h ∈ H do
  /* Enumerate linked modules' assets */
      𝒜_{h,g} ← UNROLLASSETS(A_g, h)
      let 𝒜_{h,g} ≜ {Amh_1, Amh_2, Amh_3, ...}
      DoIsec_asset_h ← DOI(𝒜_{h,g})
      let DoIsec_asset_h ≜ {DoIsec_asset_{h,1}, {DoIsec_asset_{h,2}}, ...}
  end for
  return Asset_g, DOI
```

6.2.1.2 Modular Spec. Summarization

This task is built on a retrieval-augmented generation (RAG) in-context learning-based model, where the SoC-level design specification is used as the knowledge source. For each security-relevant module, a user-guided query augments the prompt to produce a *technical summary* via LLMs, which contains all the security-relevant information both within the module and across modules in the broader SoC context, as shown in the prompt in Fig. 6.5. Overall, this *summary* acts as a concise Spec. input for the *Asset Generation* agent for the respective module.

6.2.1.3 Modular RTL Parsing

To extract the design elements, two LLM-based parsers are implemented: one for I/O ports and another for internal signals/registers, along with their types and functions at the module level. This is done to ensure that LLMs do not overlook any of the parsed design elements for asset decision, thereby reducing the chance of false negatives.

Technical Summary Generation Prompt

$Retrieved Data {Module}$

Your task is to analyze the data and generate a comprehensive, structured, and informative technical summary of this module. Your response will be used as background context for asset identification in hardware security verification. Please include the following in your response:

1. Module Introduction:

Briefly explain what the {Module} module does in the context of {IP} design.

2. Module Configurations:

- List all the sub-modules and their functionalities
- List all the ports (input ports, output ports, and inout ports) and their functionalities
- List all the internal signals' information within this {Module} module and their functionalities
- Provide a description of each of the corresponding fields' names for all the registers and their detailed functions.

If there is no explicit I/O port or register mapping information in the retrieved data for the {Module}, state this.

3. Module Theory of Operations:

- Describe in details how different sub-modules of the {Module} module as well as the module itself operates.
- Explain how different sub-modules, I/O ports, and registers of this module interact with each other, so that the {Module} module can operate properly. If the module supports multiple modes, explain how these modes work.

4. Module interfaces:

- Describe how the {Module} module is integrated in the {IP} design.

5. Module significance in the IP:

Explain how the {Module}'s functionality is used in the {IP}; that means, how the module's operation influences-

- the operational behavior of other modules/blocks in the {IP}.
- the {SoC}'s operation as a whole.

Ensure your response is **well-organized and technical**, with a focus on all the different elements (I/O ports, signals, registers' fields, etc.) of {Module} and their contribution to the {Module}'s operation.

If the Retrieved Data does not contain sufficient information on the I/O ports, signals, registers, etc. of the {Module} as a standalone module, then you will state such in the technical summary.

Fig. 6.5 Technical summary generation prompt used in the asset identification workflow

6.2.2 AI-Based Asset Identification

Using the *technical summary* and RTL as input, the few-shot in-context learning paradigm is employed for asset generation. The hardware security context is defined, complemented by the Q/A related to CIA security objectives. To further illustrate the task, several case studies are provided featuring widely recognized hardware IP blocks, e.g., AES, GPIO, and Gaussian noise generator, each annotated with asset listings and corresponding justifications explaining why some design elements are security assets and others are not. Through this step-by-step approach, the LLM is

Asset Generation Prompt

$Security Objectives Context$
$Assets Example$
$Input Design$ $Design Element$

Task:
Your task is to analyze both RTL codings and the technical specification provided above, for all the elements in the {element_list} in order to understand each of its function and significance with respect to the {Module}; and from these elements, find 3 (three) lists of assets, where the lists are based on whether they are confidentiality-critical, integrity-critical and availability-critical.

Approach:
- Conduct a thorough and systematic analysis of both the technical specification and RTL codings of the {Module} above, including the commented out lines inside the RTL, to get a detailed understanding about each of the elements; then, based on your learning about hardware security objectives (confidentiality, integrity and availability), try to find out for each of the I/O ports in the {element_list} whether it can be confidentiality-critical asset, integrity-critical asset or availability-critical asset. Do the same for the signals/registers in the {element_list}.
- Do not overlook your analysis for any element (I/O port or signal/register) as a potential asset.
- Provide precise and thorough justifications for each asset's inclusion.

Constraints:
- The {Module} would have no confidentiality-critical asset if it is not used for cryptographic relevance. Whether the {Module} is used for cryptographic purposes or not would be evident from the analysis of the above technical specification and RTL codings.
- Do not use your own knowledge to identify assets; rather, rely only on the analysis of the provided technical specification and RTL codings of the {Module}.
- Exclude structured I/O ports (typed interfaces), signals/registers from assets.

$Output Format$

Fig. 6.6 Asset generation prompt used in the asset identification workflow

systematically guided to learn and identify security assets, as shown in the prompt in Fig. 6.6.

Following this in-context learning, the conceptual assets are identified at the module level through a stepwise reasoning and appropriate constraints approach. These assets are then systematically mapped to their corresponding structural RTL references, derived from the parsed design elements—these are the *primary assets* at the module level. In addition, for each *primary asset*, the internal signals/registers that influence/violate its security objective(s) are derived—these are termed as the *secondary assets* at the module level. Asset generation steps are shown in Algorithm 6.1 (lines 5–6).

6.2.3 Asset List Refinement

When applied to security asset identification, LLMs often lean toward listing a broad set of possible security and nonsecurity assets. This tendency, while helpful in maximizing coverage, can also introduce false positives by including design elements that are hardly security relevant. Hence, a rigorous three-stage back-to-back refinement agent is designed to filter out the false positives and validate the remaining results, ensuring that the final asset list at the module level is accurately substantiated. The refinement agent comprises the following stages, also shown in Algorithm 6.1 (lines 7–17):

6.2.3.1 Attack Scenario Analysis

Each candidate primary asset is evaluated for susceptibility to seven hardware attack classes: side channel, fault injection, secure-to-nonsecure information leakage, unauthorized access, privilege escalation, hardware Trojan, and denial of service [15, 55, 59]. Using in-context training, LLMs generate high-level abstractions of plausible attack scenarios for each asset; assets without such scenarios are excluded, as shown in the prompt in Fig. 6.7. This analysis justifies asset inclusion to verification experts and, through the zero-shot reasoning capabilities of LLMs, extends beyond modular boundaries to capture system-wide security implications.

6.2.3.2 CWE Mapping

To provide additional context for the generated assets as well as refine the assets further, whether and to what extent the CWEs from the MITRE database [5] can be mapped to each primary asset is analyzed. Given the substantial size of the database, the analysis is structured to examine CWEs under each security objective one at a time, ensuring that no CWE is overlooked, as shown in the prompt in Fig. 6.7. Besides, only the security asset-relevant information in the database is used to prevent LLMs from being overwhelmed by extraneous information during the CWE mapping process. The rigorous mapping ensures that only the absolutely critical CWE(s), if any, are mapped to each asset. Finally, assets with no mappable CWE are removed.

6.2.3.3 Self-critique Revision

The asset annotations generated at the module level so far are challenged by a self-critique prompt and revised further, as shown in the prompt in Fig. 6.7. Using the input resources again, the asset relevance, RTL references, security-objective correctness, attack scenario coherence, and CWE appropriateness are rechecked, thereby improving both accuracy and reliability.

6.2.3.4 Degree of Influence (DoI) Calculation for Secondary Assets

After enumerating assets at the module level, how these assets influence one another within the design is assessed. As modular assets interact across boundaries, their interdependencies determine how local vulnerabilities may escalate into broader

Attack Analysis Prompt

$Asset Metadata$
$Hardware Security Attack Classes Context$
Task:
Your task is to assess whether the asset is truly a security asset of the {SoC} design or not. To that end, consider the following two matters:
1. How the asset can be attacked by the above-mentioned classes of hardware security attacks or other common attacks. To do this, analyze the provided RTL of the relevant RTL segments of the {module} to determine if and how such attacks could be implemented on the provided asset. Explain how the tampering of the RTLs would facilitate the attacks.
2. Whether the attack on the asset has system-level security impact on the {SoC} design.
Approach:
- Consider the following steps below in a conservative perspective. These steps guide how to reason about the provided asset.
1. Assess Attack Feasibility: Determine whether the provided asset is vulnerable to any of the specified attack classes. For each feasible attack, consider the following points:
   a. To determine whether the relevant RTLs of the asset can facilitate the corresponding attack, consider:
      - Which portion(s) of the RTL enable the attack?
      - Which specific signal, port, or FSM in the RTL is involved in the attack?
   b. Consider how any specific snippet from the provided RTL can be tampered with to implement the attack.
   c. Consider how the corresponding attack would occur at a higher abstraction level (e.g., physical access, voltage glitching, etc.).
2. Assess the System-level Impact: First, determine how significantly an attack through the asset would compromise the security of the asset's RTL segment. Then, depending on the significance of the corresponding RTL segment in the {SoC} design, determine how much the attack through the asset would impact the overall security objective of the {SoC} design.
Constraints:
- The provided asset might be attacked by multiple classes of security attacks or by no security attack at all.
- If you do not find any system-level security impact or likely attack scenario for the provided '{asset_name}' in the {IP} segment, do not consider it as a security asset anymore.
$Output Format$

CWE Mapping Prompt (Confidentiality)

$Assets List$
$Confidentiality-relevant CWE Database$
Task:
Your task is to find, for each asset whose security objective is confidentiality, whether any of the CWEs can attack the {Module} through the corresponding asset. Consider only the assets from the Initial Asset List whose 'Security Objective' field is Confidentiality. If no such asset, ignore this task.
Approach:
- Analyze the description of each of these CWEs from the Confidentiality-relevant CWE Database and provide reasoning on whether they can attack any of the confidentiality-critical assets or not. That means, consider only those CWEs that can directly attack through that corresponding asset; otherwise, exclude the moderately or less likely CWEs. Be sure to analyze each of the CWEs in the parenthesis ($CWE-IDs Lists$).
- If yes, list which of the CWE(s) attack the design through each of the considered assets. If you find any confidentiality-critical asset that not a single CWE would attack, remove that asset from consideration.
- Do this analysis under a conservative approach. The conservative approach of CWE mapping means that for each asset, only choose those CWEs from the Confidentiality-relevant CWE Database that would absolutely attack the {Module} through that asset; discard the rest of the CWEs.
$Output Format$

Self-critique Revision Prompt

$Input Design$ $Asset Metadata$
$CWE Database (concise)$
Perform the following checks for each asset:
- Relevance Check: Ensure the asset is a legitimate element of the {module} and whether the asset is declared in the RTL. Discard entries that are too generic or unrelated to the {module}.
Functionality Clarity: Ensure the functionality is clearly described and specific to the asset's role within the {module}. Vague descriptions must be flagged or rewritten.
Security Objective Accuracy: Check that the stated security objective (Confidentiality/Integrity/Availability) appropriately matches the nature and role of the asset.
Attack Analysis Coherence: Ensure that the appropriate attacks for the corresponding assets are selected, along with rational attack scenario and their explanation.
CWE pertinence: Ensure the mapped CWEs are most likely to attack the {module} through the asset. If a non-relevant CWE is mapped, exclude it.
$Output Format$

Fig. 6.7 Asset refinement prompts used in the asset identification workflow system risks. RTL analysis is used to extract hierarchical linkages among security-critical modules. Along each path, the most tamper-prone asset is designated as the *primary* asset, while the remaining assets are treated as *secondary*. Bit-level RTL connectivity is then analyzed to quantify the relationship between primary and secondary assets, and the influence is backtracked multiplicatively along each path to derive the final *DoI* metric for each secondary asset.

$$DoI = \frac{\begin{array}{c}\text{\# of primary-asset bits functionally}\\ \text{connected to the secondary asset}\end{array}}{\text{Total \# of bits of the primary asset}} \times 100\% \tag{6.1}$$

6.3 LLM-Assisted Security Asset Identification: Evaluation and Use Cases

In this section, we discuss some of the results provided by *LAsset*. We present case studies on the NEORV32 RISC-V SoC and several hardware IP blocks, under two configurations, to demonstrate the efficacy and generality of the *LAsset* workflow in identifying security assets across diverse architectures. *LAsset* uses OpenAI's state-of-the-art (SOTA) GPT-5 model for in-context learning-based asset identification. Furthermore, during the RAG-based learning in *Modular Spec. Generation* task, a chunk size of 1,000 characters with an overlap of 200 characters is used to preserve context across segments. The embedding model used during vectorization is text-embedding-ada-002 from OpenAI, and Facebook AI Similarity Search (FAISS) is employed to identify the top 20 document chunks for each retriever.

6.3.1 SoC: NEORV32 RISC-V Processor

For the SoC case study, the NEORV32 RISC-V processor is selected for evaluating the *LAsset* framework. After a full *LAsset* run, assets are identified across NEORV32 IPs, and their interconnections yield the *degree of influence* of secondary assets on their primary counterparts. For example, the primary asset *csr_rdata* in *neorv32_cpu* is linked to several secondary assets, including *csr_we_i* in *neorv32_cpu_cp_fpu*. This asset controls only 8 of 32 FPU CSR bits (*csr_fflags[4:0]*, *csr_frm[2:0]*), giving $8/32 = 25\%$. As *csr_o* maps fully to *csr_rdata*, the overall influence remains 25%. Associated threats include fault injection, unauthorized access, and hardware Trojans (CWE-1262, CWE-1247).

To validate the outputs, a reference list of security assets for each NEORV32 module is manually curated, based on established hardware security assurance principles. The manually annotated assets are further cross-validated using mainstream LLM chatbots, i.e., GPT, Gemini, and Grok. This is carried out to assess the

Table 6.1 Tabular comparison across different input configurations at SoC and IP levels. The aggregate number of true positives, false positives, false negatives, and true negatives across different IPs in the NEORV32 processor is shown, along with the accuracy compared against the reference listing. In addition, the same aggregates for the 21 IP blocks, along with the recall compared against the asset listing in Nath et al. [18] are shown

Input config	Actual class	Predicted positive	Predicted negative	Performance metric
SoC level (NEORV32)				
RTL + Spec.	Positive	272	30	Accuracy = 93.75%
	Negative	59	1064	
Only RTL	Positive	235	67	Accuracy = 91.16%
	Negative	59	1064	
IP level (21 IPs from OpenTitan, OpenCores, etc.)				
RTL + Spec.	Positive	151	11	Recall = 93.21%
	Negative	50	1735	
Only RTL	Positive	146	16	Recall = 90.12%
	Negative	50	1735	

reliability of the manual annotations against the ground truth, while also introducing an additional evaluation dimension beyond human review, which might be subject to error. The responses from *LAsset* are compared with these manual annotations, whose summary is shown in Table 6.1. Since *LAsset* deterministically computes *DoI* based on the relationships among the already validated primary and secondary assets, a separate validation for this is not performed.

6.3.2 IP: OpenTitan and OpenCores IPs

For validating *LAsset* at IP-level security asset identification, a total of 21 open-source crypto IPs, Interface-GPIO IPs, and Interface-peripheral IPs, from OpenTitan [4], OpenCores [3], and other repositories have been selected. A comprehensive primary-secondary classified assets listing for a 128-bit AES architecture is shown in Table 6.2. *LAsset*-generated asset outputs are compared with those from Nath et. al [47], as shown in Table 6.1, since both their tool-generated list of primary assets and the benchmark datasets for the same IPs are publicly available [18].

6.3.3 Performance Comparison and Analysis

In this subsection, the performance of *LAsset* is benchmarked, as shown in Fig. 6.8, to evaluate whether it aligns with the expected behavior and to analyze the underlying reasons for any observed discrepancies. A detailed, IP-wise summary of the results obtained by comparing the asset outputs with the reference list is presented in Table 6.3.

Table 6.2 Security assets for 128-bit AES design. For each of the design modules, primary-secondary assets are annotated with relevant security objectives and associated CWEs

Primary asset				
Conceptual asset	Structural asset	Security objective	Secondary asset (Design module)	CWE-ID
Key expansion logic output	Key	Confidentiality	Out-1 (expand-key-128)	1300, 1191, 1239, 1258
			Out-2 (expand-key-128)	1300, 1191, 1313
			Key (one-round)	1300, 1247, 1191, 1239, 1258, 1263
			Key (final-round)	1300, 1247, 1191, 1239, 1258, 1263
			Out (AES-128)	1247, 1319, 1384
Final encrypted output	Out	Availability	State-out (one-round)	1247, 1261, 1313
			State-out (final-round)	1247, 1261, 1313
			State	1300, 1191, 1258
Initial data processing state	State	Confidentiality	State-in (one-round)	1300, 1247, 1323, 1313, 1263
			State-in (final-round)	1300, 1247, 1323, 1313, 1263
			State (AES-128)	1300, 1191, 1258
			State (table-lookup)	1300, 1247, 1319
			P0, P1, P2, P3 (table-lookup)	1300, 1247, 1319

6.3.3.1 SoC

From Table 6.1, it is evident that the *Spec. + RTL* approach yields better results than the *only RTL* approach in SoC level. This signifies that RTL is the primary contributing resource, since RTL explicitly maps out all necessary design components and their functionalities, which LLMs can interpret effectively. However, the specifications' natural language description of operational behavior and interconnectivity helps the LLM better understand the design, more so than using RTL alone. Therefore, the takeaway from this analysis is that although an industry-standard specification alone is not enough to identify the security assets in design, it can work as a supplementary factor to the *LAsset* framework.

6.3.3.2 IP

Table 6.1 shows that the *Spec.+RTL* approach outperforms the *Only RTL* approach at the IP level. Compared with the work of Nath et al. [18], *LAsset* achieves a higher recall value of 93.21%, whereas their approach has 83.33%. These values are

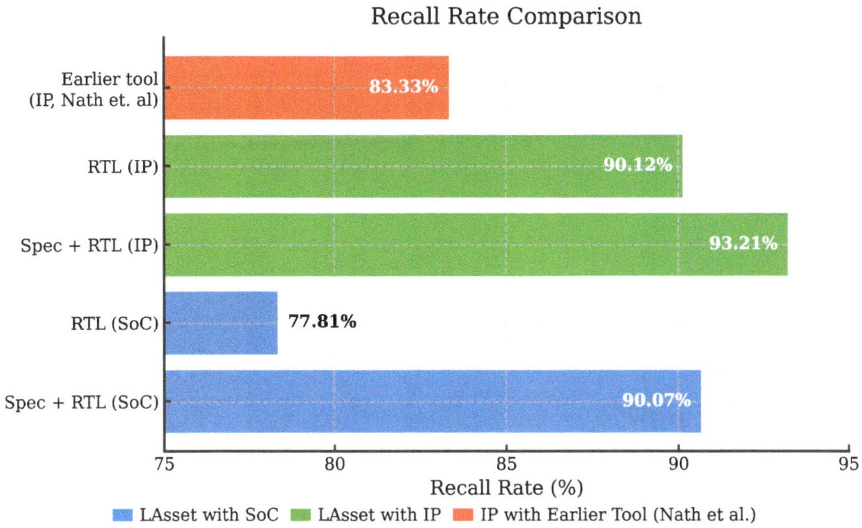

Fig. 6.8 Graphical comparison between *Spec. + RTL* and *only RTL* approaches of *LAsset* at SoC and IP levels. The *Spec. + RTL* approach performs better than *Only RTL* for both configurations. Furthermore, *LAsset* demonstrates superior performance than the one in Nath et al. [18]

Table 6.3 Summary of assets for *Spec. + RTL* vs. *RTL* only approaches across IP repositories. Each block reports reference asset list (golden), total design elements (elem.), total initial assets after *asset generation* (init.), total refined assets after *asset refinement* (ref.), true positive assets (TP), false negative assets (FN), and false positive assets (FP)

				Spec + RTL					Only RTL				
Repository	IP	Golden	Elem.	Init.	Ref.	TP	FN	FP	Init.	Ref.	TP	FN	FP
OpenTitan	hmac	11	268	15	11	8	3	3	14	12	8	3	4
	keymgr	42	688	46	44	37	5	7	44	42	36	6	6
	ot_gpio	5	114	8	7	5	0	2	7	7	5	0	2
	rc4	3	12	4	4	3	0	1	5	4	3	0	1
	robust_axi2apb	10	87	15	13	10	0	3	11	11	9	1	2
OpenCores	sha3	7	71	13	12	7	0	5	10	8	7	0	1
	simple_gpio	2	17	3	3	2	0	1	5	4	2	0	2
	tiny_aes	4	141	10	7	4	0	3	13	12	4	0	8
	ahb_m_wishbone	7	31	12	9	7	0	2	9	8	6	1	2
	CPU	14	35	15	14	12	2	2	14	13	11	3	2
	PMP (CPU)	6	23	6	6	5	1	1	7	6	4	2	2
NEORV32	Debug transport module	5	25	7	7	5	0	2	8	6	4	1	2
	TRNG	7	39	9	8	6	1	2	7	7	5	2	2
	UART	10	44	12	11	10	0	1	9	9	8	2	1
Total		133	1595	175	156	121	12	35	163	149	112	21	37

measured against their manually tailored golden asset list. Upon closely reviewing this list, however, a number of actual assets are missing. For instance, in the case of the SHA-3 IP block, the round constants `rc1` and `rc2` should be included, as tampering with them would introduce errors into the transformation steps and undermine the integrity of the algorithm. Because the golden list in Nath et al. overlooks a significant number of such critical assets, those identified by *LAsset* are incorrectly flagged as false positives, even though they must be true positives. For this reason, a direct comparison of accuracy and F1 score between *LAsset* and Nath et al. would not be a meaningful assessment; instead, recall rate provides a more appropriate metric for comparison.

6.3.4 Time-Cost-Performance Comparison Across Different LLM Models

Although GPT-5 model has been employed to obtain the results of *LAsset* presented in this chapter, the framework's model-agnostic behavior is considered as well. To this end, the end-to-end flow is executed for five sample IP blocks, i.e., AES, SHA-3, GPIO, AXI-Adapter, and AHB3LITE, and the three outcomes, recall (%), runtime, and monetary cost, are compared among five different LLM models for each IP, as shown in Fig. 6.9.

To quantify the trade-offs, a model-agnostic utility (MAU) that aggregates the normalized metrics under user-specified priorities is defined:

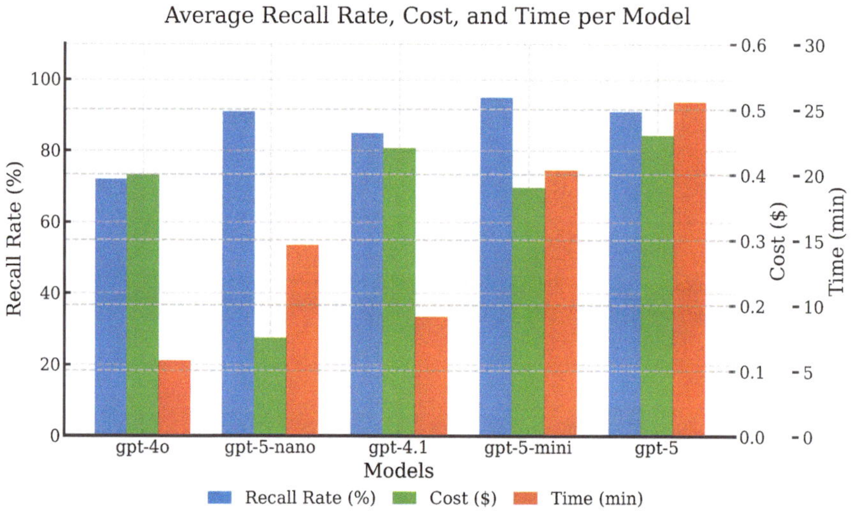

Fig. 6.9 Performance-time-cost comparison among different LLM models for *LAsset*. GPT-4o, GPT-4.1, GPT-5-nano, GPT-5-mini, and GPT-5—these five OpenAI models are considered

$$\begin{aligned}\text{MAU} &= A\,\widehat{100-\text{Rec}} + B\,\widehat{\text{Time}} + C\,\widehat{\text{Cost}},\\ \text{where,}\quad & A, B, C \geq 0,\quad A+B+C=1.\end{aligned} \tag{6.2}$$

MAU should be lower for an efficient model, as each variable should be as low as possible. Here, A, B, and C indicate the selection preference (e.g., performance-centric, latency-sensitive, or cost-constrained). For a given priority profile, the recommended model is the one that minimizes (6.2). For the experiments, performance is prioritized over cost and time, so the Eq. (6.2) is tuned with $A = 0.6$, $B = 0.1$, and $C = 0.3$. Under this configuration, *GPT-5-mini* is found to be the best candidate with the lowest (MAU).

6.4 Automated Security Property Generation Using LLM

Formal property verification (FPV) offers a more rigorous alternative, using mathematical methods to prove that a hardware design adheres to a set of specified behaviors and can offer security protection for the identified assets [25, 29]. However, the success of FPV is entirely dependent on the quality of these behavioral specifications, otherwise known as security properties. Designing a comprehensive set of properties that accurately captures all intended security behaviors is a complex task [8, 26, 27]. This manual authoring process has become a major bottleneck in the hardware design lifecycle, requiring specialized knowledge and extensive effort. Recent research has begun to explore the use of LLMs to assist in various verification tasks, such as generating formal assertions from design comments or natural language descriptions [24, 37, 39], which will be detailed in Chap. 5. While promising, these pioneering efforts often face limitations when dealing with the ambiguity of complex design documents or integrating multiple sources of information—such as threat models and register transfer level (RTL) code—into a cohesive analysis [37, 51]. The framework described in this chapter builds on existing research by creating a more structured and automated process [11, 12].

A threat model is a required structured representation of all the information that affects the security of an application. For hardware, this involves identifying potential threats and vulnerabilities. Many of these vulnerabilities can be mapped to the MITRE CWE database, which is a comprehensive catalog of known hardware and software vulnerability types [5]. The framework in this chapter focuses on four common threat categories:

Unauthorized Access: This category covers breaches where an attacker illicitly reads from or writes to protected resources, such as confidential memory locations, control registers, or critical configuration settings [42]. The goal is to compromise the confidentiality or integrity of the system.

Illegal States and Transitions: This threat involves an attacker manipulating the system to enter an undefined state or to transition between states in an unauthorized manner [43]. Such actions can disrupt normal operation, corrupt data, or create openings for further exploits.

Reset State Manipulation: This threat targets a system's reset functionality. An attacker might try to trigger a reset at a critical moment or alter the system's state during the reset sequence to bypass security mechanisms [43, 44].

Information Leakage: This category concerns the unintentional disclosure of sensitive data. Information can be leaked through software-accessible side channels, unintended physical emissions, or other covert channels that expose internal secrets like cryptographic keys or private data [41].

6.4.1 Automated Property Generation Framework

The framework was designed to analyze a hardware design by identifying its most critical components and their interactions, using this understanding to formulate applicable security requirements [11].

As illustrated in Fig. 6.10, the process is divided into three primary stages:

1. **Security Asset Identification:** The initial stage of the process involves pinpointing security-critical assets within the hardware design. The framework provides an LLM with high-level abstractions of the RTL code, the design specification document, and a description of the relevant threat model. By analyzing these inputs together, the LLM can identify which components, such as specific registers, memory blocks, or state machines, are most vital to the system's security and require protection. An example of a threat model description for secure asset identification can be seen below, where the threat model is unauthorized access:

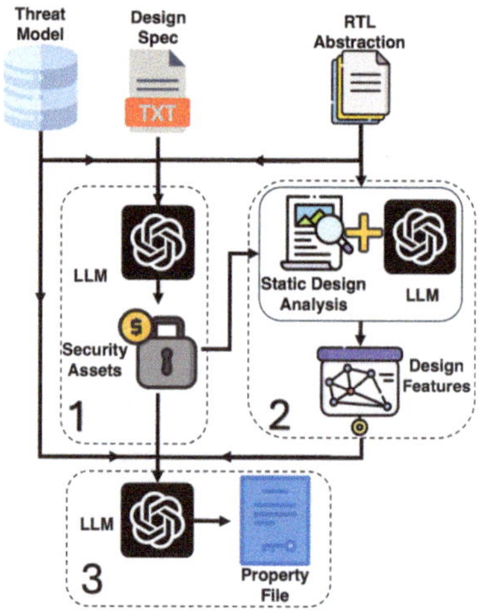

Fig. 6.10 An overview of the automated property generation framework, illustrating the flow from design inputs to natural language properties

Unauthorized Access: This threat model encompasses any scenario where an unauthorized entity (internal or external) gains access to or modifies sensitive resources within the design. This includes registers, memory locations, or configuration settings that are critical for the system's operation or that hold confidential data.

2. **Design Feature Extraction:** Following asset identification, the methodology extracts relevant design features using a two-pronged strategy. First, static analysis tools parse the RTL code to build a structural representation of the design, such as a control- and data-flow graph (CDFG). This graph reveals how data and control signals interact with the identified security assets. Concurrently, an LLM analyzes the RTL abstractions and documentation to extract more subtle, security-relevant information, such as the intended purpose of a masking feature or the logic behind a particular protocol.
3. **Natural Language Property Generation:** Lastly, the framework synthesizes the collected information. The list of security assets and their associated design features is fed back into the LLM, along with a refined and more detailed description of the threat model. This contextual input allows the LLM to generate a set of precise, actionable security properties expressed in clear, human-readable natural language. These properties define the expected secure behavior of the hardware, forming the knowledge for subsequent formal verification efforts.

6.4.2 How to Apply the Framework in Practice?

To understand the viability and usage of the described framework, an analysis was conducted on a case study for vulnerabilities in a Rivest-Shamir-Adleman (RSA) controller. The case study follows the described methodology, beginning with the definition of a relevant threat model and resulting in the discovery of a significant design flaw.

Listing 6.1 Snippet of the RTL Verilog code for the RSA FSM controller module

```
always @(posedge clk) begin
    if (start == 1'b0) begin
        FSM <= IDLE;
        round_index <= 4'h0;
        finished <= 1'b0;
    end else begin
        case (FSM)
            IDLE:
                if (start == 1'b1)
            INIT:
                FSM <= LOAD1;
            LOAD1:
                FSM <= LOAD2;
            LOAD2:
```

```
                    FSM <= MULT;
                MULT:
                    FSM <= SQR;
                SQR:
                    if (round_index == 4'b0010)
                        FSM <= RESULT;
                    else begin
                        FSM <= MULT;
                        round_index <= round_index + 4'b1;
                    end
                RESULT: begin
                    FSM <= IDLE;
                    finished <= 1'b1;
                end
            endcase
        end
end
endmodule
```

6.4.3 Threat Model Application

The analysis was guided by the "Illegal States and Transitions" threat model. This model addresses scenarios where a system could be forced into an invalid or unintended state, or transition between states in an unauthorized manner. Such behavior can disrupt normal operation, compromise data integrity, or expose vulnerabilities to an attacker. This threat model aligns with established hardware weaknesses, such as CWE-338 (insecure state management), which deals with the improper handling of system states that can lead to security bypasses or other exploits. By focusing on this threat, the analysis aims to verify that the FSM's control flow is robust and cannot be subverted.

6.4.3.1 Step 1: Security Asset Identification

With the threat model as a guiding principle, the first step in the process is to identify the security-critical assets within the hardware design. The framework analyzes an abstraction of the RTL code, the design specification, and the threat model description to pinpoint components whose compromise would directly impact the security posture of the system. For the RSA FSM controller shown in Listing 6.1, the framework identified the following key assets:

- FSM State Register: The core register holding the current state of the FSM. Its manipulation is the most direct way to violate the threat model, potentially allowing an attacker to bypass essential cryptographic steps, alter the sequence of operations, or force errors to extract sensitive information.

- `round_index` and `next_round_index`: These registers serve as counters that control the iterative process of modular exponentiation, a fundamental part of the RSA algorithm. Tampering with these counters could lead to incorrect cryptographic results, significantly weakening the encryption and making it susceptible to attack.
- `start` and `clk` Inputs: These primary control signals are also identified as assets. Unauthorized manipulation of the start signal could prematurely trigger or halt the RSA operation, leading to a denial-of-service condition. Similarly, manipulating the clock (`clk`) could induce timing errors, leading to faulty calculations and compromised security.

6.4.3.2 Step 2: Design Feature Extraction

Once the critical assets are identified, the framework performs a detailed extraction of relevant design features to understand how these assets are interconnected and controlled. This is achieved through a hybrid approach that combines the precision of static analysis with the interpretive capabilities of an LLM:

- Static Analysis: The framework utilizes a tool like *Pyverilog* to parse the RTL code and construct an abstract syntax tree (AST). By traversing this tree, the framework builds a CDFG, which provides a structured map of all data and control pathways leading to and from the identified security assets. This reveals which signals can influence the FSM state, how the round counters are updated, and what logic governs the `start` signal.
- LLM Interpretation: The LLM analyzes the RTL abstraction and design documentation to capture higher-level context and intended behavior that may not be apparent from the code structure alone. For the RSA FSM, this includes understanding the distinct roles of each FSM state (e.g., `IDLE`, `INIT`, `MULT`, `SQR`, `RESULT`) and the sequential logic of the cryptographic operation.

This approach yields a set of design features, including the defined FSM states, the legal transitions between them as specified in the RTL, and the control signals for these transitions.

6.4.3.3 Step 3: Security Property Generation and Vulnerability Discovery

The final step synthesizes the identified assets, extracted features, and a refined threat model description to generate natural language security properties. These properties are formal statements of the expected secure behavior of the design. Table 6.4 provides a consolidated summary of the key elements of the case study, linking assets to the properties generated to protect them.

The true value of this process lies in its ability to uncover vulnerabilities. The generated properties were manually converted into SystemVerilog Assertions (SVA) and tested using a pre-silicon fault-injection analysis methodology. This analysis

Table 6.4 Security properties of the RSA hardware module assets

Security asset	Rationale for selection	Associated design features	Generated security properties (examples)
FSM state register	Controls the operational flow of the RSA algorithm. Manipulation could bypass steps or force errors	States: IDLE, INIT, LOAD1, LOAD2, MULT, SQR, RESULT. Transitions: e.g., IDLE -> INIT on start signal	P1/P6: FSM shall only enter valid, defined states (e.g., 3'b000 to 3'b110). Any other state is illegal. P2: In IDLE, FSM remains in IDLE until start is asserted. P3: Transitions to states like LOAD1, MULT, etc., shall only occur from their defined preceding state. P7: The complete state sequence must follow the defined path from IDLE to RESULT and back to IDLE
round_index	Controls the number of rounds in modular exponentiation. Tampering can weaken encryption	Loop control: round_index increments in the MULT state. Transition Condition: The value of round_index determines the transition from MULT to SQR	P4: FSM remains in the SQR state for the number of rounds specified by the algorithm, as controlled by round_index
start, finished signals	Control the initiation and completion of the RSA operation	Initiation Signal: start triggers the transition from IDLE to INIT. Completion Signal: finished is asserted in the RESULT state	P2: FSM transitions from IDLE only when start goes from low to high. P5: FSM transitions from RESULT to IDLE only after finished is asserted high

revealed a critical flaw related to property P1, which asserts that the FSM must only occupy its legally defined states.

The RTL code for the FSM controller used a case statement to define state transitions but lacked a default case, leaving an unused state encoding (3'b111) unhandled. During fault simulation, an injected fault (simulating a single-event upset or a malicious bit flip) at the most significant bit (MSB) of the FSM register while in the LOAD1 state (3'b011) caused a transition to this illegal, unused state (3'b111). This immediately violated property P1.

More importantly, a subsequent fault was injected at the least significant bit (LSB) while in the illegal 3'b111 state transitioned the FSM directly to the SQR state (3'b110). This allowed the FSM to bypass several intermediate steps of the algorithm and proceed to the final RESULT state. As noted in the analysis, this state-skipping vulnerability could potentially expose intermediate encryption results. An attacker able to exploit this could use techniques described by Barenghi et al [13]. to reverse-engineer the encryption key or the original plaintext, completely compromising the security of the RSA implementation. The case study thus demonstrates a powerful connection: An abstract threat model, processed by the framework, generates a specific property that, when formally verified, uncovers a concrete and exploitable hardware vulnerability.

6.5 Practical Implications for Design and Verification Workflows

The natural language security properties generated by the framework serve as a vital bridge between different teams and stages in the SoC development lifecycle, offering significant practical value.

For design engineers, the properties provide clear, actionable insights into potential security vulnerabilities early in the design process. A property stating that a particular register should only be writable under specific conditions immediately guides the designer to check and, if necessary, harden the access control logic for that register. This proactive approach allows for security to be "designed in" rather than "bolted on" late in the cycle.

For verification engineers, the framework directly addresses the primary bottleneck in FPV: the labor-intensive and error-prone task of manually authoring security properties. By automating the generation of high-quality, SVA-ready properties, the framework significantly reduces the manual effort and specialized expertise required, thereby accelerating the overall security verification process.

Lastly, the systematic cataloging of generated properties into a database creates a powerful, reusable asset for an organization. This database facilitates future analysis, allows for the identification of recurring vulnerability patterns across different projects, and serves as a high-quality, domain-specific dataset for fine-tuning future generations of LLMs for even greater accuracy and relevance. This transforms the verification process from a series of discrete tasks into a cumulative learning system.

6.6 SoC-by-SoC Security Property Mapping

In the dynamic landscape of hardware development, the designers strive to enhance and update their existing product to meet evolving market demands by incorporating new features, address security vulnerabilities, and stay competitive. The key challenge in this scenario lies in efficiently continuing the security of the original hardware design to its upgraded version.

Traditionally, this process demands a considerable investment of time and manual effort from security engineers [7, 21, 35, 38]. Consider the intricacies involved: identifying potential threats, converting these into specific security requirements, and then expressing them as formal security properties. This manual translation process, though crucial, poses a hindrance to the agile adaptation

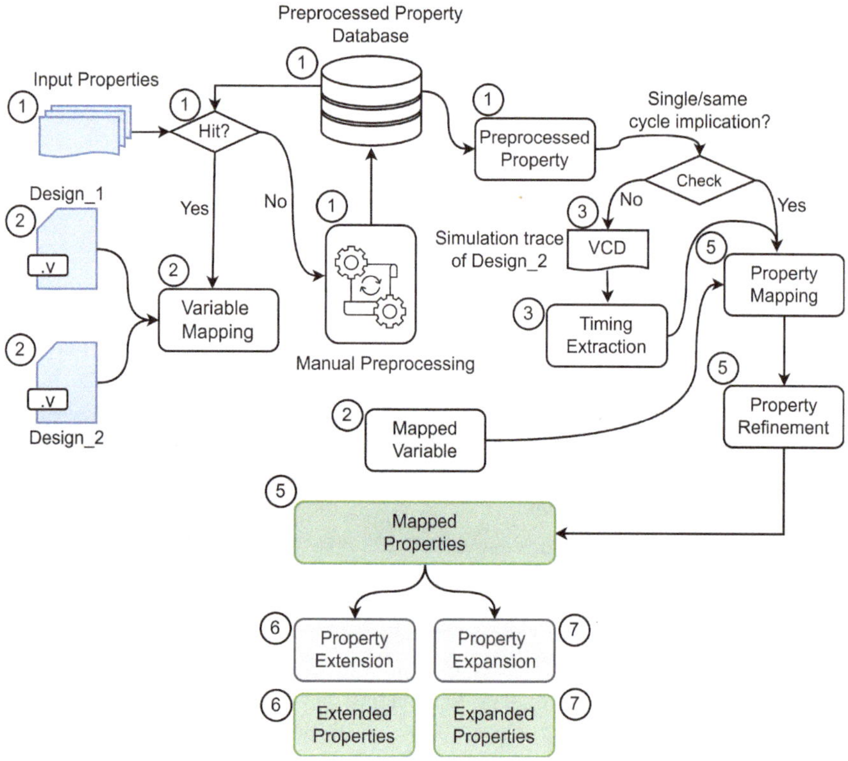

Fig. 6.11 A comprehensive depiction of the property mapping framework, organized into seven sequential steps as indicated by numerical labels in the figure. Each step further encompasses specific substeps

required in the fast-paced realm of hardware development. The complexity intensifies with each design upgrade iteration. To ensure the security continuity of the original design to the new version, the above steps should be iteratively conducted on the new updated design, creating a bottleneck that impedes the rapid deployment of upgraded products to the market.

To address this issue, a property mapping framework is presented in Fig. 6.11, offering a broad perspective. This framework operates on the RTL implementation of both the original and updated design versions, along with a set of security properties from the legacy design intended for mapping into the newer version of the design. The process unfolds through a sequence of seven pivotal steps, namely: (1) property preprocessing, (2) variable mapping, (3) timing extraction, (4) property mapping, (5) property refinement, (6) property extension, and (7) property expansion.

Each step plays a crucial role in the overarching property mapping methodology. The framework integrates a database containing preprocessed properties. Initially, the input security properties are examined within this database. If analogous properties are identified (hit), they undergo automated preprocessing. In instances where no match is found, manual preprocessing is executed, and the processed properties are appended to the database. During this phase, LLM assistance is utilized to interpret informal security specifications and generate preliminary formal property templates. The output of this stage is a set of preprocessed properties slated for mapping into the updated design version. Next, the variable mapping step identifies equivalent variables in the updated design. This process combines graph-based structural analysis with LLM-assisted semantic interpretation of RTL signals and available design documentation to improve mapping accuracy. Subsequently, a timing extraction phase derives the clock cycle information embedded within the properties while avoiding the propagation of potential design vulnerabilities. Lastly, the property refinement step conducts the fine-tuning of the properties for any additional antecedent variables, finalizing the mapped property. The ensuing sections elaborate on each step of the framework.

6.6.1 Property Preprocessing

The framework initiates by preprocessing the input property set, pursuing a two-fold objective. Firstly, it seeks to optimize properties for efficient memory consumption during formal verification. Secondly, it simplifies the property mapping process. The forthcoming discussion highlights the crucial aspect of property preprocessing, elucidated through the examination of two initial properties and their respective preprocessed forms.

Property P1 If *req* is asserted, *gnt* will eventually be asserted.

This is a common liveness property ensuring that a request is followed by a grant. For formal verification, the property is expressed in SystemVerilog as follows.

```
property p1;
@(posedge clk) req |-> s_eventually(gnt);
endproperty
assert property(p1);
```

Consider Fig. 6.12, illustrating a sample trace for Property P1. In this trace, the *req* signal rises at cycle 1 and remains high until the *gnt* signal is observed at cycle *n*,

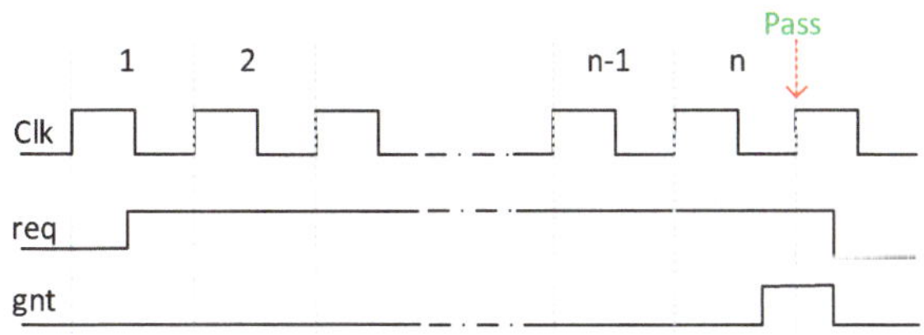

Fig. 6.12 Example trace for property P1, where a request signal rises and stays high up to and including the clock cycle at which gnt is asserted

validating Property P1. In the context of formal verification, at each positive edge of the *clk* signal, the presence of a property's antecedent necessitates the occurrence of the consequent at the specified clock cycle. For Property P1, each clock cycle where *req* is true requires the occurrence of *gnt*. If *gnt* is absent, a copy of the property is stored in memory.

Consider a scenario where *req* goes high and remains so for an extended period, say 1000 clock cycles, before *gnt* is asserted. In this case, 1000 copies of the property are created and stored in memory. All these copies are satisfied when a single assertion of *gnt* occurs, rendering this property memory intensive. While the intent is correct, the formulation is inefficient in terms of memory consumption. To address this, the property can be rephrased as follows.

```
property p1_1;
@(posedge clk) $rose(req) |-> s_eventually(gnt);
endproperty
assert property(p1_1);
```

In this modification, the antecedent of the property is revised from `req` to `$rose(req)`, signifying that the assertion of the *gnt* signal is contingent upon the observation of the rising edge of the *req* signal. In this instance, despite the *req* signal remaining high for an extended 1000 clock cycles, it experiences a single rising edge, serving as the antecedent for the property. Consequently, in contrast to the previous formulation, 1000 copies of the property are not stored in memory. Instead, a single copy is retained and resolved after the assertion of *gnt*.

This adjustment renders the property notably more memory efficient than its predecessor. Nonetheless, it still necessitates the storage of a single copy of the property until the assertion of *gnt*, spanning a duration of 1000 clock cycles in this instance. To enhance efficiency further, this property can undergo additional preprocessing as follows.

```
property p1_2;
@(posedge clk) req && !gnt |=> req;
endproperty
assert property(p1_2);
```

Property P1_2 addresses the same design aspect as properties P1 and P1_1, yet it excels in several key aspects:

- **Memory Efficiency**: P1_2 operates as a single-cycle implication property. Unlike P1 and P1_1, it eliminates the need to store copies of the property in memory for prolonged durations.
- **Ease of Debugging**: The complexity of debugging increases with the length of a counterexample (CEX). As a single-cycle implication, P1_2 facilitates straightforward debugging for verification engineers when encountering property failures.
- **Dependency on SystemVerilog Reference Manual Version (SV LRM)**: Reusability is a critical factor for properties, and their dependency on specific SystemVerilog versions can hinder this. Notably, the SystemVerilog keyword

Table 6.5 Comparison among original and preprocessed properties in terms of meeting design requirements, memory efficiency, debugging, and SystemVerilog language reference manual version dependency

Property	Meets requirement?	Memory efficiency	Easy to debug?	Depends on SV LRM version?
P1	Yes	Very low	No	Yes
P1_1	Yes	High	No	Yes
P1_2	Yes	Very high	Yes	No

`s_eventually` is exclusively supported from SV09 onward. In contrast, Property P1_2 demonstrates independence from any particular SystemVerilog version, ensuring its broad reusability.

Table 6.5 provides a comparison of the three properties, evaluating their efficacy in meeting requirements, memory efficiency, ease of debugging, and dependency on SystemVerilog Language Reference Manual (SV LRM).

Now let us consider another property **P2** and its preprocessing.

Property P2 *b* has to be stable, while *a* is true and remains stable until and including the clock cycle when *a* falls to "0."

Verification engineers have the flexibility to depict **P2** through different formulations in SystemVerilog assertions. Two alternative representations for **P2** in SystemVerilog assertions are outlined below.

```
property p2_1;
@(posedge clk) a |=> strong($stable(b)
[*0 $] ## 1 ($stable(b) && !a));
endproperty
assert property(p2_1);

property p2_2;
@(posedge clk) a |-> ($stable(b) s_until_with !a);
endproperty
assert property(p2_2);
```

While both SystemVerilog assertion representations of **P2** are equivalent and align with the requirement specified in the natural language property, they exhibit certain drawbacks. Primarily, they lack clarity and ease of understanding, both of which are crucial characteristics for effective property specifications. The complexity of these representations makes it challenging to discern their intended purpose at a glance. Moreover, these formulations are not inherently reusable as they are contingent on specific versions of the SystemVerilog Language Reference Manual (SV LRM). This dependency arises from the usage of *strong*, and *s_until_with*, which may not be universally supported across all SV LRM versions. Additionally, these representations represent infinite-length properties, involving [*0:$] and s_until_with, posing challenges for formal verification tools. Furthermore, in the event of property failures within this infinite-length sequence, debugging becomes a nontrivial task.

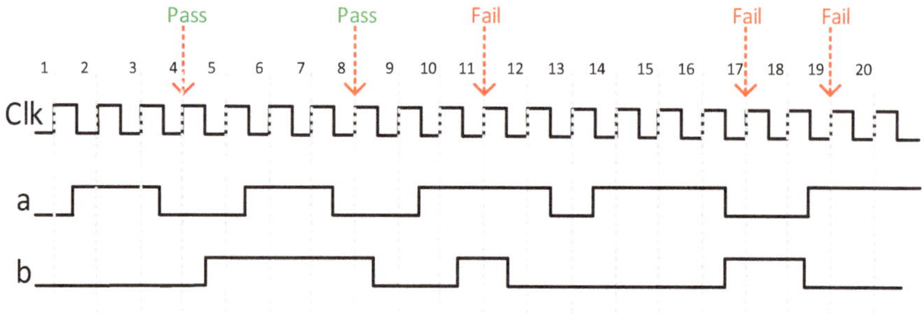

Fig. 6.13 Example trace for property P2, where *b* has to be stable, while *a* is true and remains stable until and including the clock cycle when *a* falls to "0"

To overcome these challenges, a fundamental shift in mindset is imperative when formulating properties. It is crucial to acknowledge that formal tools evaluate properties independently. Thus, a strategic approach involves decomposing complex properties into smaller, more straightforward components, mitigating the aforementioned difficulties. Illustratively, examining the trace for property **P2** in Fig. 6.13 reveals instances of both pass (at cycles 4 and 8) and failure (at cycles 11, 17, and 19) of the properties. Consequently, **P2** can be effectively disaggregated into the following two distinct properties.

```
property p2_3;
  @(posedge clk) a |=> $stable(b);
endproperty

assert property(p2_3);
```

```
property p2_4;
@(posedge clk) $fell(a) |-> $stable(b);
endproperty

assert property(p2_4);
```

Together, properties **p2_3** and **p2_4** form a cohesive solution that fulfills the specifications outlined for property **p2**. These properties are characterized by their simplicity, clarity, and reusability. Importantly, they employ a single-cycle implication, a feature that greatly facilitates the formal tool verification process. This simplified decomposition of the property not only enhances comprehensibility but also simplifies debugging, presenting a significant advancement over the previous representation, particularly in scenarios involving property failures.

As part of the property mapping framework, a database of manually preprocessed properties is maintained. This database consists of an initial set of original properties and their corresponding preprocessed versions. Every time a new property is fed into the framework, it looks into the database and searches for a property with a similar pattern. If there is a match, the associated preprocessed property is retrieved

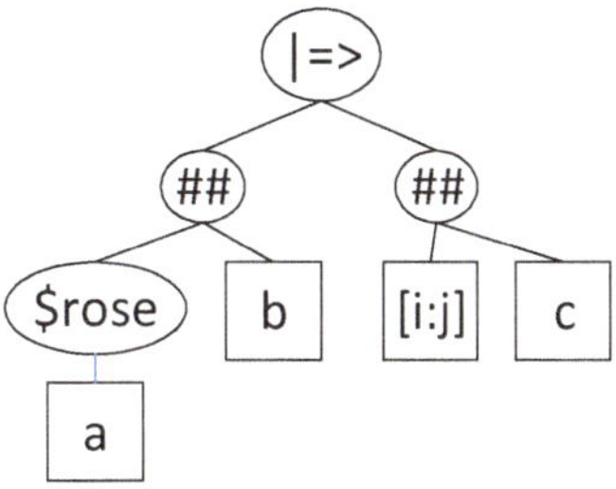

Fig. 6.14 Example of property graph

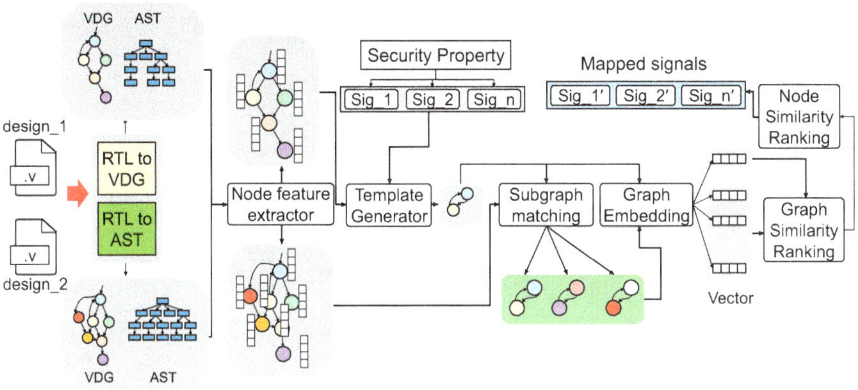

Fig. 6.15 Overall workflow of variable mapping

and used to guide the preprocessing of the new property. To enable efficient pattern matching and retrieval, both the original and preprocessed properties are represented and stored in graph form. An example of the graph format of the following property is shown in Fig. 6.14.

$rose(a) ## b|⇒ ## [$i$: j] $rose(c)

In the graph representation of the properties, registers, wires, I/O ports and constant values are represented as leaf nodes, while operators are considered non-leaf nodes. In this case, a, b, c, i, and j are leaf nodes.

6.6.2 *Variable Mapping*

Before mapping security properties from one design to another, it is crucial to identify the corresponding variables in the newer version that match those in the original design. However, this can be a challenging task as the equivalent counterpart may not have the same naming convention or exhibit the exact functional behavior as the original variable. Nonetheless, the equivalent variable should demonstrate a similar pattern of behavior relative to the overall design as the target variable in the original design. Figure 6.15 depicts the workflow for variable mapping. It consists of six major components: graph conversion, node feature extractor, template generator, subgraph matching module, graph-embedding module, graph similarity ranking module, and the node similarity ranking module. In the following section, all of these components are discussed in detail.

6.6.2.1 Graph Conversion

This module converts both the legacy and the newer version of the RTL design into two graphs: variable dependency graph (VDG) and abstract syntax tree (AST). VDG is used to represent the semantic behavior of the design, while the AST provides the syntactic representation of the design.

Variable Dependency Graph (VDG) The VDG [50] is a graphical representation used to depict the dependencies between variables in an RTL design. In VDG, each variable in the RTL design is represented as a node, and the dependencies between them are represented as edges. The VDG is a helpful tool for capturing the semantic behavior of an RTL design. On the other hand, AST is a syntactic representation of the design.

Definition 6.1 VDG is a weighted directed graph and is defined as G(**V**,**E**,**W**), where **V**, **E**, **W** denote the set of nodes, edges, and weights of the edges. Each node $v \in \mathbf{V}$ represents a variable in the RTL design. Each edge $e_{ij} \in \mathbf{E}$ denotes the dependency of variable v_i on variable v_j. Each $w_{ij} \in \mathbf{W}$ represents the weight of the edge $e_i j$, which considers both the data and control dependency between variable v_i and v_j.

Abstract Syntax Tree (AST) AST is the graphical representation of the lexical information of a program. In this case, AST is generated from the RTL implementation of the design to extract the syntactic features of the variables in the RTL design.

6.6.2.2 Node Feature Extractor

The variable mapping module utilizes the graph neural network (GNN) to find the equivalent variables and utilizes the graph- and node-level embedding, which necessitates the extraction of important node features. Toward this goal, two types of features are extracted: *semantic and structural features* and *syntactic features* of each node. As discussed in Sect. 6.6.2.1, the VDG provides a semantic representation of the RTL design. Therefore, semantic and structural features are extracted from the VDG, while syntactic features are derived from the corresponding AST.

Semantic and Structural Features A set of semantic and graph-level structural features is extracted for each node in the VDG. These features represent the characteristics of a particular node in the VDG in terms of functional dependency and structural information. Under this category, the following features are extracted by the node feature extractor:

- **In-Degree Centrality**: It is defined as fraction of nodes connected to the incoming edges of a node.
- **Out-Degree Centrality**: It is defined as fraction of nodes connected to the outgoing edges of a node.
- **Closeness Centrality**: This is a measure of how close a particular node is to the entire network [30].
- **Betweenness Centrality**: This centrality is originally introduced to measure the importance of a node as a channel of information flow in a network [17]. It is calculated as the fraction of shortest paths that pass through a given node.
- **Percolation Centrality**: It is a measure of relative impact of nodes based on their topological connectivity [53].
- **Page Rank**: Page rank is originally introduced by Google to rank the web pages [49]. However, this concept is widely used in graph theory to measure the importance of nodes of a graph. The page rank of a particular node depends on the importance of its incoming edges.
- **Trophic Levels**: Trophic levels represent a hierarchical measure assigned to nodes in a directed graph, indicating the order in which a quantity would flow through the graph. It demonstrates that the flow of a quantity would primarily pass from nodes at lower trophic levels before reaching nodes at higher trophic levels.
- **Relative Variable Complexity (RVC)**: RVC is introduced in [50]. It is a measure of the temporal and spatial distance of a particular variable with respect to a target variable. It also evaluates how difficult it is for a human analyst to understand the execution path between the source and target variables.
- **Relative Variable Importance**: For each variable, a relative importance score is computed with respect to all output ports of the design. A higher relative importance score of a variable with respect to the target variable indicates that the variable captures more subtle behaviors and has a greater impact on the target variable.

Syntactic Features A set of syntactic features of each node is also extracted to achieve a better representation of the variables of the RTL design. The following syntactic features are extracted from the AST of the RTL design:

- **Variable Type**: It indicates the type of the variable, i.e., wire, reg.
- **Variable Width**: This is the length of a multibit variable.
- **Port Type**: Indicates if the variable is input, output or in/out port.
- **If Frequency**: This is defined as the fraction of the usage of a variable in the if statement over the total number of if statement. It is a measure of how much the flow of the operation depends on a particular variable. A variable with higher value indicates the larger dependency of the routing of the operation on that variable.

- **Case Frequency**: The case frequency of a variable is the proportion of case statements in which the variable appears out of the total number of case statements. A variable with a high case frequency is likely to be a critical component in the design and used in many decision-making processes.
- **Always/Process Frequency**: This is defined as the proportion of the always/process statements in which a variable appears out of the total number of always/process statements. A variable with high always/process frequency may have a greater impact on the functionality of a design.

6.6.2.3 Template Generator

Template generator takes VDG of the original design and the variables of the original property as inputs. It then generates subgraph template for each of the variables. For each of the variables, it also requires a list of metadata as follows:

- **Root**: This should be set true, if the variable should be at the root/source of the subgraph template.
- **Leaf**: This field indicates whether the variable should be a leaf of the subgraph template.
- **Center**: If this is set true, then a subgraph template will be generated around the variable as a center of the subgraph.
- **Depth**: This is an integer value, which represents the depth of the subgraph template to be generated. If the variable is set to be root/leaf, depth indicates the upper limit of the hops the template generator needs to traverse. On the other hand, if the variable is set to be at the center, depth is treated as the radius of the subgraph.

6.6.2.4 Region of Interest (ROI) Selector

The ROI selector module plays a critical role in this approach, identifying candidate regions/subgraphs in the newer version graph that match the template subgraph obtained from the previous module. These candidate subgraphs represent potential regions where the equivalent variable of the target variable is likely located. By selecting these regions of interest (ROIs), the search space is reduced effectively, addressing the scalability challenges in larger designs. The interaction between the template generator and the ROI selector is depicted in Fig. 6.16. Algorithm 6.2 outlines the steps involved in the ROI selection process. For subgraph matching, the VF2 algorithm [20] is employed. The ROI selection process follows an iterative approach. It begins with an initial template subgraph of lower depth and identifies all possible subgraphs in the newer design that match the template. This iterative process continues, refining the template subgraph by incrementally increasing its depth and searching for matched subgraphs in the newer design graph. The process terminates when a fixed point is reached, indicating that the number of matched

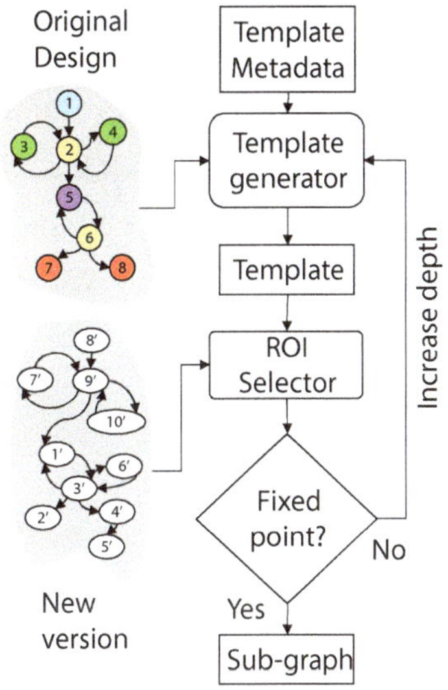

Fig. 6.16 Iterative ROI selection

subgraphs no longer changes. This approach effectively minimizes the number of matched subgraphs and reduces the search space for identifying the equivalent variables.

Algorithm 6.2 ROI selection

Input: Template subgraph T, initial depth d, newer version graph G
Output: Set of matched subgraphs M

$depth \leftarrow d$
$template \leftarrow T$
$matched \leftarrow \{\}$
repeat
  $prevMatched \leftarrow matched$
  $depth \leftarrow depth + 1$
  $template \leftarrow$ GENERATENEWTEMPLATE($template, depth$)
  $matched \leftarrow$ FINDMATCHEDSUBGRAPHS($template, G$)
until $matched = prevMatched$
return $matched$

6.6.2.5 Subgraph Ranking

Once the fixed point is obtained based on the number of matched subgraphs found, it becomes imperative to rank the candidate subgraphs in order to identify the most similar subgraph among the set of candidates. To achieve this, a state-of-the-art approach is employed using graph neural networks (GNN) for graph embedding, facilitating the ranking of subgraphs. This module leverages GNN to perform graph embedding, converting all subgraphs in the candidate set, as well as the subgraph template from the original design, into fixed-length vectors. Subsequently, the similarity between the template and each subgraph in the candidate list is computed

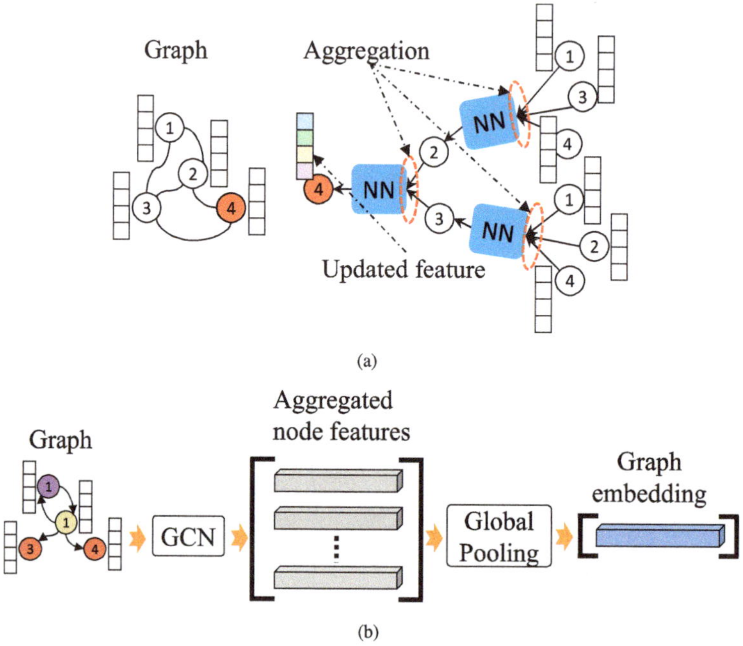

Fig. 6.17 (**a**) GNN layer (**b**) Graph embedding

using the cosine similarity metric. The node-embedding and graph-embedding workflows are depicted in Fig. 6.17.

For node embedding, the GraphSAGE architecture of GNN [32] is adopted due to its flexibility in selecting aggregation functions and considering self-embedding from the previous layer during message transformation, thereby mitigating information loss. This is particularly advantageous in this case, where the input graphs, consisting of the template subgraph and candidate subgraph found in the newer version of the design, are relatively small in size. Preserving vital information during embedding is of paramount importance, and GraphSAGE effectively accomplishes this by facilitating efficient message transformation and aggregation functions. The node-embedding process, illustrated in Fig. 6.17a, entails constructing a computation tree for each node, transforming node features/messages using a linear function, followed by order-invariant feature aggregation. Subsequently, the aggregated feature vectors undergo a nonlinear activation function, such as ReLU.

The message transformation and aggregation of GraphSAGE can be defined as follows:

$$h_v^0 = X_v \tag{6.3}$$

$$h_v^k = \sigma(W_k.AGG(h_u^{k-1}, \forall u \in N(v)) + B_k.h_v^{k-1}) \tag{6.4}$$

$$AGG = \sum_{u \in N(v)} \frac{h_u^{k-1}}{|N(v)|}, \tag{6.5}$$

where h_v^k represents the embedding of node v at the kth layer of the GNN, $\sigma(.)$ denotes the nonlinear activation function, AGG indicates the aggregation function, $N(v)$ represents the set of neighboring nodes of node v, and W_k and B_k are the weight matrices for message transformation of the target node's neighbors and itself, respectively. The weight matrices include the learnable parameters, which get updated during the training by learning the important element of the neighboring nodes' features. In this case, the weight matrices are randomly initialized for the linear transformation of node features. As the framework does not perform node-level prediction, the primary objective is to generate vector representations of nodes that facilitate graph embedding and subsequent graph similarity computation.

After the node embedding, the embedded node features are utilized to achieve the graph-level embedding. The graph-embedding steps are shown in Fig. 6.17b. To get the graph-level embedding, a pooling layer is added to the graph convolutional neural (GCN) network. Both max and mean pooling are applied in this work, and the final embedding is achieved by the concatenation of these two pooling. Finally, the cosine distance is calculated between the embedding of the template subgraph and all other candidate subgraphs to rank the subgraphs based on their similarity with the template graph. The cosine similarity between two vectors A and B is calculated as follows:

$$Similarity(A, B) = \cos(\theta) = \frac{A.B}{||A||||B||}, \tag{6.6}$$

where θ is the angle between the vectors A and B. $||A||$ and $||B||$ are the length of the vectors A and B.

6.6.2.6 Node Similarity Ranking

This module plays a crucial role in identifying the most similar variable within the candidate subgraphs. Upon obtaining the ranked subgraphs, the distance between the target variable in the template subgraph and all nodes within each candidate subgraph is measured. This is achieved by calculating the cosine similarity metric between the feature vectors of the nodes, as described in Eq. (6.6). While it is highly probable that the equivalent variable resides within the candidate subgraph with the highest rank, node similarity is calculated for all candidate subgraphs. This is done to ensure accurate identification of the equivalent variable. Specifically, if the highest-ranked subgraph fails to exhibit a similarity score exceeding 90%, it suggests a potential misranking. In such cases, the node similarity is calculated for the next highest-ranked candidate subgraph, thereby enhancing the precision of the variable identification process.

6.6.3 Timing Extraction

Once the equivalent variables are identified in the new design, this module focuses on establishing the temporal relationships among them. To achieve this, a hybrid approach is employed that combines formal verification and simulation techniques to determine the temporal distance among the mapped variables. When the antecedent variables represent peripheral signals, simulation-based methods are employed, applying test patterns to fulfill the antecedent conditions. Conversely, for antecedent variables representing internal signals, the hybrid approach is utilized. To identify the antecedent activation scenario, a property representing the negation of the antecedent condition is constructed and formally verified. The expected verification failure produces a counterexample that satisfies the original antecedent condition. This counterexample facilitates the creation of targeted test patterns, significantly reducing the time required for test pattern generation. These test patterns are then utilized to simulate both the original and new design versions. The simulation results are recorded in value change dump (VCD) files, which are subsequently parsed to extract the temporal distance among the mapped variables.

6.6.4 Property Mapping

Integration of the obtained equivalent variables and their temporal distance is the final step in achieving the translated security property. In this step, a template-based approach is employed. Initially, the original property is converted into a graph-based template as shown in Fig. 6.14. The template graph represents variables and constant values as leaf nodes, while operators are considered nonleaf nodes. Leaf nodes are assigned tags to indicate whether they represent variables or constant values, as well as their association with the antecedent or consequent of the security property. Leveraging the variable mapping and temporal distance information obtained from previous steps, node replacement is performed within the property template. This replacement involves substituting the original leaf nodes with their respective mapped counterparts, resulting in the translated security property.

6.6.5 Property Refinement

During this phase, the mapped properties undergo refinement through the potential addition of new terms in their antecedents, as necessitated. For instance, an original property formulated as `A |-> B` could be mapped to any of the following properties in the updated design:

1. `A_mapped |-> B_mapped`
2. `A_mapped` $\vee$ `C |-> B_mapped`
3. `A_mapped` $\wedge$ `C |-> B_mapped`

In this context, `A_mapped` and `B_mapped` represent variables in the revised design, analogous to the variables `A` and `B` in the original design. The term `C` is an additional component that must be incorporated into the antecedent to fully characterize the property in the new design. The extraction of term `C` is facilitated by leveraging the goldmine open-source assertion mining tool [1]. Goldmine utilizes the simulation trace of a design and its RTL implementation to mine assertions relative to a target variable.

Goldmine also necessitates information on the clock cycle distance between the antecedent (`A_mapped`) and the consequent (`B_mapped`), a parameter obtained during the timing extraction phase of the framework. For instance, in the initial mapped property `A_mapped |-> B_mapped`, the clock cycle distance between them is 1, as denoted by the single-cycle implication (`|->`). In this scenario, the target variable is represented by the variable in the consequent, `B_mapped`. In the process of refining the property, Goldmine is supplied with the RTL design, simulation trace (Value Change Dump File), the target variable (`B_mapped`), and the clock cycle distance of 1. Subsequently, Goldmine generates a comprehensive set of possible assertions for `B_mapped` within the specified clock cycle distance. To illustrate, let us assume Goldmine extracts the following assertions for the target variable `B_mapped`:

1. `C |-> B_mapped`
2. `D |-> B_mapped`

In this instance, the variable `B_mapped` is contingent on either `C` or `D`, and this dependency can be consolidated as `C ∨ D` to form a unified antecedent. Subsequently, this unified antecedent is combined with the original antecedent in the mapped property through a logical AND operation, producing the following result.

Refined mapped property: `A_mapped ∧ (C ∨ D) |-> B_mapped`

6.6.6 *Property Extension*

The concept of property extension, initially introduced in [10], plays a crucial role in extending properties from one abstraction level to another. A similar approach is applied across different design versions. Property extension proves to be a vital concept as it enables the formulation of security properties that can adapt to additional attack surfaces, thereby identifying new vulnerabilities in the updated design version. While property mapping ensures the security continuity of the legacy design in the new version, it falls short in identifying novel vulnerabilities introduced during design enhancement and optimization. Property extension enhances this capability by identifying new vulnerabilities on top of maintaining the security continuity of the legacy design.

To illustrate the concept of property extension, consider Fig. 6.18, which depicts the interfaces of two versions of AES. In Fig. 6.18a, the interfaces include *plaintext*

Fig. 6.18 (**a**) Legacy AES (**b**) New version of AES

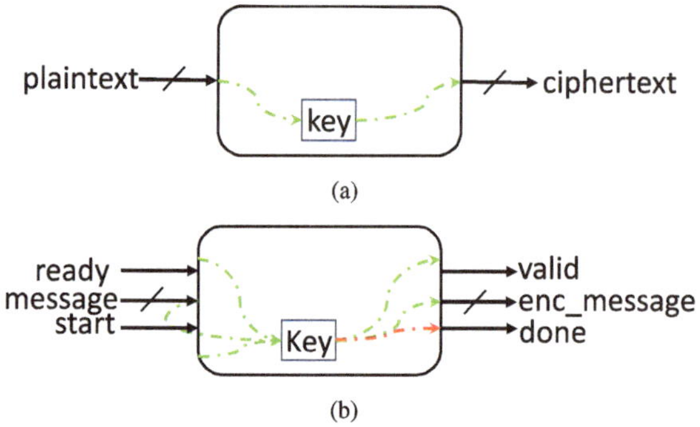

and *ciphertext*, whereas the new version of AES in Fig. 6.18b is augmented with a valid-ready protocol and additional handshaking signals such as *start*, *ready*, *valid*, and *done*.

The *key* register is a critical asset in both designs requiring protection against unauthorized access. In the legacy design, the information flow property below is employed to verify any potential *key* leakage to the specified target ports.

```
check_spv -from key -to plaintext ciphertext
```

Upon automatically mapping this property to the new version of AES, the resulting mapped property becomes

```
check_spv -from key -to message enc_message
```

While this mapped property effectively verifies potential information leakage in the newer version through ports (*message, enc_message*) that are equivalent to those in the legacy design (*plaintext, ciphertext*), it falls short of verifying information leakage through other ports. Property extension is employed to identify potential attack surfaces, prompting an extension of the mapped property to ensure verification against all possible vulnerabilities in the updated design version. An example of an extended property for this mapped property is as follows:

```
check_spv -from key -to message enc_message ready start valid
done
```

This illustration demonstrates a property extension specifically tailored for the vulnerability of *information leakage*. Notably, this approach is versatile and applicable to a spectrum of other vulnerabilities, including but not limited to timing side channel, access control violation, fault injection, and more.

6.6.7 *Property Expansion*

Property expansion, introduced as a technique in [10], was originally devised to ensure the persistence of a security goal across abstraction levels by generating

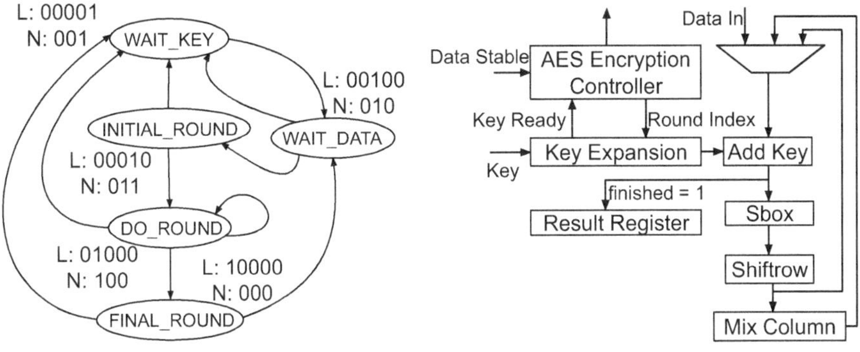

(a) (b)

Fig. 6.19 (**a**) Controller circuit of AES encryption algorithm (**b**) Data path of AES encryption algorithm

additional security properties. This concept is applied to maintain the continuity of the security goal across different design versions. To illustrate the concept of property expansion, consider Fig. 6.19, presenting both the controller and data path of the AES encryption module [3]. The state machine encompasses five states: **WAIT_KEY**, **WAIT_DATA**, **INITIAL_ROUND**, **DO_ROUND**, and **FINAL_ROUND**. In the newer version, the controller is enhanced in terms of area efficiency by transitioning from a one-hot state encoding (**L**) in the legacy design to a binary state encoding (**N**), requiring only three flip-flops and thus achieving area savings. Notably, **key** and *plaintext* are loaded during the **WAIT_KEY** and **WAIT_DATA** states, respectively. At the **INITIAL_ROUND**, *key* and *plaintext* are processed, while **DO_ROUND** orchestrates iterative operations such as *SubBytes*, *ShiftRows*, *MixColumns*, and *AddRoundKey*, contingent on the round number. Upon completing all rounds, **FINAL_ROUND** is reached, where the control signal *finished* is set high, and the encrypted result is stored in the result registers, as depicted in Fig. 6.19b.

The state **FINAL_ROUND** is designated as a protected state due to the risk that, if an attacker manages to circumvent intermediate states and gain access to this state, the prematurely encrypted result may be revealed in the output register. This could potentially assist attackers in analyzing premature cryptographic operations and extracting the cryptographic key. The following subsection delves into a plausible attack scenario that exploits a poorly implemented state encoding, resulting in access control violations within the design.

Fault-Injection Attack The vulnerability of a weakly implemented finite state machine (FSM) state encoding to fault-injection attacks has been demonstrated in prior research [23, 45]. Specifically, the fault-injection attack proposed in [46] highlights potential security risks associated with binary state encoding. Consider the enhanced version of the controller circuit featuring a three-bit state encoding, as depicted in Fig. 6.19a. After the logic synthesis of the register-transfer-level (RTL)

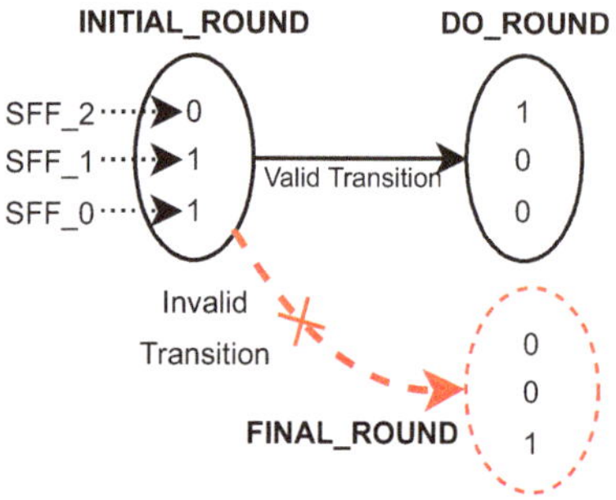

Fig. 6.20 Fault-injection attack during the transition from INITIAL_ROUND to DO_ROUND to gain unauthorized access to FINAL_ROUND

design, the three-bit encoding undergoes a transformation into three-state flip-flops (SFF) in the gate-level design.

Let us focus on a valid transition in the FSM, such as the transition from INITIAL_ROUND to DO_ROUND. It is important to note that no valid transition exists from INITIAL_ROUND to FINAL_ROUND. During this transition (from INITIAL_ROUND to DO_ROUND), SFF_2, SFF_1, and SFF_0 undergo changes from 0 to 1, 1 to 0, and 1 to 0, respectively (refer to Fig. 6.20). Exploiting the encoding illustrated in Fig. 6.20, an attacker can access FINAL_ROUND from INITIAL_ROUND by manipulating SFF_2 and SFF_0 to retain their values (0 and 1, respectively), while causing SFF_1 to transition from 1 to 0. If an attacker can compromise the setup time of SFF_2 and SFF_0 during the normal transition from INITIAL_ROUND to DO_ROUND, these flip-flops will remain unchanged (0 and 1, respectively), allowing SFF_1 to change its value to 0. This fulfills the requirement to align the state flip-flops' values with FINAL_ROUND (001), enabling unauthorized access to FINAL_ROUND directly from INITIAL_ROUND.

Property Expansion to Verify Access Control Violation This subsection explores the expansion of a security property to address access control violation vulnerabilities induced by fault-injection attacks. Consider the security property designed to ascertain the absence of access control violations in the AES controller module.

Property P3 FINAL_ROUND can only be accessed from DO_ROUND.
Assuming that this property is formally verified during the assessment of the legacy design, ensuring its resilience against such vulnerabilities, its counterpart in the enhanced version (binary-encoded controller) is established through the property mapping steps outlined in Sect. 6.6.4. However, formal verification of this mapped property in the new design does not automatically guarantee the absolute absence of access control violation vulnerabilities. This arises due to potential opportunities for attackers introduced by area optimization through binary encoding in the enhanced controller, as discussed in the preceding section. During valid transitions, fault injection can lead to the exploration of states not legitimately authorized for access during that specific transition. Therefore, supplementary properties are requisite to ensure comprehensive mitigation of access control vulnerabilities.

In this scenario, an additional property must be formulated alongside the mapped property, formally examining that the modified state encoding is not exploitable to breach access control. A binary metric named fault-injection (FI) feasibility metric, developed in [10], aids in creating such a property. Given the state encoding of a finite state machine, where S_x and S_y denote the present and next states of a valid transition, and S_p is the protected state, the FI feasibility metric is defined as follows:

$$\textbf{FI feasibility} = (\prod_{i=0}^{n-1}((\overline{S_{yi} \oplus S_{pi})} \parallel \overline{(S_{xi} \oplus S_{pi})})).$$

Here, i represents the i_{th} bit of an n-bit state encoding. An FSM exhibiting an FI feasibility value of 0 signifies the absence of fault-injection vulnerability within that particular FSM. Leveraging the FI feasibility metric, the SystemVerilog property for the three-bit binary state encoding of the controller depicted in Fig. 6.19a can be formulated as follows:

```
property p4;
  @(posedge clk) !(state == FINAL_ROUND) && !($past(state) ==
     FINAL_ROUND) |->
  !((!(state[0] & FINAL_ROUND[0]) | !($past(state[0]) & FINAL_ROUND[0]))
    &&
    (!(state[1] & FINAL_ROUND[1]) | !($past(state[1]) & FINAL_ROUND[1]))
    &&
    (!(state[2] & FINAL_ROUND[2]) | !($past(state[2]) & FINAL_ROUND[2])));
endproperty

assert property(p4);
```

This property analyzes all valid transitions within the finite state machine, excluding the strongly connected transition involving the protected state (FINAL_ROUND), and assesses the FI feasibility metric's value. In the event that the metric returns a true value, signifying vulnerability, the property fails. This failure implies that the state encoding is prone to a fault-injection attack. This additional property is formulated as a part of the property expansion methodology, ensuring a thorough verification against access control violations stemming from fault-injection attacks.

6.7 LLM-Assisted Property Mapping: Use Cases

To evaluate the effectiveness of the proposed property mapping framework, experiments were conducted on three widely used cryptographic benchmarks: AES, RSA, and SHA-256. Two versions of each design were obtained from OpenCores [3] and an open-source GitHub repository, with one designated as the legacy version and the other as the enhanced version. The enhanced designs featured optimizations such as binary state encoding, supplemental handshaking signals for SoC integration, and incorporation of a purpose-built malicious implant. This implant, active under specific trigger conditions, was intended to assess the framework's ability to identify vulnerabilities, a notable improvement over prior methodologies [61]. To comprehensively assess the framework, a security verification plan was conducted to verify the following vulnerabilities in the design: *timing side channel (TSC), information flow through primary input output ports (IL(PIO)), information flow through handshaking signals (IL(HS)), malicious implant (MI), FSM state encoding (FSE), fault injection (FI)*, and *denial of service (DOS)*. Consequently, 18 properties were formulated for these cryptographic designs, detailed in Table 6.6. Each property addresses specific security aspects, contributing to a holistic evaluation.

AES.P1: Validates that the "Finished" signal only asserts at the conclusion of the AES encryption, averting premature encrypted data storage.

Table 6.6 List of properties for AES, RSA, and SHA-256 used to evaluate the framework

Design	Property no.	Property
AES	AES.P1	Finished signal should be high only at `FINAL_ROUND`
	AES.P2	Finished signal should be high 10 cycles after the `INITIAL_ROUND`
	AES.P3	Key register should not flow to `data_stable`, `key_ready`, `finished`, and `round_type_sel`
	AES.P4	`next_FSM` register should not be observable or controllable by `data_stable`, `key_ready`, `finished`, and `round_type_sel`
	AES.P5	`FINAL_ROUND` should only be accessed from `DO_ROUND`
	AES.P6	`DO_ROUND` must occur before the `FINAL_ROUND`
	AES.P7	If the AES encryption starts, it eventually ends
RSA	RSA.P1	Finished signal should be high at `RESULT` state
	RSA.P2	`RESULT` state can only be accessed from `SQR` state
	RSA.P3	Both the `MULT` and `SQR` states must occur before reaching the `RESULT` state
	RSA.P4	Finished signal should be high nine cycles after the `INIT` state
	RSA.P5	`next_FSM` register should not be observable or controllable by the `start` and `finished` signals
	RSA.P6	If the RSA encryption starts, it eventually generates the encrypted data
SHA256	SHA.P1	done signal should be high only at the `st_sha_data_valid` state
	SHA.P2	`Data_input` state must occur before reaching the `st_sha_data_valid` state
	SHA.P3	`st_sha_data_valid` state should only be accessed from `st_sha_blk_nxt` state
	SHA.P4	`hash_control_st_reg` register should not be observable or controllable by the peripheral ports
	SHA.P5	If hashing starts, it eventually produces the hashed output

AES.P2: Ensures a constant clock cycle count for complete AES encryption, preventing timing side-channel vulnerabilities.

AES.P3: An information flow property confirming the encryption key remains secure through specified peripheral ports.

AES.P4: Verifies that the state register is neither observable nor controllable from untrusted ports.

AES.P5: Ensures protected state access is limited to the "DO_ROUND," averting potential key leakage.

AES.P6: Validates that encryption mandates progression through "DO_ROUND" for essential operations.

AES.P7: Confirms the design's availability, guarding against deadlocks. RSA Properties:

RSA.P1: Checks for premature encrypted data in the output register to prevent information leakage.

RSA.P2: Validates that only the "SQR" state can access the "RESULT" state, preventing access control violations.

RSA.P3: Ensures that crucial operations like Montgomery modular multiplication in "MULT" and "SQR" states are not bypassed.

RSA.P4: Verifies the absence of timing side channels.

RSA.P5: Confirms the internal state register is secure from potential untrusted port access.

RSA.P6: Ensures design availability by guarding against deadlocks. SHA-256 Properties:

SHA.P1: Verifies that no intermediate hashed data leaks to the output during the "st_sha_data_valid" state.

SHA.P2: Ensures the hashed output is collision-free by preventing bypasses in the "data_input" state.

SHA.P3: Confirms adherence to access control policies.

SHA.P4: Validates that the state register cannot be manipulated for unauthorized access.

SHA.P5: Property ensuring the availability of the design.

The outcomes of variable mapping, timing mapping, and formal verification for original, mapped, extended, and expanded properties are presented in Table 6.7. The verification process leveraged Cadence JasperGold [2], yet the methodology remains adaptable to various formal verification tools and property languages.

Variable Mapping Result In the variable mapping domain, the property translation exhibited a commendable accuracy, with only three out of 18 properties featuring an incorrect mapping for a single variable. Specifically, the "Next State" register was inaccurately mapped as the "Current State" for these instances. It is noteworthy that despite these discrepancies, the misassigned variables share strikingly similar functionalities within the design context.

Timing Mapping Result The timing mapping analysis demonstrated precise synchronization of clock cycle information across all three designs. This accuracy is exemplified through properties such as AES.P2, AES.P7, RSA.P4, RSA.P6, and

Table 6.7 Results of property translation (variable/timing mapping), extension, and expansion with detected vulnerabilities

		Property mapping							
		Var. map		Timing map		Detected vuln.		Property exp.	
Design	Property no.	Correct	Wrong	Orig.	Map.	Orig. prop.	Map. prop.	Detected vuln.	Detected vuln.
AES	AES.P1	3/3	0/3	–	–	–	–	–	FSE, FI
	AES.P2	3/3	0/3	10	10	–	MI	–	FSE, FI
	AES.P3	5/5	0/5	–	–	–	–	IL(HS)	–
	AES.P4	4/5	1/5	–	–	–	–	IL(HS)	–
	AES.P5	3/3	0/3	–	–	–	–	–	FSE, FI
	AES.P6	3/3	0/3	–	–	–	–	–	FSE, FI
	AES.P7	2/2	0/2	1	1	–	–	–	DOS
RSA	RSA.P1	3/3	0/3	–	–	–	–	–	FSE, FI
	RSA.P2	3/3	0/3	–	–	–	–	–	FSE, FI
	RSA.P3	4/4	0/4	–	–	–	–	–	FSE, FI
	RSA.P4	3/3	0/3	9	9	–	MI, TSC	–	FSE, FI
	RSA.P5	2/3	1/3	–	–	–	–	IL(HS)	–
	RSA.P6	2/2	0/2	1	1	–	–	–	DOS
SHA256	SHA.P1	3/3	0/3	–	–	–	–	–	FSE, FI
	SHA.P2	3/3	0/3	–	–	–	–	–	FSE, FI
	SHA.P3	3/3	0/3	–	–	–	–	–	FSE, FI
	SHA.P4	14/15	1/15	–	–	IL(PIO)	–	IL(HS)	–
	SHA.P5	2/2	0/2	1	1	–	–	–	DOS

SHA.P5. Notably, the framework effectively discerned and excluded the malicious implant from the scope of properties during the mapping process. This capability signifies the framework's resilience in not inadvertently incorporating malicious elements into the verified properties.

Design Vulnerability/Bug A significant limitation in prior methodologies [1, 61] resides in their susceptibility to incorporating design vulnerabilities or bugs as integral components of mapped properties during translation. To illustrate, let us examine the AES controller module depicted in Fig. 6.19a. In its standard configuration, each encryption operation adheres to a 10-clock cycle completion. However, envision an enhanced version harboring a malicious implant capable of bypassing all intermediate states when a specific trigger condition is met, accomplishing the operation in a mere three cycles. The conventional approach, during property mapping, would erroneously project the legacy clock cycle information onto a range [3–10], thus validating the nefarious behavior within the property itself. To surmount this inherent flaw, a simulation-based strategy is adopted. By subjecting the design to simulation with a restricted set of randomly generated plaintext inputs, the risk of triggering the rare condition is proactively mitigated. Given the rare activation likelihood of the trigger during simulation, especially with a limited input set, the enhanced versions of all three designs, each containing the concealed malicious implant, evaded getting picked up in the initial property mapping phase. This precautionary simulation step effectively defers the discovery of malicious functionality to the subsequent formal verification stage. As a result, the proposed approach enhances the resilience of the verification process against the inadvertent inclusion of design vulnerabilities during property translation.

Vulnerability Identification The efficacy of mapping, extension, and expansion is meticulously assessed through the verification of mapped, extended, and expanded properties. These properties, scripted in SystemVerilog, undergo verification utilizing the Cadence JasperGold formal verification tool [2]. The findings reveal that solely the legacy SHA256 design contains an information leakage vulnerability, identified by the original legacy property (SHA.P4). However, the mapped properties identified a vulnerability (MI) in the new version of AES and two vulnerabilities (MI, TSC) in the updated RSA version. In contrast, the extended properties unveil a total of four information leakage vulnerabilities across the three designs, as identified by AES.P3, AES.P4, RSA.P5, and SHA.P4. Remarkably, the expanded properties contribute to the identification of 14 vulnerabilities collectively for all benchmark applications. The properties, along with their expanded versions and the corresponding identified vulnerabilities, are as follows: AES.P1 (FSE, FI), AES.P2 (FSE, FI), AES.P5 (FSE, FI), AES.P6 (FSE, FI), AES.P7 (DOS), RSA.P1 (FSE, FI), RSA.P2 (FSE, FI), RSA.P3 (FSE, FI), RSA.P4 (FSE, FI), RSA.P6 (DOS), SHA.P1 (FSE, FI), SHA.P2 (FSE, FI), SHA.P3 (FSE, FI), SHA.P5 (DOS).

References

1. https://sites.google.com/view/goldmine-illinois/home.
2. https://www.cadence.com/en_US/home/tools/system-design-and-verification/formal-and-static-verification.html.
3. OpenCores Projects. https://opencores.org/projects. OpenCores. [Online]. Accessed: 2024-02-04.
4. OpenTitan: Open Source Silicon Root of Trust (RoT). https://opentitan.org/. Accessed: 2024-02-04.
5. CWE - Common Weakness Enumeration. https://cwe.mitre.org/index.html, 2024. Accessed: 2024-02-04.
6. Accellera Systems Initiative. Security Annotation for Electronic Design Integration Standard. https://www.accellera.org/images/downloads/standards/Accellera_SA-EDI_Standard_v10.pdf. Apr. 2021. [Online]. Accessed: 2024-02-04.
7. Sohrab Aftabjahani, Ryan Kastner, Mark Tehranipoor, Farimah Farahmandi, Jason Oberg, Anders Nordstrom, Nicole Fern, and Alric Althoff. Special session: Cad for hardware security-automation is key to adoption of solutions. In *2021 IEEE 39th VLSI Test Symposium (VTS)*, pages 1–10. IEEE, 2021.
8. Sohrab Aftabjahani, Mark Tehranipoor, Farimah Farahmandi, Bulbul Ahmed, Ryan Kastner, Francesco Restuccia, Andres Meza, Kaki Ryan, Nicole Fern, Jasper Van Woudenberg, et al. Special session: Cad for hardware security-promising directions for automation of security assurance. In *2023 IEEE 41st VLSI Test Symposium (VTS)*, pages 1–10. IEEE, 2023.
9. Baleegh Ahmad, Hammond Pearce, Ramesh Karri, and Benjamin Tan. LASHED: LLMs And Static Hardware Analysis for Early Detection of RTL Bugs. *arXiv preprint arXiv:2504.21770*, 2025.
10. Bulbul Ahmed, Fahim Rahman, Nick Hooten, Farimah Farahmandi, and Mark Tehranipoor. Automap: Automated mapping of security properties between different levels of abstraction in design flow. In *2021 IEEE/ACM International Conference On Computer Aided Design (ICCAD)*, pages 1–9. IEEE, 2021.
11. Avinash Ayalasomayajula, Nusrat Farzana Dipu, Mark M Tehranipoor, and Farimah Farahmandi. Automatic asset identification for assertion-based SoC security verification. *IEEE Transactions on Computer-Aided Design of Integrated Circuits and Systems*, 43(10):3264–3277, 2024.
12. Avinash Ayalasomayajula, Rui Guo, Jingbo Zhou, Sujan Kumar Saha, and Farimah Farahmandi. LASP: LLM assisted security property generation for SoC verification. In *Proceedings of the 2024 ACM/IEEE International Symposium on Machine Learning for CAD*, pages 1–7, 2024.
13. Alessandro Barenghi, Luca Breveglieri, Israel Koren, and David Naccache. Fault injection attacks on cryptographic devices. Proceedings of the IEEE, 100(11):3056–3076, 2012.
14. Abhishek Basak, Swarup Bhunia, and Sandip Ray. A flexible architecture for systematic implementation of soc security policies. In *2015 IEEE/ACM International Conference on Computer-Aided Design (ICCAD)*, pages 536–543. IEEE, 2015.
15. Md Kawser Bepary, Tao Zhang, Jingbo Zhou, Fahim Rahman, Farimah Farahmandi, and Mark Tehranipoor. Towards efficient gate-level electromagnetic side-channel leakage modeling and vulnerability assessment. *Journal of Hardware and Systems Security*, 9(1-2):24–37, 2025.
16. Mohammad Mahdi Bidmeshki, Yunjie Zhang, Monir Zaman, Liwei Zhou, and Yiorgos Makris. Hunting security bugs in SoC designs: Lessons learned. *IEEE Design & Test*, 38(1):22–29, 2020.
17. Ulrik Brandes. A faster algorithm for betweenness centrality. *Journal of mathematical sociology*, 25(2):163–177, 2001.
18. CalgaryISH. Asset Dataset for Crypto, GPIO, and Peripheral IPs using Partial Keyword and Signal Category-based Automatic Tool. https://github.com/CalgaryISH/Asset_Dataset_using_PKG. GitHub. [Online]. Accessed: 2024-02-04.

19. Gustavo K Contreras, Adib Nahiyan, Swarup Bhunia, Domenic Forte, and Mark Tehranipoor. Security vulnerability analysis of design-for-test exploits for asset protection in SoCs. In *2017 22nd Asia and South Pacific Design Automation Conference (ASP-DAC)*, pages 617–622. IEEE, 2017.
20. Luigi P Cordella, Pasquale Foggia, Carlo Sansone, and Mario Vento. A (sub) graph isomorphism algorithm for matching large graphs. *IEEE transactions on pattern analysis and machine intelligence*, 26(10):1367–1372, 2004.
21. Calvin Deutschbein, Andres Meza, Francesco Restuccia, Ryan Kastner, and Cynthia Sturton. Isadora: Automated information flow property generation for hardware designs. In *Proceedings of the 5th Workshop on Attacks and Solutions in Hardware Security*, pages 5–15, 2021.
22. Qingxiu Dong, Lei Li, Damai Dai, Ce Zheng, Jingyuan Ma, Rui Li, Heming Xia, Jingjing Xu, Zhiyong Wu, Tianyu Liu, et al. A survey on in-context learning. *arXiv preprint arXiv:2301.00234*, 2022.
23. Carson Dunbar and Gang Qu. Designing trusted embedded systems from finite state machines. *ACM Transactions on Embedded Computing Systems (TECS)*, 13(5s):1–20, 2014.
24. Zhiyuan Yan, Wenji Fang, Mengming Li, Min Li, Shang Liu, Zhiyao Xie, and Hongce Zhang. AssertLLM: Generating Hardware Verification Assertions from Design Specifications via Multi-LLMs. In Proceedings of the 30th Asia and South Pacific Design Automation Conference, ASPDAC'25, pages 614–621, New York, NY, USA, 2025. Association for Computing Machinery.
25. Farimah Farahmandi, Yuanwen Huang, and Prabhat Mishra. Formal approaches to hardware trust verification. In *The Hardware Trojan War: Attacks, Myths, and Defenses*, pages 183–202. Springer, 2018.
26. Farimah Farahmandi, M Sazadur Rahman, Sree Ranjani Rajendran, and Mark Tehranipoor. *CAD for hardware security*. Springer, 2023.
27. Farimah Farahmandi, Ankur Srivastava, Giorgio Di Natale, and Mark Tehranipoor. Introduction to the special issue on cad for security: Pre-silicon security sign-off solutions through design cycle, 2023.
28. Nusrat Farzana, Avinash Ayalasomayajula, Fahim Rahman, Farimah Farahmandi, and Mark Tehranipoor. SAIF: Automated Asset Identification for Security Verification at the Register Transfer Level. In *2021 IEEE 39th VLSI Test Symposium (VTS)*, pages 1–7, 2021.
29. Nusrat Farzana, Fahim Rahman, Mark Tehranipoor, and Farimah Farahmandi. SoC security verification using property checking. In *2019 IEEE International Test Conference (ITC)*, pages 1–10, 2019.
30. Linton C Freeman et al. Centrality in social networks: Conceptual clarification. *Social network: critical concepts in sociology. Londres: Routledge*, 1:238–263, 2002.
31. Yunfan Gao, Yun Xiong, Xinyu Gao, Kangxiang Jia, Jinliu Pan, Yuxi Bi, Yixin Dai, Jiawei Sun, Meng Wang, and Haofen Wang. Retrieval-augmented generation for large language models: A survey. *arXiv preprint arXiv:2312.10997*, 2(1), 2023.
32. Will Hamilton, Zhitao Ying, and Jure Leskovec. Inductive representation learning on large graphs. *Advances in neural information processing systems*, 30, 2017.
33. Khan Thamid Hasan, Md Ajoad Hasan, Nashmin Alam, Md Touhidul Islam, Upoma Das, and Farimah Farahmandi. AI-assisted hardware security verification: A survey and AI accelerator case study. *arXiv preprint arXiv:2604.01572*, 2026.
34. Md Ajoad Hasan, Dipayan Saha, Khan Thamid Hasan, Nashmin Alam, Azim Uddin, Sujan Kumar Saha, Mark Tehranipoor, and Farimah Farahmandi. LAsset: An LLM-assisted security asset identification framework for system-on-chip (SoC) verification. *arXiv preprint arXiv:2601.02624*, 2026.
35. Wei Hu, Alric Althoff, Armaiti Ardeshiricham, and Ryan Kastner. Towards property driven hardware security. In *2016 17th International Workshop on Microprocessor and SOC Test and Verification (MTV)*, pages 51–56. IEEE, 2016.
36. IEEE P3164 Working Group. Asset Identification for Electronic Design IP. https://ieeexplore.ieee.org/document/10496567. Apr. 2024. [Online]. Accessed: 2024-02-04.

37. Rahul Kande, Hammond Pearce, Benjamin Tan, Brendan Dolan-Gavitt, Shailja Thakur, Ramesh Karri, and Jeyavijayan Rajendran. Llm-assisted generation of hardware assertions. *arXiv preprint arXiv:2306.14027*, 2023.
38. Hareesh Khattri, Narasimha Kumar V Mangipudi, and Salvador Mandujano. HSDL: A security development lifecycle for hardware technologies. In *2012 IEEE International Symposium on Hardware-Oriented Security and Trust*, pages 116–121. IEEE, 2012.
39. Bhabesh Mali, Karthik Maddala, Vatsal Gupta, Sweeya Reddy, Chandan Karfa, and Ramesh Karri. ChIRAAG: ChatGpt informed rapid and automated assertion generation. In *2024 IEEE Computer Society Annual Symposium on VLSI (ISVLSI)*, pages 680–683. IEEE, 2024.
40. Andres Meza and Ryan Kastner. Information flow coverage metrics for hardware security verification. *arXiv preprint arXiv:2304.08263*, 2023.
41. MITRE. CWE-200: Information Exposure. https://cwe.mitre.org/data/definitions/200.html, 2024. Accessed: 2024-05-21.
42. MITRE. CWE-284: Improper Access Control. https://cwe.mitre.org/data/definitions/284.html, 2024. Accessed: 2024-05-21.
43. MITRE. CWE-338: Insecure State Management. https://cwe.mitre.org/data/definitions/338.html, 2024. Accessed: 2024-05-21.
44. MITRE. CWE-614: Sensitive Cookie in HTTPS Session Without 'Secure' Attribute. https://cwe.mitre.org/data/definitions/614.html, 2024. Accessed: 2024-05-21.
45. Adib Nahiyan, Farimah Farahmandi, Prabhat Mishra, Domenic Forte, and Mark Tehranipoor. Security-aware FSM design flow for identifying and mitigating vulnerabilities to fault attacks. *IEEE Transactions on Computer-aided design of integrated circuits and systems*, 38(6):1003–1016, 2018.
46. Adib Nahiyan, Kan Xiao, Kun Yang, Yeir Jin, Domenic Forte, and Mark Tehranipoor. AVFSM: A framework for identifying and mitigating vulnerabilities in FSMs. In *Proceedings of the 53rd Annual Design Automation Conference*, pages 1–6, 2016.
47. Subroto Kumer Deb Nath and Benjamin Tan. Toward automated potential primary asset identification in verilog designs. *arXiv preprint arXiv:2502.04648*, 2025.
48. Livinus Obiora Nweke and Stephen Wolthusen. A review of asset-centric threat modelling approaches. 2020.
49. Lawrence Page, Sergey Brin, Rajeev Motwani, and Terry Winograd. The pagerank citation ranking: Bringing order to the web. Technical report, Stanford infolab, 1999.
50. Debjit Pal, Spencer Offenberger, and Shobha Vasudevan. Assertion ranking using RTL source code analysis. *IEEE Transactions on Computer-Aided Design of Integrated Circuits and Systems*, 39(8):1711–1724, 2019.
51. Sudipta Paria, Aritra Dasgupta, and Swarup Bhunia. Divas: An llm-based end-to-end framework for soc security analysis and policy-based protection. *arXiv preprint arXiv:2308.06932*, 2023.
52. Eric Peeters. SoC security architecture: Current practices and emerging needs. In *Proceedings of the 52Nd Annual Design Automation Conference*, pages 1–6, 2015.
53. Mahendra Piraveenan, Mikhail Prokopenko, and Liaquat Hossain. Percolation centrality: Quantifying graph-theoretic impact of nodes during percolation in networks. *PloS One*, 8(1):e53095, 2013.
54. Juan Portillo, Eugene John, and Seetharam Narasimhan. Building trust in 3pip using asset-based security property verification. In *2016 IEEE 34th VLSI Test Symposium (VTS)*, pages 1–6. IEEE, 2016.
55. Pantha Protim Sarker, Tianze Kan, Jingchen Liang, Ozgur Tuncer, Bo He, Zelin Lu, Sudarshan Mallu, Lang Lin, Norman Chang, Riku Hasegawa, et al. ML-EMFI: A machine learning-driven pre-silicon electromagnetic fault injection security evaluation for robust IC design. In *2025 IEEE International Symposium on Hardware Oriented Security and Trust (HOST)*, pages 57–66. IEEE, 2025.
56. Max Schäfer, Sarah Nadi, Aryaz Eghbali, and Frank Tip. An empirical evaluation of using large language models for automated unit test generation. *IEEE Transactions on Software Engineering*, 50(1):85–105, 2023.

57. Patanjali Slpsk, Jonathan Cruz, Sandip Ray, and Swarup Bhunia. Protects: Secure provisioning of system-on-chip assets in untrusted testing facility. In *2023 IEEE International Test Conference India (ITC India)*, pages 1–6. IEEE, 2023.
58. Shams Tarek, Dipayan Saha, Khan Thamid Hasan, Sujan Kumar Saha, Mark Tehranipoor, and Farimah Farahmandi. Assertain: Automated security assertion generation using large language models. *arXiv preprint arXiv:2604.01583*, 2026.
59. Mark Tehranipoor, N Nalla Anandakumar, and Farimah Farahmandi. Hardware Trojan insertion. In *Hardware Security Training, Hands-on!*, pages 73–92. Springer, 2023.
60. Jan von der Assen, Jamo Sharif, Chao Feng, Christian Killer, Gérôme Bovet, and Burkhard Stiller. Asset-centric threat modeling for ai-based systems. *arXiv:2403.06512*, 2024.
61. Rui Zhang and Cynthia Sturton. Transys: Leveraging common security properties across hardware designs. In *2020 IEEE Symposium on Security and Privacy (SP)*, pages 1713–1727. IEEE, 2020.

Chapter 7
LLM-Driven IP Protection Strategies for SoCs

Hadi Kamali and Kimia Azar

Over the years, hardware intellectual property (IP) in modern system-on-chip (SoC) designs faces escalating security threats, including reverse engineering, IP piracy, overproduction, and hardware Trojan insertion. To counter these risks, design-time protection techniques such as logic locking and hardware obfuscation introduce key-based functionality and structural transformations that prevent unauthorized use. At the same time, recent advances in large language models (LLMs) have demonstrated strong capabilities in code understanding, generation, and reasoning, creating new opportunities, and new challenges, for hardware security. This chapter presents a comprehensive treatment of LLM-enabled IP protection in SoCs, with a primary focus on hardware obfuscation and logic locking. We first review the SoC IP threat model and established countermeasures and then introduce the key capabilities of LLMs that are relevant to hardware design and security workflows. Building on this, we examine how LLMs can enable IP protection across the SoC design flow, from automated insertion of locking and obfuscation structures, to security-aware design transformations, to verification and resilience assessment. We further discuss the integration of LLMs with electronic design automation (EDA) tools for adversarial security evaluation, i.e., attacks implementation, design interpretation, Trojan insertion and detection, and robustness assessment of protected designs. A detailed case study illustrates an LLM-driven framework for circuit interpretation and logic locking that strengthens resistance against widely used SAT-based de-obfuscation. Finally, we analyze the risks and security concerns associated with deploying LLMs in IP leakage, model misuse, and trustworthiness issues and outline mitigation strategies. We conclude by highlighting open research challenges and emerging opportunities, such as hardware-security-specialized LLMs, secure training pipelines, and LLM-assisted verification frameworks.

H. Kamali (✉) · K. Azar
University of Central Florida, Orlando, FL, USA
e-mail: hadi.mardanikamali@ucf.edu

F. Farahmandi et al. (eds.), *LLM for SoC Design and Security*,
https://doi.org/10.1007/978-3-032-34116-7_7

7.1 Background: IP Protection in SoCs

Modern system-on-chip (SoC) designs incorporate substantial amounts of hardware intellectual property (IP), often assembled from multiple third-party cores and fabricated through globally distributed supply chains [28, 43]. Consequently, a single silicon die typically integrates IP blocks originating from diverse vendors located across different geographic regions. For example, the power management circuitry of an IC may be sourced from an analog IP provider in Texas, while the cryptographic core may be licensed from a vendor in Europe [141]. From an abstraction standpoint, hardware IPs are commonly categorized into three classes: *soft IP*, *firm IP*, and *hard IP*. Soft IP is typically delivered at the RTL level and described using hardware description languages (HDLs) such as Verilog. Firm IP is provided at the gate level after logic synthesis, while hard IP corresponds to layout-level implementations or hardened macros (e.g., embedded SRAM blocks) that are tightly coupled to specific process technologies.

While this horizontally integrated design model significantly reduces cost and accelerates time to market (TTM) [104, 127], it also introduces multiple trust boundaries across the SoC lifecycle [1, 76, 77]. As illustrated in Fig. 7.1, modern design houses routinely depend on external IP vendors [46, 98, 109], offshore foundries [140], third-party verification services, and contract manufacturers [34, 78, 132]. Any untrusted entity within this distributed ecosystem may attempt to extract, clone, modify, or overproduce integrated circuits [47, 89]. Reported industry losses due to semiconductor IP theft and cloning reach billions of dollars annually, highlighting the significant economic and security implications of these threats [63, 103]. Common attack vectors include (see Fig. 7.2):

1. ***IP piracy, cloning, and overproduction:*** Theft of RTL or gate-level netlists for illegal reuse or resale [31, 64, 75]. This can include unauthorized fabrication of

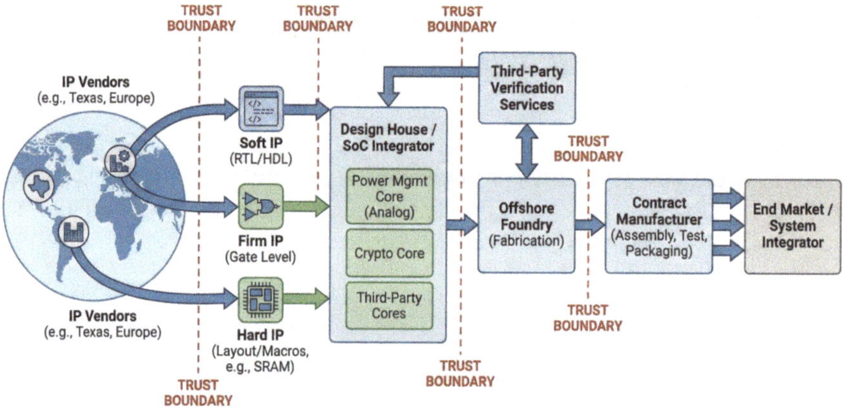

Fig. 7.1 The main steps of IC supply chain in horizontal model

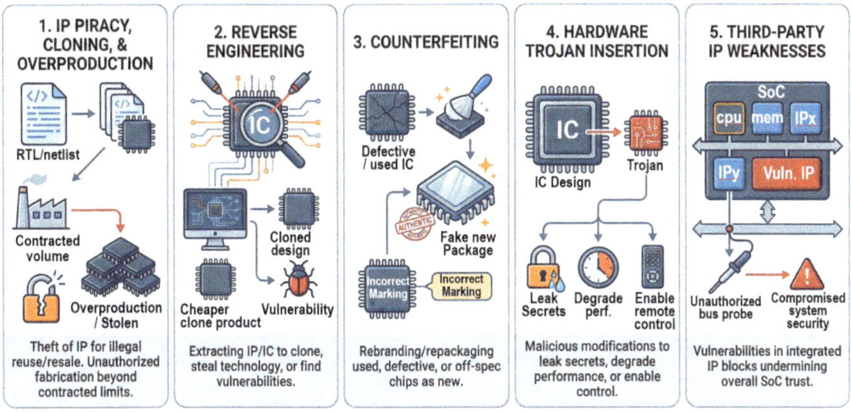

Fig. 7.2 Threats in IC supply chain, locations, and aims

extra chips beyond contracted volumes [47, 105, 121, 126]. With access to the IP at different layers of abstraction in different stages of the IC supply chain, an adversary could use the IP illicitly as is (e.g., by claiming its ownership and overproducing it).

2. ***Reverse engineering:*** It aims for the unauthorized party to retrieve the IP or the entire IC (i) to manufacture and package an identical product, which can be sold at a cheaper price by illegally rebranding it; (ii) to be used in their own IC products without compensating the rightful IP vendors; (iii) to steal the technology and integrate their own technology on top for bringing an advanced one faster into the market; and (iv) to find vulnerabilities (e.g., weak security primitives) [15, 49, 57, 89, 110].
3. ***Counterfeiting:*** Rebranding or repackaging cloned chips as authentic parts [31, 47, 50, 121]. A counterfeit component (i) is an unauthorized copy; (ii) does not conform to OCM design, model, and/or performance standards; (iii) is not produced by the OCM or is produced by unauthorized contractors; (iv) is an off-specification, defective, or used OCM product sold as new or working; or (5) has incorrect or false markings and/or documentation.
4. ***Hardware Trojan insertion:*** Malicious circuit modifications (tampering) that leak secrets, degrade performance, or enable remote control [16, 23, 26, 106]. Tampering can occur in both targeted and untargeted attacks. In a targeted attack, an adversary who has partial or complete knowledge of the hardware design under-attack may insert a Trojan designed to achieve a specific malicious effect [24, 25, 123]. In an untargeted attack, on the other side, the adversary's goal is sabotage or denial-of-service attack by exploiting specific portions of the design such as the power supply net or clock pin, without complete knowledge of the underlying design [51, 93].
5. ***Third-party IP weaknesses:*** Vulnerabilities in integrated IP blocks that undermine overall SoC trustworthiness [22, 48, 82, 116]. As an instance, integrating a

malicious IP can bring the potential of eavesdropping in an IC by probing (in an unauthorized way) the accessed bus data. This is more critical, when other IPs are unaware of such vulnerabilities and transmit security-critical data. Because SoCs integrate many heterogeneous IP cores, a weakness in a single component can compromise the security and value of the entire system.

7.1.1 Countermeasures Landscape

These risks have motivated a range of design-time protection techniques that preserve correct functionality for authorized users while degrading usability and reproducibility for adversaries [41]. Classical IP protection approaches include logic locking and hardware obfuscation [18], as well as hardware watermarking [36, 107, 122] (which embeds identifiable signatures to prove ownership) and layout camouflaging [95] (which uses structurally similar but functionally different standard cells to hinder reverse engineering). These representative techniques are summarized below (see Fig. 7.3):

- ***Gate camouflaging:*** IC camouflaging replaces conventional standard cells with specially designed gates whose physical layouts obscure their true functionality [95]. A widely used implementation relies on true versus dummy contacts [138], enabling multiple logical functions to share an indistinguishable physical footprint. This structural ambiguity increases reverse engineering effort because an adversary cannot reliably infer gate functionality from layout inspection alone [95]. However, camouflaging often incurs nontrivial area, power, and delay overheads.

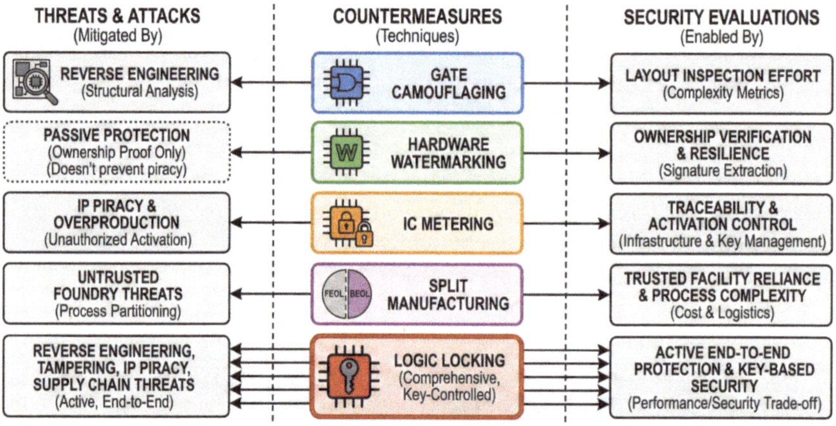

Fig. 7.3 Security threats in IC supply chain, countermeasures against these threats, and measuring the reliability and evaluation of countermeasures

- ***Hardware watermarking:*** Hardware watermarking embeds a unique signature into an IP core without altering its intended functionality, enabling postfabrication ownership verification [14, 53]. An effective watermark should be easy to embed and extract, incur minimal overhead, and remain resilient against removal or modification attacks [14, 52, 94, 122]. Despite extensive research, watermarking is generally considered a *passive* protection technique because it primarily supports ownership claims rather than preventing IP piracy, cloning, or overproduction [35].
- ***IC metering:*** Hardware (IC) metering enables IP vendors to control the activation and distribution of fabricated chips [67]. Passive metering provides device uniqueness and traceability, whereas active metering extends this capability by incorporating on-chip control logic (e.g., locks or monitors) that can enable or disable functionality. While effective for supply chain control, IC metering typically requires additional infrastructure for key management and activation.
- ***Split manufacturing:*** Split manufacturing partitions fabrication steps between trusted and untrusted facilities [54]. Typically, the front-end-of-line (FEOL) layers are produced in an untrusted high-end foundry, while the back-end-of-line (BEOL) interconnects are completed in a trusted facility. This approach primarily mitigates threats originating at untrusted foundries but introduces manufacturing complexity, cost overhead, and reliance on trusted partners.

Among these defenses, *logic locking* [4, 18, 85, 90, 91, 139] has emerged as a particularly comprehensive and flexible protection mechanism for modern IC designs. As SoC supply chains become increasingly distributed, threats arise across multiple lifecycle stages, rendering single-point solutions insufficient [108, 142]. Traditional IP encryption mainly protects early front-end stages (e.g., design exchange and synthesis) and remains vulnerable to side-channel leakage of decryption keys [69]. In contrast, logic locking can be inserted across multiple abstraction levels, from RTL [86] through gate-level [58] and even layout [84], and maintains postfabrication control, while allowing designers to selectively protect high-value modules.

Compared with watermarking [14, 35, 122] and fingerprinting [145], which are largely passive and may be removed without affecting functionality, logic locking actively perturbs circuit behavior through key-controlled structures, thereby providing stronger resistance to reverse engineering [89] and tampering [83] without exposing explicit ownership metadata. Relative to IC metering [67], which often depends on physical measurements and may incur higher validation overhead, logic locking achieves protection through structural modification with typically lower system-level impact [20]. Finally, although split manufacturing [126] addresses threats from untrusted foundries, it targets a narrower threat model and introduces significant process complexity. Logic locking, by contrast, can provide comparable protection at the layout level, while simultaneously addressing broader supply chain threats without requiring multiple fabrication facilities or major process changes.

7.1.2 Threat Landscape

In the context of IP protection (focusing on logic locking), de-obfuscation attacks can be broadly classified based on the adversary's assumptions and capabilities [57, 143]. From the attacker's perspective, attacks that require access to an activated reference chip (serving as an oracle) are referred to as *oracle-guided attacks*, whereas those that operate without such access are known as *oracle-less attacks*. Although much of the early literature focused on oracle-guided models, obtaining a functional oracle is often impractical in real-world settings, making oracle-less threats increasingly important [57]. Many proposed attacks are *invasive*, as they rely on acquiring the chip's netlist. An adversary may reconstruct the netlist through physical reverse engineering [89] of a purchased IC or extract it from design artifacts such as GDSII at the foundry. In both scenarios, the recovered implementation is typically the locked design without the correct key. Even within the design house, most implementation teams operate on keyless locked netlists during standard flows, with the correct key restricted to trusted entities [116].

In contrast, semi-invasive and noninvasive attacks are less common but remain technologically significant. These approaches often employ optical probing techniques [39, 55, 92], such as electro-optical probing (EOP) or electro-optical frequency mapping (EOFM), to directly observe sensitive internal signals, thereby posing a technology-agnostic threat to logic locking. The presence of design-for-testability (DFT) infrastructure, particularly scan chains, further expands the attack surface [21, 60]. When scan access is fully enabled, attackers can control and observe internal states, effectively decomposing the de-obfuscation problem into smaller and more tractable subproblems. Even when scan access is restricted, prior studies show that key recovery may still be feasible through primary input/output analysis [17, 97]. Additionally, a powerful adversary at the foundry may manipulate layout data (e.g., GDSII) to insert stealthy hardware Trojans that disable protection mechanisms or leak the secret key after activation [32, 70]. Collectively, these capabilities define a realistic and multifaceted threat landscape that modern logic-locking defenses must address.

Over the past decade, logic locking has been the focus of an intense arms race between defenders and attackers [19]. Under the threat models described above, numerous attack strategies have been developed to recover secret keys or remove locking logic [10, 11, 17, 19, 20, 56, 97, 112, 133]. The most influential class is the SAT attack [118], which combines Boolean satisfiability solving with oracle queries to iteratively eliminate incorrect key candidates. Many early locking schemes were shown to be vulnerable to SAT attacks, often within minutes. In response, researchers proposed SAT-resilient primitives such as Anti-SAT [130], SARLock [136], and SFLL variants [134], which introduce carefully crafted structures intended to frustrate solver convergence.

Beyond SAT-based methods, several additional attack classes have emerged:

- **Removal and sensitization attacks:** Structural analysis techniques that identify and eliminate likely key gates [112, 137]

- **Functional and structural heuristics:** Methods based on signal probability, controllability, observability, or dependency analysis [17, 59, 112]
- **Machine learning attacks:** Classifiers that detect obfuscation patterns [30] or predict key-gate locations [11, 12]
- **Approximate and oracle-less attacks:** Techniques that recover high-quality approximate functionality without requiring a working oracle [111]

While stronger locking schemes improve resilience, they often incur higher area, power, or performance overheads. A persistent challenge is therefore achieving robust, attack-resistant locking with minimal design cost. This tension motivates the development of more adaptive and intelligent locking strategies. Prior to the emergence of LLMs, researchers explored conventional machine learning to guide hardware obfuscation and IP protection [113, 115]. These approaches typically relied on task-specific models and handcrafted features. Representative examples include learning-guided obfuscation (LeGO) [9], an iterative framework that inserts locks, evaluates them against simulated attacks, and incrementally reinforces vulnerable regions using ML guidance. Other studies demonstrated that lightweight neural networks can generate circuit perturbations, such as buffer insertion, inverter addition, or logic reordering, to increase SAT solver complexity while preserving functionality [66, 114]. In parallel, graph-based IP detection methods (e.g., GNN-based approaches) have been proposed to identify reused or stolen IP subcircuits through structural learning [13, 38].

Collectively, these efforts demonstrated that AI techniques can help navigate the large design space of obfuscation transformations and attack responses. However, most prior approaches were narrow in scope, required domain-specific training, and lacked general reasoning capability. With the growing complexity of SoC designs and the increasingly adaptive nature of attacks, these limitations motivate the exploration of more general and reasoning-capable automation. Large language models introduce new capabilities, such as code understanding, generation, transformation, and cross-domain reasoning, that can potentially support both the design of stronger protection mechanisms and the systematic evaluation of their resilience.

7.2 LLM Foundations for Hardware Security

Large language models (LLMs) were originally developed for natural language processing; however, their rapid advances in code understanding, generation, and multistep reasoning have made them increasingly relevant to hardware design workflows [71, 120, 125]. As shown in Fig. 7.4, this relevance stems from the fact that hardware description languages (HDLs), such as Verilog and VHDL, are fundamentally structured code representations [73]. In recent years, both academia and industry have begun to treat LLMs as potential codesign agents capable of assisting with RTL generation, verification collateral creation, and, increasingly, security analysis and assurance (e.g., LLM-as-a-judge paradigms) [8, 101]. This

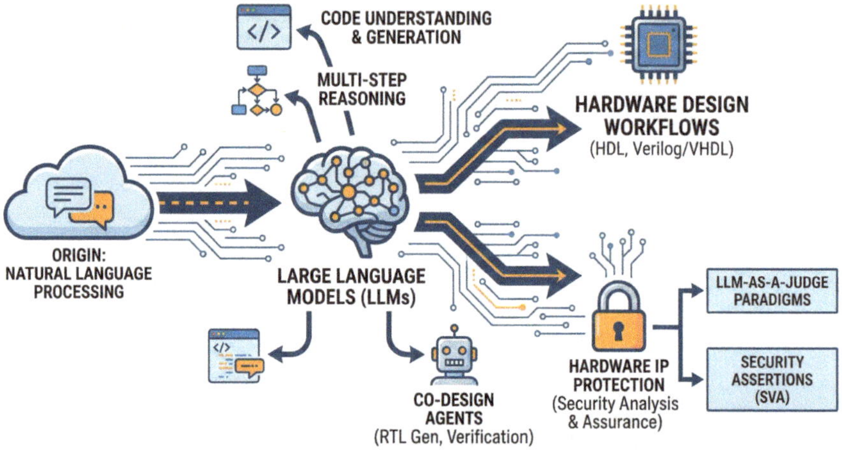

Fig. 7.4 Trend of usage of LLM: from NLP to hardware design, verification, and security

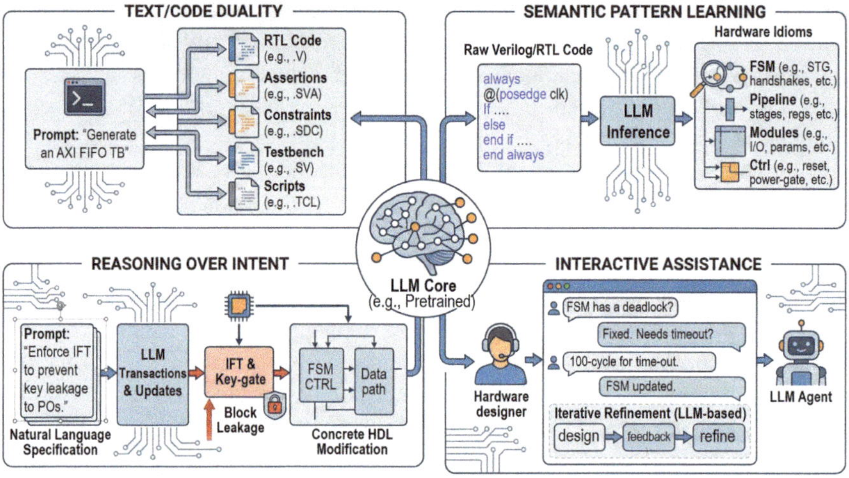

Fig. 7.5 Mapping LLM for hardware tasks: a perfect match

section summarizes the core capabilities and architectural properties that make LLMs promising for hardware IP protection.

7.2.1 LLM Transfer to Hardware

As discussed in related chapters, and as shown in Fig. 7.5, several intrinsic properties of LLMs map naturally onto hardware design workflows:

- **Text/code duality:** RTL, assertions, constraints, testbenches, build scripts, and EDA tool directives are all textual artifacts that LLMs can generate, transform, and analyze.
- **Semantic pattern learning:** Pretraining on large-scale code corpora enables LLMs to internalize common hardware idioms, including finite-state machines (FSMs), handshake protocols, pipelining structures, reset strategies, and parameterized module patterns.
- **Reasoning over intent:** LLMs can translate natural language specifications and threat models into concrete HDL modifications (e.g., inserting trigger-payload Trojans, enforcing information flow constraints, or introducing key gates, while preserving functionality).
- **Interactive assistance:** Conversational interfaces support iterative refinement, which is particularly valuable in hardware design where correctness constraints are strict and one-shot generation is rarely sufficient.

These capabilities do not guarantee functional correctness; however, they position LLMs as powerful intermediaries between human design intent and HDL/EDA artifacts [27, 124]. Early hardware-focused LLM research followed a common trajectory: starting from strong general-purpose code models and adapting them to hardware domains through curated datasets [29, 146], fine-tuning [6, 33, 147], agentic approaches [148], and reasoning [5, 128]. Research prototypes have demonstrated that fine-tuned transformer models can generate Verilog modules from prompts describing behavior, interfaces, and design constraints. Empirical studies further show that domain-adapted models, ranging from hundreds of millions to tens of billions of parameters, achieve improved syntactic correctness and functional plausibility compared with untuned general-purpose models on Verilog-centric tasks.

Because correctness is paramount in hardware design, the community has introduced dedicated benchmarks and datasets to evaluate HDL-specific capabilities. Representative examples include VerilogEval [72] and curated task suites such as RTLLM [74], which assess syntax validity, functional behavior (via simulation or formal analysis), and code quality. These benchmarks are critical for quantifying both the strengths of current models (e.g., boilerplate RTL generation and common structural patterns) and their limitations (e.g., long-range dependencies, subtle corner cases, and timing-sensitive control logic). In parallel, EDA vendors and semiconductor companies have begun integrating AI copilots into design environments to support code generation, verification assistance, and script automation [27, 71]. Domain-specific initiatives, such as chip-design assistants and EDA script generators, reflect a broader industry shift toward productizing LLMs as engineering productivity tools. This trend has direct implications for security-oriented applications, including automated IP protection, vulnerability discovery, and adversarial robustness evaluation.

7.2.2 Foundational Security-Relevant Capabilities of LLM

Once an LLM can reliably parse and generate HDL, a range of security-centric capabilities becomes feasible. Importantly, IP protection is not a monolithic task; rather, it spans design transformation, verification, and adversarial evaluation. Modern LLMs have demonstrated the potential to support each of these pillars, as summarized below:

- ***Vulnerability and Trojan Insertion (Offense as Evaluation):*** Multiple studies indicate that, when guided by high-level malicious objectives, LLMs can generate RTL modifications that implement stealthy trigger, i.e., payload behaviors, such as covert data leakage or functionality corruption [42, 65, 102]. This capability is inherently dual use: (i) *Risk:* It lowers the barrier for attackers to craft hardware Trojans or subtle backdoors; and (ii) *Defensive value:* It enables automated red-teaming by generating candidate Trojans or insider-like mistakes for validating detection pipelines. This duality is particularly relevant to IP protection, where robust locking and obfuscation must be evaluated against adaptive adversaries.
- ***Vulnerability Detection and Security Code Review:*** LLMs can function as security-oriented reviewers of RTL and related design artifacts [3, 7]. Given a module, specification, or policy description, an LLM can (i) highlight suspicious constructs (e.g., hidden always blocks or rarely activated conditions), (ii) flag risky coding patterns (e.g., uninitialized signals, unsafe default cases, incomplete privilege checks), and (iii) identify potential attack surfaces such as debug logic, test modes, scan chains, or undocumented interfaces. Although these assessments are not guaranteed to be sound, they can effectively triage designer attention and generate hypotheses for formal or simulation-based validation [61].
- ***Policy Compliance Checking and Information Flow Reasoning:*** A particularly powerful capability is translating natural language security policies into code-level analyses [81]. Example policies include: (i) no combinational path from key storage registers to primary outputs, (ii) debug mode must remain disabled unless an authenticated sequence occurs, and (iii) secret-dependent control must be confined to trusted domains. LLMs can scan RTL for potential violations and, importantly, synthesize concrete verification artifacts such as assertions, checklists, or testbench scenarios [62]. This translation is valuable because security requirements are often written in prose and inconsistently enforced across teams and IP vendors.
- ***Formal Property and Assertion Generation:*** Emerging work shows that LLMs can propose candidate assertions and invariants directly from RTL descriptions [62], which can then be validated using formal engines [131] (e.g., model checking, SMT solving, or mutation testing—see Fig. 7.6). This is significant because property authoring is often the primary bottleneck in scalable hardware assurance. LLM assistance can accelerate (i) generation of candidate safety properties, (ii) construction of environment assumptions and constraints, and (iii) coverage-driven refinement prompts (e.g., identifying missing properties exposed

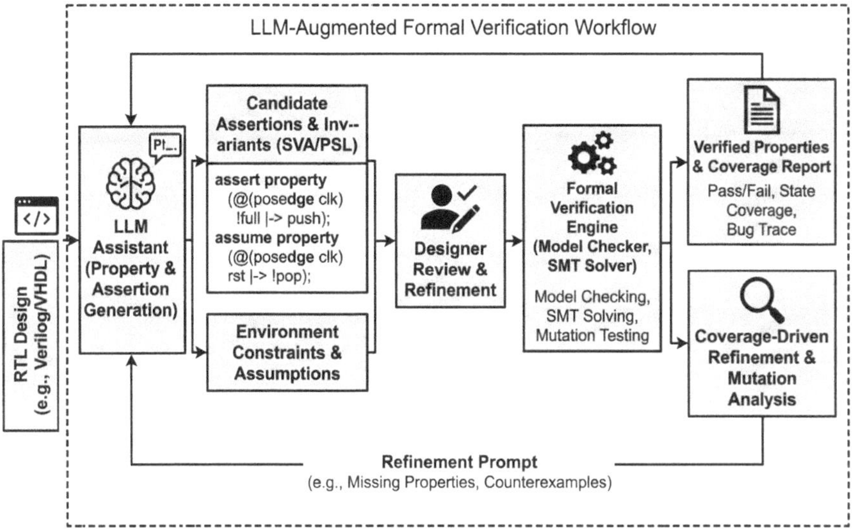

Fig. 7.6 An example of LLM-assisted formal property and assertion generation flow

by mutation analysis) [119]. This capability becomes particularly important when integrating LLMs with EDA-based security evaluation flows.

- ***Repair and Mitigation:*** Beyond detection, LLMs can propose candidate mitigations, such as inserting guards, tightening reset logic, strengthening privilege gating, or restructuring vulnerable datapaths [119]. For third-party IP, where documentation may be incomplete and formal models unavailable, LLMs can provide valuable first-pass remediation suggestions that human engineers subsequently validate [62].

Collectively, these capabilities, HDL understanding and generation, semantic reasoning, property synthesis, and adversarial code manipulation, form the foundation for LLM-assisted IP protection (see Fig. 7.7). They enable two complementary operational roles:

1. **LLMs as defenders:** generating or transforming designs to incorporate protection (e.g., locking or obfuscation), producing verification collateral, and assisting engineers in reasoning about security trade-offs
2. **LLMs as attackers (red-team agents):** automating attack reasoning, generating de-obfuscation strategies, inserting evaluation Trojans, and stress-testing protection mechanisms

Crucially, the strengths of LLMs also expose important limitations in hardware contexts. First, *context length and scale* remain significant challenges, as real SoCs and even mid-sized IP blocks may exceed practical prompt windows, necessitating hierarchical summarization and tool-assisted retrieval. Second, LLM outputs lack formal correctness guarantees; plausible-looking RTL modifications

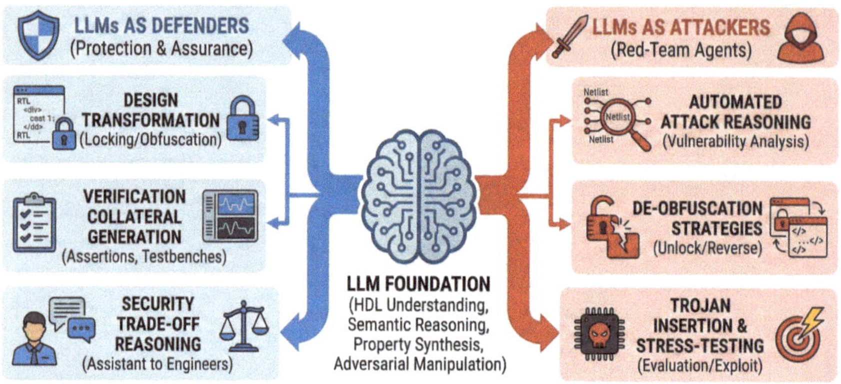

Fig. 7.7 LLM in hardware security: for both red- and blue-teaming

may still violate functional, timing, or power constraints. Third, the *cost of errors* in hardware is exceptionally high—postsilicon fixes are expensive or infeasible, requiring that LLM-generated artifacts be tightly coupled with simulation, linting, and formal verification. With these foundations established, we next examine how these capabilities translate into practical LLM-assisted IP protection across the SoC design flow, and how LLMs can be integrated with EDA tools to both strengthen defenses and evaluate resilience under realistic attack models.

7.3 LLM-Assisted IP Protection Across the SoC Flow

The application of LLMs to hardware IP protection can span multiple stages of the SoC design lifecycle, from early specification analysis to post-obfuscation security evaluation. In this section, we survey representative studies and illustrate how LLMs can be leveraged at different points in the flow to assist designers in protecting IP, particularly through logic locking and obfuscation, and to validate the robustness of those protections against advanced attacks.

One promising use of LLMs is the analysis of high-level system specifications to automatically infer security requirements and potential vulnerabilities. This capability is particularly valuable for IP protection because effective key-gate insertion typically depends on identifying security-critical paths and control points. For example, DIVAS is an LLM-driven framework that consumes user-defined SoC specifications (e.g., descriptions of bus architectures and protected assets) and automatically identifies potential security weaknesses [88]. The framework maps detected issues to common weakness enumerations (CWEs) [2] and generates corresponding security assertions and policy checks.

In practice, such a system enables workflows in which an LLM reads a specification and produces actionable guidance, e.g., *"The design does not mention*

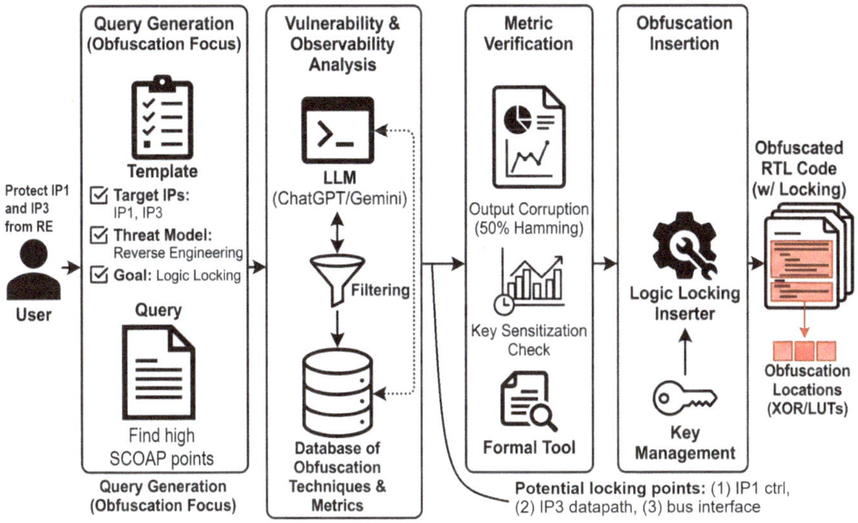

Fig. 7.8 An exemplary view of automated framework for determining hardware obfuscation points [88]

authentication on the debug interface; this corresponds to CWE-306 (Missing Authentication); insert a lock (or blocking module) here." Using models such as ChatGPT, DIVAS demonstrates how natural language specifications can be translated into concrete security instrumentation. This early-stage guidance is particularly valuable because it allows designers to plan obfuscation and locking strategies before RTL implementation. A generalized version of this pipeline is illustrated in Fig. 7.8. By identifying potential IP exposure points at the specification stage, designers can proactively embed protection mechanisms (e.g., key-locking configuration registers of a DSP core that control sensitive functionality).

Beyond specification analysis, recent work has explored using LLMs directly to identify optimal key-gate locations within gate-level or RTL designs. One notable example is GLLaMoR [99], which leverages the reasoning capabilities of LLMs to guide logic locking. As shown in Fig. 7.9, GLLaMoR parses Verilog netlists into graph-based representations and presents structured adjacency information to the LLM. The framework uses prompt-guided reasoning (e.g., DFS-style traversal instructions) to identify nodes that lie on a large number of input-to-output paths and are therefore attractive-locking candidates.

While initial zero-shot and few-shot experiments struggled with large circuit complexity, the study demonstrates that fine-tuning on curated netlist datasets significantly improves the model's ability to detect security-critical nodes. Experimental results show high true-positive rates in identifying important locking locations, along with notable speedups compared with traditional Python-based analysis tools. Overall, GLLaMoR suggests that LLM-guided node selection can

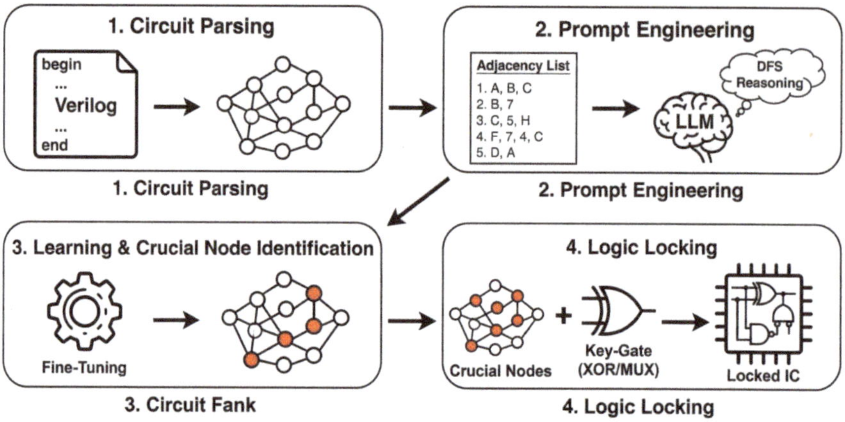

Fig. 7.9 Main steps in GLLaMoR: graph-enhanced LLM-driven approach for identifying logic-locking elements' location [99]

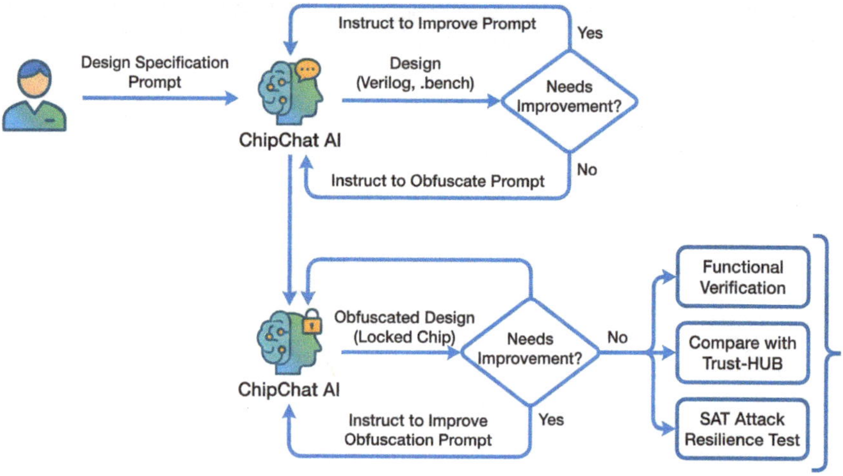

Fig. 7.10 Overview of Obfus-Chat: from prompt to obfuscation [68]

improve the end-to-end robustness of locked designs against both SAT-based and functional analysis (FALL) attacks.

Perhaps the most direct application of LLMs to IP protection is the automated insertion of logic locks and obfuscation structures. Traditionally, designers either developed custom (heuristic) algorithms or manually selected signals to protect, which is a process that is time-consuming and error-prone. LLM-driven tools aim to automate this workflow. One of the earliest examples is Obfus-Chat [68], an LLM-based logic obfuscation framework. As shown in Fig. 7.10, Obfus-Chat takes an HDL netlist and a target key size as input and automatically produces an obfuscated version of the design.

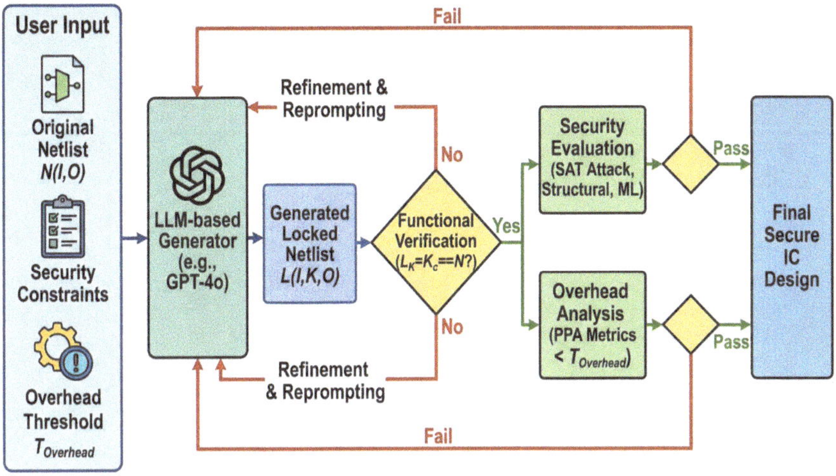

Fig. 7.11 Iterative LLM-driven logic-locking methodology [44]

In this framework, the LLM is guided via structured prompts to insert key gates, multiplexers, LUTs, and related transformations while preserving functional equivalence. The generated obfuscated HDL can then be integrated into the standard design flow. Evaluation on Trust-Hub benchmark circuits demonstrates functional correctness and improved resistance to SAT-based attacks within the evaluated threat model, indicating that LLM-guided transformations can capture meaningful obfuscation strategies. These results highlight the potential of LLMs to learn from existing locking heuristics and reduce manual design effort.

Beyond Obfus-Chat, Gandhi et al. [44] proposed an iterative LLM-driven logic-locking methodology. As illustrated in Fig. 7.11, their framework employs pretrained transformers (including GPT-4-class models) in a feedback loop that progressively refines locked netlists. Each iteration evaluates functional equivalence, resilience to structural and SAT attacks, and compliance with power, performance, and area (PPA) constraints. Experimental results on benchmark circuits show that iterative prompting improves context retention and locking quality. However, the study also reports that achieving fully verified locked netlists remains challenging, highlighting the need for tighter integration between LLM generation and formal verification.

The emergence of multiple independent efforts over the past two years strongly suggests that LLMs are capable of understanding and implementing logic locking at the RTL level. These models can analyze circuit structure, identify heuristic or optimization-driven lock locations, and generate the corresponding HDL required to implement obfuscation constructs. An important extension of this idea is the use of coordinated multiagent LLM pipelines. The LockForge framework [100] exemplifies this direction by addressing a long-standing reproducibility challenge in hardware security: Many published locking schemes lack publicly available implementations. LockForge employs a team of specialized LLM agents to read

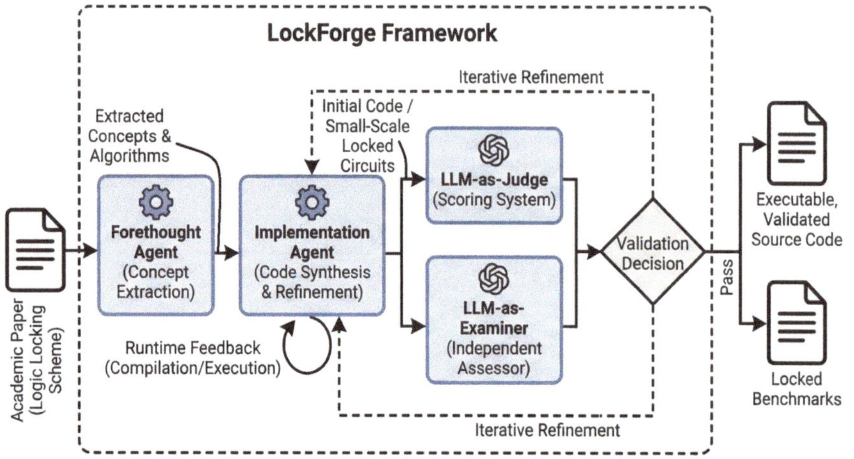

Fig. 7.12 An overview of multiagent framework for paper-to-code automation in LockForge [100]

logic-locking papers and automatically generate corresponding HDL implementations.

An overview of the multiagent framework in LockForge is shown in Fig. 7.12. The pipeline consists of four stages: *Forethought*, where an agent extracts key concepts from the paper; *Implementation*, where another agent generates and inserts locking logic; *Refinement*, where the implementation is cross-checked against the paper's described mechanisms; and *Validation*, where independent LLM agents act as judge and examiner. Using this approach, LockForge successfully implemented 10 previously unpublished locking schemes. This result demonstrates that LLMs can interpret complex security concepts and translate them into executable hardware transformations. In the longer term, such pipelines could enable designers to specify custom security policies in natural language and automatically synthesize compliant hardware protections.

Incorporating IP protection is only one side of the problem; the other is verifying that protection mechanisms are both effective and functionally safe. LLMs are beginning to assist in this phase as well. For example, after obfuscation, designers often need to verify properties such as: *"Under incorrect keys, the circuit should produce well-corrupted or random outputs."* Writing such formal properties manually is labor-intensive. Recent work shows that LLMs can generate security-oriented systemVerilog assertions (SVAs) directly from design specifications. Frameworks such as AssertLLM [40] and ChIRAAG [79] demonstrate that multistage prompting can produce meaningful assertions suitable for formal tools. For instance, such capabilities can enable us to use models to synthesize formal properties from plain-text specifications and validate them using the solver (e.g., Z3 solver), automatically generating a substantial fraction of the required security properties. Integrating LLMs into this stage enables automated checking that obfuscation has not inadvertently introduced new vulnerabilities (e.g., exposed debug paths or bypass conditions), effectively creating an AI-augmented security sign-off process.

In summary, across the SoC design flow, LLMs can (1) identify where and how to apply protection, (2) automatically insert obfuscation and locking logic, and (3) generate verification collateral to validate security properties. Together, these capabilities are forming an emerging end-to-end pipeline for LLM-assisted IP protection. While current approaches remain nascent and require human oversight—particularly for functional validation—they point toward a future in which significant portions of secure hardware design can be augmented by intelligent automation.

7.4 LLM + EDA Integration for Attack and Security Evaluation

Ironically, the same LLM capabilities that assist defenders can also empower attackers, or be deliberately harnessed by defenders in an adversarial evaluation mode. Integrating LLMs into the EDA toolchain for security assessment enables AI-driven simulation and execution of advanced attack strategies, allowing designers to identify vulnerabilities before real adversaries exploit them. A recently explored threat scenario involves using LLMs to perform full hardware IP piracy. LLMPirate [45] is the first framework that leverages LLMs to generate functionally equivalent pirated versions of hardware from a locked or watermarked design. The core challenge is enabling the LLM to understand circuit functionality and then rewrite the implementation in a sufficiently different form to evade IP protection tools while preserving behavioral equivalence.

As shown in Fig. 7.13, LLMPirate addresses this challenge through automated prompt engineering and hierarchical partitioning. The framework converts Verilog

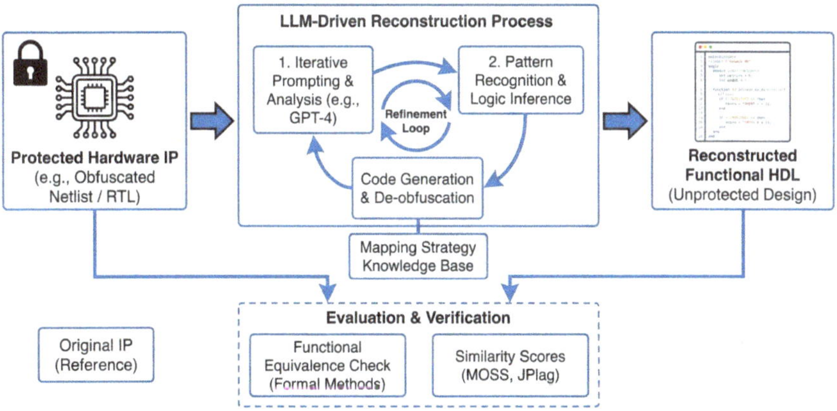

Fig. 7.13 LLM-assisted hardware IP theft in LLMPirate [45]

designs into LLM-friendly representations (e.g., Boolean expressions or structured pseudocode) to fit within context limits, prompts the model to regenerate or refactor the design, and iteratively refines the output to maintain functional equivalence. Experimental evaluation across eight LLMs, including both open-source models and GPT-4-class systems, demonstrated that the generated variants evaded state-of-the-art piracy detection tools in all tested cases. In particular, similarity-based detectors (e.g., graph-based IP matching tools) failed to identify the rewritten netlists as derivatives of the originals. Case studies on nontrivial designs such as the IBEX RISC-V core and MOR1KX OpenRISC processor further confirmed the practicality of the approach.

These results suggest that LLMs can effectively act as automated pirate engineers, capable of refactoring stolen IP into structurally distinct yet functionally equivalent implementations. For the hardware security community, this represents a significant warning: Protection mechanisms that rely primarily on structural similarity or watermark detection may become increasingly vulnerable to AI-assisted transformations. From a defensive perspective, however, LLMPirate-like techniques can be repurposed as robustness evaluators. Designers can attempt automated AI-driven cloning of protected designs; if the process succeeds without key knowledge, the protection scheme likely requires strengthening. In this sense, LLM-driven piracy can serve as an automated red-team capability analogous to exploit generation in software security.

Another critical attack vector is hardware Trojan insertion, in which stealthy modifications are introduced to leak information or disrupt functionality under rare trigger conditions. Designing an effective Trojan, small, covert, and difficult to detect, is itself a complex task. Recent work by Kokolakis et al. [65] demonstrates that LLMs can assist in this process (see Fig. 7.14). In their framework, GPT-4 is used to analyze large HDL codebases and identify high-impact insertion points.

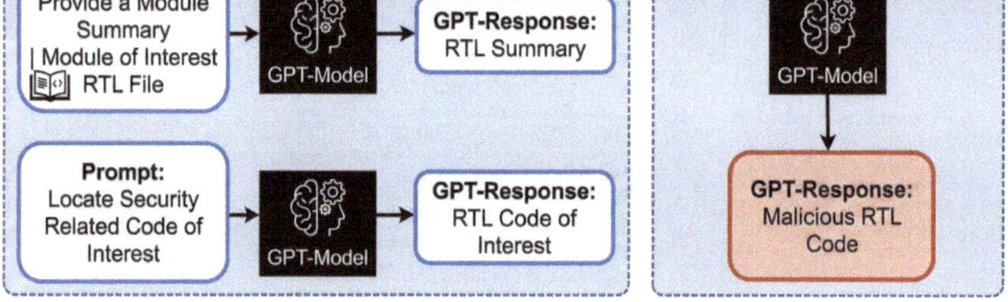

Fig. 7.14 LLM-assisted hardware Trojan insertion [65]

Rather than processing the entire design monolithically, the approach employs context filtering to focus the model on relevant modules or signal groups, thereby addressing prompt-length limitations. The resulting pipeline enables a general-purpose LLM to (i) scan RTL, (ii) propose candidate Trojan insertion locations, and (iii) outline trigger—payload logic. A proof of concept demonstrated end-to-end Trojan insertion in a complex RISC-V processor, producing functional backdoors capable of secret leakage.

From a defensive standpoint, such AI-assisted attack generation can be highly valuable. Integrating similar workflows into verification flows enables automated probing for Trojan-prone regions. If an LLM repeatedly identifies a specific module (e.g., an unprotected bus controller) as a high-value target, designers can proactively reinforce that component using monitors, redundancy, or hardened logic.

LLMs can also act as powerful simplifiers for hardware reverse engineering. A long-standing challenge in this domain is reconstructing high-level RTL from flattened gate-level netlists, which is a process often referred to as *netlist decompilation* or *functional reverse engineering*. During synthesis, HDL descriptions undergo aggressive optimizations that remove hierarchy and semantic structure, making reconstruction difficult [117]. Traditional techniques rely heavily on pattern matching against known subcircuit libraries, which does not scale well to modern optimized designs. As shown in Fig. 7.15, hybrid frameworks can combine graph neural networks (GNNs) with LLMs to address this limitation. In a representative pipeline, the gate-level netlist is first converted into a directed graph, after which a graph attention network (GAT) classifies gates into functional categories (e.g., adder, multiplier, comparator). These semantic hints are then embedded into prompts for a fine-tuned LLM, which generates higher-level RTL abstractions. The synthesized RTL is subsequently validated via formal equivalence checking.

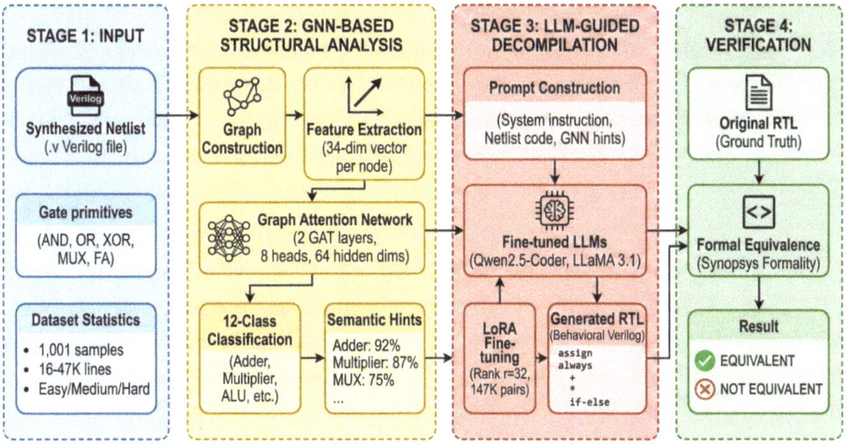

Fig. 7.15 LLM-assisted simplifier for netlist decompilation: an exemplary framework

Such approaches demonstrate how LLMs can reduce the manual burden of reverse engineering and improve analysis of untrusted third-party IP.

From a practical EDA perspective, these advances point toward a future "security evaluation mode" within CAD tools. When enabled, such a mode could automatically invoke multiple LLM-assisted analyses on the design under test, including:

- An LLM-based piracy simulation (LLMPirate-style refactoring)
- An LLM-guided Trojan vulnerability scan
- A holistic LLM-driven vulnerability audit

Because LLMs can be guided using high-level adversarial prompts (e.g., "identify how this module could leak secret keys"), they effectively act as expert red-team analysts. In early exploratory systems such as Chip-chat [27] interactive designer, LLM security reviews, while other studies demonstrate LLMs operating as both attacker and auditor within the same workflow.

This dual-use nature underscores both the promise and the risk of LLM integration. On one hand, it lowers the technical barrier for attackers by enabling automated generation of sophisticated hardware exploits. On the other hand, it provides defenders with powerful new tools for proactive vulnerability discovery and mitigation. The key challenge for the community is ensuring that defensive capabilities evolve at least as rapidly as offensive ones. By incorporating LLM-based adversarial testing into standard chip validation flows, designers can identify blind spots earlier and harden systems more effectively. In this emerging paradigm, the LLM becomes a new member of the verification team—one capable of reasoning like an attacker and uncovering weaknesses that traditional analyses may miss.

7.5 One Case Study: LLM-Assisted FSM Logic Locking

To demonstrate how this concept works, this section demonstrates steps of a case study of how an LLM can support *circuit interpretation* and *logic-locking design* to improve both automation and robustness against de-obfuscation attacks. Consider a baseline design (e.g., a microprocessor controller) that we wish to protect using logic locking such that common attack classes, most notably SAT-based attacks (either combinational or sequential) and structural/functional analyses, become impractical. In traditional flows, an engineer may select a set of "critical" control signals and insert key gates (e.g., XOR/XNOR gates, key-controlled MUXes, or LUT-based blocks) and then validate security under a chosen attack model. While such approaches can be effective for specific settings, selecting high-impact lock locations and designing resilient locking structures remains challenging and often depends on expert intuition. Here, we illustrate how an LLM can enhance this workflow by (i) identifying security-critical locking points and (ii) synthesizing a nontrivial locking strategy that increases adversarial effort.

Target Design and Threat Model Assume the target is a finite-state machine (FSM) controller in a processor pipeline, which is a critical locking target because it governs global sequencing and control decisions. We consider an attacker who can obtain a locked netlist (e.g., through reverse engineering or supply chain leakage) but does not possess the secret key. The adversary attempts either (i) oracle-guided de-obfuscation (e.g., SAT-based key search when a functional oracle is available) or (ii) oracle-less bypass strategies that exploit structural patterns and functional reasoning about the FSM. The defender's objective is to enforce that incorrect keys induce deceptive behavior that is difficult to disentangle, while the correct key preserves full functional equivalence. As shown in Fig. 7.16, this can be done in four main steps:

Step 1—LLM-assisted identification of locking candidates: We first provide the LLM with the RTL (Verilog) description of the controller along with a structured prompt requesting identification of security-critical states, transitions, and control

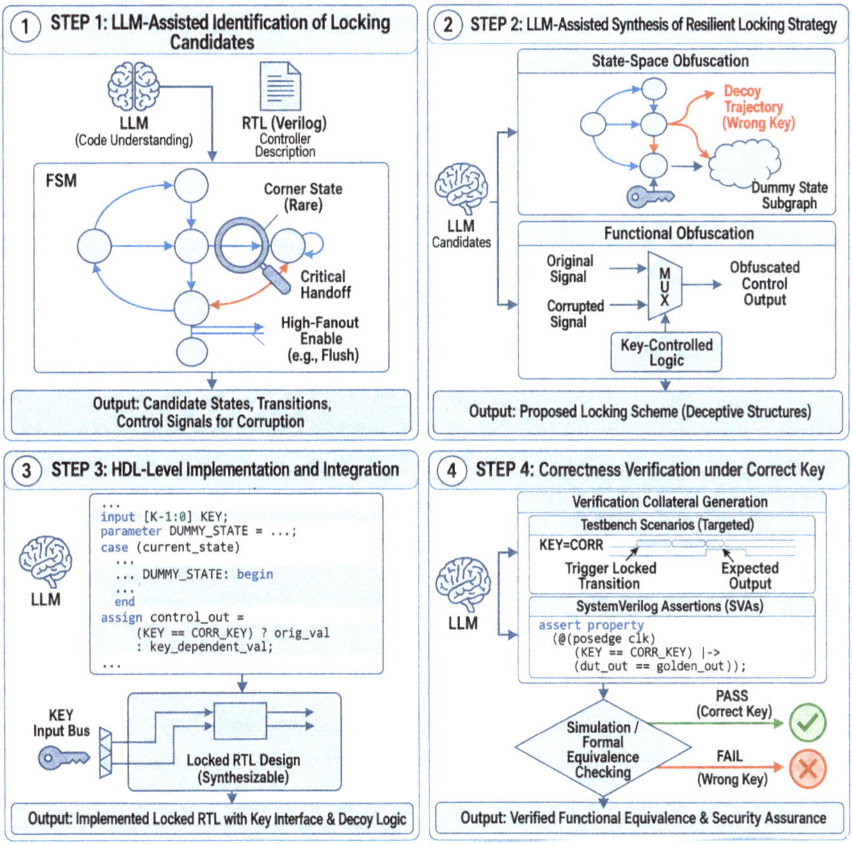

Fig. 7.16 A case study of FSM obfuscation using LLM understanding of deep states

signals. A representative prompt is: *"Given this FSM RTL, identify state variables, transitions, or control outputs whose corruption would maximally disrupt correct operation while remaining difficult to diagnose via local reasoning. Prioritize rare transitions and high-fanout control points."* Using its code understanding capability, the LLM can highlight candidate elements such as (i) corner or rarely visited states, (ii) transitions that gate progress (e.g., initialization-to-execution handoffs), (iii) high-fanout control enables (e.g., stall/flush, mode enables), or (iv) signals that activate infrequent modes (e.g., debug/test microsequences). These candidates are attractive because small perturbations can induce large behavioral divergence, forcing deeper adversarial analysis.

Step 2—LLM-assisted synthesis of a resilient locking strategy: Next, we prompt the LLM to propose a locking scheme that focuses on the identified candidates while incorporating deceptive structures. This could be done by prompting like *"Propose a logic-locking scheme that uses a key to preserve correct functionality under the right key while inducing plausible but incorrect behaviors under wrong keys. Include decoy logic and state-space obfuscation. Ensure the scheme is implementable at RTL."* A common LLM-proposed pattern is to combine multiple obfuscation tactics: (i) *state-space obfuscation*, where key-dependent transitions divert the FSM into one or more "dummy" states (or deep subgraphs of states) under wrong keys; and (ii) *functional obfuscation*, where key-controlled multiplexers or LUT logic selectively corrupt high-impact control outputs. For example, the LLM may propose key-conditioned perturbations of the next-state function such that, under an incorrect key, specific legitimate transitions are redirected into a decoy trajectory that is internally consistent but ultimately incorrect. Similarly, key-controlled MUXes can swap or invert selected control signals, producing behavior that is not trivially distinguishable from a benign design bug.

Step 3—HDL-level implementation and integration: We then task the LLM with implementing the proposed scheme in Verilog. This includes adding dummy states, updating the state-transition logic, inserting key-dependent gates/MUXes, and defining the key interface (e.g., an input bus `KEY` or a key register loaded during secure provisioning). A well-structured LLM-generated modification updates the state encoding (or state parameters), extends the `case` statement to include new states, and ensures that dummy-state logic is quiescent under the correct key. The goal is to produce a locked RTL that can be synthesized and integrated into the standard flow without manual rework beyond review and verification.

Step 4—Correctness verification under the correct key: A critical requirement is that the locked design remains functionally equivalent to the original design when the correct key is applied. While the LLM can provide a first-pass reasoning check, correctness must ultimately be established using simulation and/or formal equivalence checking. In practice, an LLM can assist by generating verification collateral, such as: (i) testbench scenarios targeting locked transitions and corner states, and (ii) SystemVerilog assertions (SVAs) encoding equivalence expectations (e.g., "if `KEY` is correct, then outputs match the golden design for all reachable states under valid environments"). These artifacts can then be validated using standard simulation, bounded model checking, or equivalence tools.

Why can this LLM-guided locking be more robust than naive key-gate insertion? First, by focusing locks on high-leverage semantic points (rare transitions, high-fanout controls), the approach increases the coupling between key correctness and global behavior. Second, the addition of decoy state subgraphs and key-dependent redirections increases ambiguity for structural reasoning and can increase solver workload in oracle-guided settings by complicating distinguishing input generation. More broadly, when the locking introduces carefully designed "plausible" incorrect behaviors (rather than obviously broken outputs), it may reduce the attacker's ability to quickly isolate the key logic via heuristic pruning.

In a typical evaluation, a baseline XOR-based lock (similar to what has been done in Obfus-Chat [68]) may be rapidly defeated by SAT-based attacks in oracle-guided settings, whereas a more structured FSM-centric lock can significantly increase attack complexity. Importantly, such claims should be validated under explicit attack assumptions and benchmark settings. Prior work on LLM-assisted obfuscation reports improved resistance to SAT-based attacks under their experimental conditions, suggesting that LLM-guided synthesis can capture meaningful defensive structure. In our case study, the intent is not to claim universal SAT-hardness, but to illustrate a concrete workflow in which LLMs help design stronger, semantically informed locks and produce the corresponding HDL transformations. This case study demonstrates how LLMs can interpret design intent, identify high-impact locking points, and synthesize implementable locking strategies that combine state-space and functional obfuscation. While we emphasized an FSM controller for concreteness, the same workflow can generalize to other IP blocks. For example, in cryptographic modules, LLM-guided locking may target key schedule control or mode sequencing; in signal-processing blocks, it may target configuration paths or control-dependent datapaths. In all cases, LLMs should be treated as productivity and reasoning aids whose outputs must be coupled with rigorous verification and threat-model-specific evaluation.

7.6 Risks and Security Concerns of Using LLMs

While LLMs open exciting frontiers for hardware security, their use is not without significant risks and concerns. It is important to recognize these challenges to use LLMs responsibly in the context of IP protection. One immediate concern is that feeding proprietary hardware designs into large public LLMs (e.g., via an API to ChatGPT) could inadvertently leak sensitive information [80, 135]. Unless carefully sandboxed, the queries we pose to an LLM might be retained on servers or even surface in model outputs to other users [87]. Companies have become wary of this. For instance, some chip design firms banned employees from pasting source code into external LLM tools for fear of IP exposure [96]. The broader community echoes this; indeed, building trust into using LLMs for chip design is seen as essential. Solutions could include on-premises or fine-tuned LLMs that are maintained internally (so data never leaves) [37, 129], and robust data handling policies [144]. But until those are standard, the confidentiality of IP in the age of LLMs is a paramount concern. This extends to the model's training data too: If an LLM was trained on public HDL repositories, there is a possibility that proprietary snippets (if leaked online) could be embedded in the training distribution and later reproduced by the model [80, 129]. Ensuring that LLM-generated designs do not inadvertently replicate someone else's IP (thus causing legal and security issues) will be an important part of future IP management.

Beyond explicit data exposure through prompts, recent work highlights a more subtle but critical risk: *memorization-driven IP leakage* from LLMs themselves. Even when proprietary RTL is not directly provided at inference time, large code models may inadvertently reproduce structurally or functionally similar hardware blocks that were memorized during training. This risk is particularly acute in hardware design because RTL exhibits strong structural regularities, and functionally equivalent implementations can appear syntactically distinct, while still revealing proprietary design intent. Unlike natural language, where approximate paraphrasing may be acceptable, small structural similarities in RTL can directly expose microarchitectural strategies or optimization techniques. Recent work such as CircuitGuard [80] systematically studies this phenomenon in the context of HDL generation (see Fig. 7.17). The authors identify multiple leakage modes, including verbatim reproduction, structural cloning, and functional replication of proprietary circuits. Their analysis further reveals that memorization behavior in transformer models tends to concentrate in a relatively small subset of latent features located in deeper network layers. This observation suggests that hardware IP leakage may occur even when surface-level similarity checks fail. Importantly, the study demonstrates that conventional lexical similarity metrics are insufficient for RTL

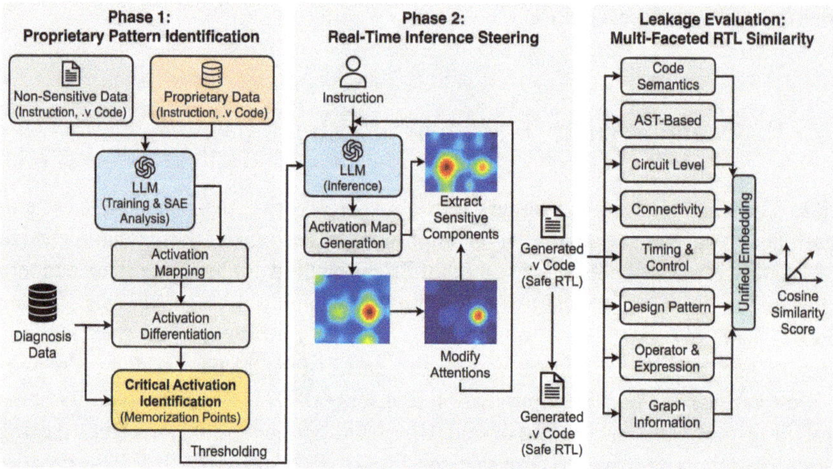

Fig. 7.17 Overview of CircuitGuard [80]: from identifying memorization-critical activations using sparse autoencoders to editing activations during inference

and instead motivates multifaceted similarity measures that incorporate syntax, circuit topology, connectivity, and functional patterns.

To mitigate this risk, CircuitGuard [80] proposes an inference-time activation steering mechanism that suppresses memorization-critical latent directions without retraining the model. Experimental results report substantial reductions (up to approximately 80%) in semantic similarity to proprietary designs, albeit with a measurable trade-off in functional correctness. These findings introduce a new dimension to hardware IP protection: Even trusted internal use of LLM-based RTL generation may require dedicated memorization safeguards to prevent inadvertent IP leakage. From an SoC security perspective, this work expands the traditional threat model. Historically, IP protection focused on post-design threats such as reverse engineering, piracy, and de-obfuscation. However, the increasing use of LLMs in RTL generation introduces a *design-time leakage channel* in which the AI assistant itself may reproduce sensitive structures. This suggests that future secure hardware workflows must consider both circuit-level protections (e.g., logic locking, obfuscation) and model-level protections (e.g., memorization mitigation, controlled deployment, and similarity screening).

LLMs, especially without domain-specific tuning, are prone to hallucinate—that is, produce outputs that are plausible-sounding but incorrect [6]. In hardware, a hallucination could equate to a design error that ruins a chip or a security flaw that opens a backdoor [8]. For example, if asked to fix a bug, an LLM might generate a change that appears to resolve the issue but subtly violates a timing constraint or a design assumption, thereby introducing a new vulnerability. In the context of logic locking, an LLM could mistakenly create a lock that is bypassable (e.g., inserting a key gate while unintentionally leaving a debug override path). Thus, relying blindly on LLM outputs is dangerous. As emphasized in Obfus-Chat [68], even with LLM automation, functional verification remains essential; prior studies employed simulation and formal checks to ensure the obfuscated design remained consistent with the original design. Any use of LLMs in hardware must therefore include rigorous verification steps [72]. Moreover, unlike software where a bug might be patched with an update, a hardware bug, especially postsilicon, is far more costly. This raises the stakes for LLM accuracy. Techniques such as chain-of-thought prompting [5] may help developers better interpret AI reasoning. In safety-critical IP, LLMs may be best positioned as advisory tools, with human experts vetting every change.

When integrating an LLM into a secure workflow, we must also consider that the model effectively becomes part of the trusted computing base. An important question is whether an attacker could manipulate the LLM (via prompt injection, poisoned documentation, or corrupted training data) to produce insecure designs. This scenario is plausible: If an adversary knows a company relies on LLM-assisted design, they might craft specifications or comments intended to bias the model toward weaker implementations. Furthermore, LLMs may exhibit blind spots if certain attack classes were underrepresented in training data. Ensuring the trustworthiness of LLM-powered design, including robustness to misinformation, bias, and adversarial prompting, remains an open research challenge. This suggests that additional safeguards such as multimodel cross-checking, retrieval-grounded prompting, and formal back-end validation may be necessary in high-assurance flows.

Perhaps the biggest ethical concern is the dual-use nature of LLMs. The same model that helps protect IP can also help attack it. For example, we discussed LLMPirate [45] and LLM-aided Trojan insertion [65]; these effectively turn LLMs into automated adversarial tools. Once such techniques are public, there is a risk of misuse by malicious actors. The community must therefore develop countermeasures. One approach is to "fight fire with fire," using LLMs to detect AI-modified or pirated designs. Future work may include training models to recognize machine-generated HDL or suspicious structural patterns. On the policy side, organizations may embed watermarks or fingerprints into AI-generated HDL to enable provenance tracking and misuse detection.

A less technical but still significant concern is over-reliance on LLM tools. If engineers routinely delegate security decisions to AI (e.g., "AI, lock my design"), they may gradually lose deep understanding of the underlying protection mechanisms. This mirrors historical concerns with excessive reliance on high-level synthesis or automated compilers. LLMs should therefore augment, not replace, human expertise. The most robust workflow is likely a human-AI synergy in which engineers specify security intent and constraints, while AI assists with candidate implementations and cross-checking. Maintaining strong education and training in hardware security fundamentals remains essential. There is also a practical concern regarding deployment cost. State-of-the-art LLMs are resource-intensive; running GPT-4-class models locally may be infeasible for many EDA environments without significant infrastructure investment. This introduces both financial cost and potential dependence on external providers. Encouragingly, recent work on domain-specialized models (e.g., ChipNeMo [71]) suggests that smaller, hardware-focused LLMs can achieve competitive performance at lower cost. It is plausible that future "hardware security LLMs" will be compact enough for secure on-premises deployment, mitigating both privacy and cost concerns. Until such solutions mature, however, practical adoption may be constrained by trust, cost, and infrastructure considerations.

In summary, the introduction of LLMs into hardware IP workflows must be approached with caution. The community increasingly recognizes the need to build trust in these AI-assisted design tools. Emerging best practices include keeping humans in the loop, using private or fine-tuned models for sensitive data, applying memorization safeguards, and rigorously verifying AI-generated designs. LLMs should be treated as powerful but fallible assistants rather than infallible oracles. By proactively addressing these risks, designers can harness the benefits of LLMs, while minimizing potential harm to IP security.

7.7 Research Opportunities and Open Problems

The confluence of LLMs and hardware IP protection is a young and rapidly evolving field. As such, many open research questions and opportunities remain. We highlight several key directions where further advances are needed to fully realize the potential of LLMs in hardware security:

1. Specialized LLM Architectures for Hardware Security: Most current efforts rely on general-purpose LLMs (e.g., GPT-class models) or lightly fine-tuned variants for hardware code. There is a significant opportunity to design architectures specifically tailored to hardware security workflows. This may include pretraining on large corpora of hardware designs and security annotations, incorporating domain priors into model structure, or developing hybrid models that jointly reason over text and circuit graphs. Prior surveys have already highlighted the need for domain-specific enhancements to improve reliability on RTL tasks.

   One promising direction is the development of models capable of native structural reasoning, e.g., architectures that ingest netlist graphs, FSM representations, or timing constraints alongside HDL text. Concepts similar to emerging "large circuit models" that fuse graph learning with language modeling could be particularly impactful. Such specialized hardware security LLMs may detect subtle vulnerabilities that generic models overlook while also improving computational efficiency. Additionally, ensemble and evolutionary approaches, where multiple specialized models (e.g., code, formal, graph) collaborate, represent a promising but largely unexplored design space.
2. Improved Datasets and Benchmarks: Progress in hardware-focused LLMs is currently constrained by limited availability of large, diverse, and security-labeled HDL datasets. Unlike natural language domains, publicly available RTL corpora are relatively small, and datasets containing annotated vulnerabilities, obfuscation transformations, or attack outcomes are even rarer. Developing high-quality benchmarks would significantly accelerate research. Potential dataset directions include: (i) large collections of circuits with labeled security flaws and patches, (ii) paired original/obfuscated designs, (iii) Trojan-inserted benchmarks with ground truth triggers, and (iv) oracle-guided vs. oracle-less attack datasets. Existing resources such as RTLLM and Trust-Hub provide important starting points but do not yet cover the full design-attack-defense spectrum. Researchers have increasingly called for continuous evaluation frameworks for LLM-in-EDA systems. Synthetic data generation via simulation and formal tools also represents a promising avenue, for example, generating traces under incorrect keys to help models learn behavioral divergence patterns. Community-driven repositories of hardware vulnerabilities (analogous to CVE databases in software) would further strengthen this ecosystem.
3. Integration with Formal Methods and Guarantees: A central open problem is bridging the gap between the probabilistic nature of LLM outputs and the strict correctness guarantees required in hardware design. Future research must explore tighter integration between LLMs and formal engines such as SAT solvers, model

checkers, and equivalence tools. One promising paradigm is a closed-loop secure synthesis flow: The LLM proposes a locking or mitigation strategy, formal tools evaluate it, counterexamples are fed back, and the model iteratively refines the design.

Early work combining LLM-generated assertions with SMT solvers demonstrates the feasibility of this direction, but much remains to be done. Reinforcement learning with formal feedback, correctness-by-construction prompting, and proof-aware generation are particularly promising avenues. Ultimately, increasing trust in LLM-assisted hardware design will require workflows where AI-generated security claims are systematically backed by formal evidence.

4. Human-AI Collaboration Interfaces: The optimal interaction model between hardware designers and LLM systems remains an open question. Possible paradigms range from continuous copilot-style assistants embedded in EDA environments to batch analysis tools that generate structured security reports. The conversational paradigm (e.g., interactive design dialogue systems) is especially promising because it supports iterative refinement and human oversight. Closely related is the challenge of explainability. For high-assurance hardware security tasks, users must understand *why* an LLM recommends a particular lock or mitigation. Developing systems that produce structured rationales (e.g., identifying critical signals, attack surfaces, and expected threat reduction) remains an important research direction. Improved human-AI interfaces will likely play a major role in practical adoption.
5. Defending Against AI-Powered Adversaries: As attackers begin leveraging LLMs, hardware security faces a new AI-driven arms race. This creates opportunities for adversarially robust design methodologies. Potential research directions include adversarial training using AI-generated attacks, automated detection of LLM-synthesized Trojans or pirated IP, and robustness evaluation pipelines that explicitly model AI-capable adversaries. Another important open problem is watermark robustness under AI-driven refactoring. Traditional watermarking and similarity detection techniques may degrade when faced with LLM-based semantic rewriting. Designing watermarks and fingerprints that survive AI transformations, possibly leveraging analog behaviors, physical effects, or cross-layer invariants, remains largely unexplored. Co-evolution of AI-based attacks and defenses is likely to become a defining theme of future hardware security research.
6. Scaling to Large Systems and Hierarchical Designs: Most current demonstrations operate on individual modules or small-scale SoCs. Scaling LLM-assisted protection to industrial-scale designs with millions of gates presents significant challenges. Context window limitations, hierarchical dependencies, and cross-module interactions all complicate direct application of current models.

   Promising directions include hierarchical reasoning pipelines, modular analysis frameworks, retrieval-augmented design understanding, and multiagent systems in which specialized agents analyze different portions of the design. Incremental security analysis, where LLM recommendations evolve alongside the design, also remains an open problem. Achieving robust scalability will be critical for real-world deployment.

7. Benchmarking and Metrics for LLM-Protected Designs: Evaluating the effectiveness of LLM-assisted protection remains an open methodological challenge. While traditional metrics (e.g., SAT attack runtime, key search complexity, PPA overhead) remain relevant, new metrics may be required to capture AI-specific effects. For example, measuring semantic diversity of obfuscated designs, resilience to AI-based refactoring, or human effort reduction could provide more holistic evaluation.

   Standardized evaluation suites and community competitions could play an important role in this space. Quantifying productivity gains alongside security improvements will be particularly important for industrial adoption.
8. Ethical, Legal, and Policy Frameworks: Beyond technical challenges, the community must develop clear guidelines for responsible use of LLMs in hardware security. Questions of liability, IP ownership of AI-assisted designs, training data provenance, and acceptable deployment contexts remain largely unresolved. Establishing best practices—such as disclosure of AI involvement, dataset transparency, and third-party validation of AI-generated security mechanisms—may become increasingly important as adoption grows.

   The intersection of LLMs and hardware IP protection represents fertile ground for innovation. Early successes in automated obfuscation, AI-assisted vulnerability discovery, and adversarial evaluation suggest a potential paradigm shift in SoC security. However, substantial research is still required to achieve scalable, trustworthy, and formally grounded AI-assisted protection workflows. Continued progress across the directions outlined above will be essential to ensure that generative AI ultimately strengthens, rather than weakens, the security posture of future hardware systems.

References

1. A. Yeh. Trends in the Global IC Design Service Market. *DIGITIMES*, 2012.
2. Baleegh Ahmad, Wei-Kai Liu, Luca Collini, Hammond Pearce, Jason M Fung, Jonathan Valamehr, Mohammad Bidmeshki, Piotr Sapiecha, Steve Brown, Krishnendu Chakrabarty, et al. Don't CWEAT it: Toward CWE analysis techniques in early stages of hardware design. In *Proceedings of the 41st IEEE/ACM International Conference on Computer-Aided Design*, pages 1–9, 2022.
3. Baleegh Ahmad, Shailja Thakur, Benjamin Tan, Ramesh Karri, and Hammond Pearce. On hardware security bug code fixes by prompting large language models. *IEEE Transactions on Information Forensics and Security*, 19:4043–4057, 2024.
4. Bulbul Ahmed, Sazadur Rahman, Kimia Zamiri Azar, Farimah Farahmandi, Fahim Rahman, and Mark Tehranipoor. SeeMLess: Security evaluation of logic locking using machine learning oriented estimation. In *Proceedings of the Great Lakes Symposium on VLSI 2024*, pages 489–494, 2024.
5. Mohammad Akyash, Kimia Azar, and Hadi Kamali. DecoRTL: A run-time decoding framework for RTL code generation with LLMs. In *2025 IEEE/ACM International Conference On Computer Aided Design (ICCAD)*, pages 1–9. IEEE, 2025.

6. Mohammad Akyash, Kimia Azar, and Hadi Kamali. RTL++: Graph-enhanced LLM for RTL code generation. In *2025 IEEE International Conference on LLM-Aided Design (ICLAD)*, pages 44–50. IEEE, 2025.
7. Mohammad Akyash and Hadi Mardani Kamali. Self-HWDebug: Automation of LLM self-instructing for hardware security verification. In *2024 IEEE Computer Society Annual Symposium on VLSI (ISVLSI)*, pages 391–396. IEEE, 2024.
8. Mohammad Akyash and Hadi M Kamali. Evolutionary large language models for hardware security: A comparative survey. In *Proceedings of the great lakes symposium on VLSI 2024*, pages 496–501, 2024.
9. Abdulrahman Alaql, Saranyu Chattopadhyay, Prabuddha Chakraborty, Tamzidul Hoque, and Swarup Bhunia. LeGO: A learning-guided obfuscation framework for hardware IP protection. *IEEE Transactions on Computer-Aided Design of Integrated Circuits and Systems*, 41(4):854–867, 2021.
10. Abdulrahman Alaql, Md Moshiur Rahman, and Swarup Bhunia. Scope: Synthesis-based constant propagation attack on logic locking. *IEEE Transactions on Very Large Scale Integration (VLSI) Systems*, 29(8):1529–1542, 2021.
11. Lilas Alrahis, Satwik Patnaik, Faiq Khalid, Muhammad Abdullah Hanif, Hani Saleh, Muhammad Shafique, and Ozgur Sinanoglu. GNNUnlock: Graph neural networks-based oracle-less unlocking scheme for provably secure logic locking. In *2021 Design, Automation & Test in Europe Conference & Exhibition (DATE)*, pages 780–785. IEEE, 2021.
12. Lilas Alrahis, Satwik Patnaik, Muhammad Shafique, and Ozgur Sinanoglu. OMLA: An oracle-less machine learning-based attack on logic locking. *IEEE Transactions on Circuits and Systems II: Express Briefs*, 69(3):1602–1606, 2021.
13. Lilas Alrahis, Abhrajit Sengupta, Johann Knechtel, Satwik Patnaik, Hani Saleh, Baker Mohammad, Mahmoud Al-Qutayri, and Ozgur Sinanoglu. GNN-RE: Graph neural networks for reverse engineering of gate-level netlists. *IEEE Transactions on Computer-Aided Design of Integrated Circuits and Systems*, 41(8):2435–2448, 2021.
14. N Nalla Anandakumar, M Sazadur Rahman, Mridha Md Mashahedur Rahman, Rasheed Kibria, Upoma Das, Farimah Farahmandi, Fahim Rahman, and Mark M Tehranipoor. Rethinking watermark: Providing proof of IP ownership in modern SoCs. *Cryptology ePrint Archive*, 2022.
15. Md Tanvir Arafin, Andrew Stanley, and Praveen Sharma. Hardware-based anti-counterfeiting techniques for safeguarding supply chain integrity. In *2017 IEEE International Symposium on Circuits and Systems (ISCAS)*, pages 1–4. IEEE, 2017.
16. Giuseppe Ateniese, Aggelos Kiayias, Bernardo Magri, Yiannis Tselekounis, and Daniele Venturi. Secure outsourcing of cryptographic circuits manufacturing. In *International Conference on Provable Security*, pages 75–93. Springer, 2018.
17. Kimia Zamiri Azar, Hadi Mardani Kamali, Farimah Farahmandi, and Mark Tehranipoor. Warm up before circuit de-obfuscation? an exploration through bounded-model-checkers. In *2022 IEEE International Symposium on Hardware Oriented Security and Trust (HOST)*, pages 13–16. IEEE, 2022.
18. Kimia Zamiri Azar, Hadi Mardani Kamali, Farimah Farahmandi, and Mark Tehranipoor. *Understanding logic locking*. Springer, 2024.
19. Kimia Zamiri Azar, Hadi Mardani Kamali, Houman Homayoun, and Avesta Sasan. SMT attack: Next generation attack on obfuscated circuits with capabilities and performance beyond the sat attacks. *IACR Transactions on Cryptographic Hardware and Embedded Systems*, pages 97–122, 2019.
20. Kimia Zamiri Azar, Hadi Mardani Kamali, Houman Homayoun, and Avesta Sasan. NNgSAT: Neural network guided sat attack on logic locked complex structures. In *Proceedings of the 39th International Conference on Computer-Aided Design*, pages 1–9, 2020.
21. Kimia Zamiri Azar, Hadi Mardani Kamali, Houman Homayoun, and Avesta Sasan. From cryptography to logic locking: A survey on the architecture evolution of secure scan chains. *IEEE Access*, 9:73133–73151, 2021.

22. Abhishek Basak, Swarup Bhunia, Thomas Tkacik, and Sandip Ray. Security assurance for system-on-chip designs with untrusted IPs. *IEEE Transactions on Information Forensics and Security*, 12(7):1515–1528, 2017.
23. Kanad Basu, Samah Mohamed Saeed, Christian Pilato, Mohammed Ashraf, Mohammed Thari Nabeel, Krishnendu Chakrabarty, and Ramesh Karri. Cad-base: An attack vector into the electronics supply chain. *ACM Transactions on Design Automation of Electronic Systems (TODAES)*, 24(4):1–30, 2019.
24. Georg T Becker, Francesco Regazzoni, Christof Paar, and Wayne P Burleson. Stealthy dopant-level hardware trojans. In *Cryptographic Hardware and Embedded Systems-CHES 2013: 15th International Workshop, Santa Barbara, CA, USA, August 20-23, 2013. Proceedings 15*, pages 197–214. Springer, 2013.
25. Shivam Bhasin, Jean-Luc Danger, Sylvain Guilley, Xuan Thuy Ngo, and Laurent Sauvage. Hardware Trojan horses in cryptographic IP cores. In *2013 Workshop on Fault Diagnosis and Tolerance in Cryptography*, pages 15–29. IEEE, 2013.
26. Shivam Bhasin and Francesco Regazzoni. A survey on hardware Trojan detection techniques. In *2015 IEEE International Symposium on Circuits and Systems (ISCAS)*, pages 2021–2024. IEEE, 2015.
27. Jason Blocklove, Siddharth Garg, Ramesh Karri, and Hammond Pearce. Chip-chat: Challenges and opportunities in conversational hardware design. In *2023 ACM/IEEE 5th Workshop on Machine Learning for CAD (MLCAD)*, pages 1–6. IEEE, 2023.
28. Clair Brown, Greg Linden, and Jeffrey T Macher. Offshoring in the semiconductor industry: A historical perspective [with comment and discussion]. In *Brookings trade forum*, pages 279–333. JSTOR, 2005.
29. Paul E Calzada, Zahin Ibnat, Tanvir Rahman, Kamal Kandula, Danyu Lu, Sujan Kumar Saha, Farimah Farahmandi, and Mark Tehranipoor. VerilogDB: The largest, highest-quality dataset with a preprocessing framework for LLM-based RTL generation. *arXiv preprint arXiv:2507.13369*, 2025.
30. Zhiqian Chen, Lei Zhang, Gaurav Kolhe, Hadi Mardani Kamali, Setareh Rafatirad, Sai Manoj Pudukotai Dinakarrao, Houman Homayoun, Chang-Tien Lu, and Liang Zhao. Deep graph learning for circuit deobfuscation. *Frontiers in big Data*, 4:608286, 2021.
31. Gustavo K Contreras, Md Tauhidur Rahman, and Mohammad Tehranipoor. Secure split-test for preventing IC piracy by untrusted foundry and assembly. In *2013 IEEE International symposium on defect and fault tolerance in VLSI and nanotechnology systems (DFTS)*, pages 196–203. IEEE, 2013.
32. Jonathan Cruz, Pravin Gaikwad, and Swarup Bhunia. Analysis of hardware Trojan resilience enabled through logic locking. In *2022 Asian hardware oriented security and trust symposium (AsianHOST)*, pages 1–6. IEEE, 2022.
33. Fan Cui, Chenyang Yin, Kexing Zhou, Youwei Xiao, Guangyu Sun, Qiang Xu, Qipeng Guo, Yun Liang, Xingcheng Zhang, Demin Song, et al. OriGen: Enhancing RTL code generation with code-to-code augmentation and self-reflection. In *Proceedings of the 43rd IEEE/ACM International Conference on Computer-Aided Design*, pages 1–9, 2024.
34. Pinchen Cui, Julie Dixon, Ujjwal Guin, and Daniel Dimase. A blockchain-based framework for supply chain provenance. *IEEE Access*, 7:157113–157125, 2019.
35. Upoma Das, Md Rafid Muttaki, Mark M Tehranipoor, and Farimah Farahmandi. ADWIL: A zero-overhead analog device watermarking using inherent IP features. In *2022 IEEE International Test Conference (ITC)*, pages 155–164. IEEE, 2022.
36. Upoma Das, Mohammad Rahman, Akshay Kulkarni, Mark Tehranipoor, and Farimah Farahmandi. PSCMark: Power side channel-based watermarking for SoC IPs using clock gates. *ACM Transactions on Design Automation of Electronic Systems*, 30(5):1–27, 2025.
37. Zehang Deng, Ruoxi Sun, Minhui Xue, Wanlun Ma, Sheng Wen, Surya Nepal, and Yang Xiang. Hardening LLM fine-tuning: From differentially private data selection to trustworthy model quantization. *IEEE Transactions on Information Forensics and Security*, 20:7211–7226, 2025.

38. Ziad El Sayed, Zeng Wang, Hana Selmani, Johann Knechtel, Ozgur Sinanoglu, and Lilas Alrahis. Graph neural networks for integrated circuit design, reliability, and security: Survey and tool. *ACM Computing Surveys*, 58(4):1–44, 2025.
39. Susanne Engels, Max Hoffmann, and Christof Paar. The end of logic locking? a critical view on the security of logic locking. *Cryptology ePrint Archive*, 2019.
40. Wenji Fang, Mengming Li, Min Li, Zhiyuan Yan, Shang Liu, Zhiyao Xie, and Hongce Zhang. AssertLLM: Generating and evaluating hardware verification assertions from design specifications via multi-LLMs. *arXiv preprint arXiv:2402.00386*, 2024.
41. Farimah Farahmandi, Ankur Srivastava, Giorgio Di Natale, and Mark Tehranipoor. Introduction to the special issue on cad for security: Pre-silicon security sign-off solutions through design cycle, 2023.
42. Md Omar Faruque, Peter Jamieson, Ahmad Patooghy, and Abdel-Hameed A Badawy. Unleashing ghost: An LLM-powered framework for automated hardware Trojan design. *Electronics*, 14(23):4745, 2025.
43. Elena Jean Folgo. *Accelerating time-to-market in the global electronics industry*. PhD thesis, Massachusetts Institute of Technology, 2008.
44. Jugal Gandhi, Diksha Shekhawat, M Santosh, Jaya Dofe, and Jai Gopal Pandey. Large language model driven logic locking: A generative approach to secure IC design. In *2024 IEEE 33rd Asian Test Symposium (ATS)*, pages 1–4. IEEE, 2024.
45. Vasudev Gohil, Matthew DeLorenzo, Veera Vishwa Achuta Sai Venkat Nallam, Joey See, and Jeyavijayan Rajendran. LLMPirate: LLMs for black-box hardware IP piracy. *arXiv preprint arXiv:2411.16111*, 2024.
46. Peter C Grindley and David J Teece. Managing intellectual capital: licensing and cross-licensing in semiconductors and electronics. *California management review*, 39(2):8–41, 1997.
47. Ujjwal Guin, Ke Huang, Daniel DiMase, John M Carulli, Mohammad Tehranipoor, and Yiorgos Makris. Counterfeit integrated circuits: A rising threat in the global semiconductor supply chain. *Proceedings of the IEEE*, 102(8):1207–1228, 2014.
48. Ujjwal Guin, Qihang Shi, Domenic Forte, and Mark M Tehranipoor. Fortis: a comprehensive solution for establishing forward trust for protecting IPs and ICs. *ACM transactions on design automation of electronic systems (TODAES)*, 21(4):1–20, 2016.
49. Ujjwal Guin, Ziqi Zhou, and Adit Singh. Robust design-for-security architecture for enabling trust in IC manufacturing and test. *IEEE Transactions on Very Large Scale Integration (VLSI) Systems*, 26(5):818–830, 2018.
50. Ke Huang, John M Carulli, and Yiorgos Makris. Counterfeit electronics: A rising threat in the semiconductor manufacturing industry. In *2013 IEEE international test conference (ITC)*, pages 1–4. IEEE, 2013.
51. Zhao Huang, Quan Wang, Yin Chen, and Xiaohong Jiang. A survey on machine learning against hardware Trojan attacks: Recent advances and challenges. *IEEE Access*, 8:10796–10826, 2020.
52. Zahin Ibnat, M Sazadur Rahman, Mridha Mashahedur Rahman, Hadi Mardani Kamali, Mark Tehranipoor, and Farimah Farahmandi. ActiWate: Adaptive and design-agnostic active watermarking for IP ownership in modern SoCs. In *2023 60th ACM/IEEE design automation conference (DAC)*, pages 1–6. IEEE, 2023.
53. Zahin Ibnat, Mrida Md Mashahedur Rahman, Sazadur Rahman, Jingbo Zhou, and Farimah Farahmandi. Redefining tradition: An active watermarking approach for IP protection in SoCs. *IEEE Transactions on Computer-Aided Design of Integrated Circuits and Systems*, 2025.
54. Frank Imeson, Ariq Emtenan, Siddharth Garg, and Mahesh Tripunitara. Securing computer hardware using 3D integrated circuit ({IC}) technology and split manufacturing for obfuscation. In *22nd {USENIX} Security Symposium ({USENIX} Security 13)*, pages 495–510, 2013.
55. Ayush Jain, Ujjwal Guin, M Tanjidur Rahman, Navid Asadizanjani, Danielle Duvalsaint, and RD Shawn Blanton. Special session: Novel attacks on logic-locking. In *2020 IEEE 38th VLSI Test Symposium (VTS)*, pages 1–10. IEEE, 2020.

56. Ayush Jain, Ziqi Zhou, and Ujjwal Guin. TAAL: tampering attack on any key-based logic locked circuits. *ACM Transactions on Design Automation of Electronic Systems (TODAES)*, 26(4):1–22, 2021.
57. Hadi Mardani Kamali, Kimia Zamiri Azar, Farimah Farahmandi, and Mark Tehranipoor. Advances in logic locking: Past, present, and prospects. *Cryptology ePrint Archive*, 2022.
58. Hadi Mardani Kamali, Kimia Zamiri Azar, Houman Homayoun, and Avesta Sasan. Full-lock: Hard distributions of sat instances for obfuscating circuits using fully configurable logic and routing blocks. In *Proceedings of the 56th Annual Design Automation Conference 2019*, pages 1–6, 2019.
59. Hadi Mardani Kamali, Kimia Zamiri Azar, Houman Homayoun, and Avesta Sasan. Interlock: An intercorrelated logic and routing locking. In *Proceedings of the 39th International Conference on Computer-Aided Design*, pages 1–9, 2020.
60. Hadi Mardani Kamali, Kimia Zamiri Azar, Houman Homayoun, and Avesta Sasan. On designing secure and robust scan chain for protecting obfuscated logic. *arXiv preprint arXiv:2005.04262*, 2020.
61. Rahul Kande, Vasudev Gohil, Matthew DeLorenzo, Chen Chen, and Jeyavijayan Rajendran. LLMs for hardware security: Boon or bane? In *2024 IEEE 42nd VLSI Test Symposium (VTS)*, pages 1–4. IEEE, 2024.
62. Rahul Kande, Hammond Pearce, Benjamin Tan, Brendan Dolan-Gavitt, Shailja Thakur, Ramesh Karri, and Jeyavijayan Rajendran. (Security) assertions by large language models. *IEEE Transactions on Information Forensics and Security*, 19:4374–4389, 2024.
63. Simon Klix, Nils Albartus, Julian Speith, Paul Staat, Alice Verstege, Annika Wilde, Daniel Lammers, Jörn Langheinrich, Christian Kison, Sebastian Sester-Wehle, et al. Stealing Maggie's secrets-on the challenges of IP theft through FPGA reverse engineering. In *Proceedings of the 2024 on ACM SIGSAC Conference on Computer and Communications Security*, pages 3391–3405, 2024.
64. Johann Knechtel, Satwik Patnaik, and Ozgur Sinanoglu. Protect your chip design intellectual property: An overview. In *Proceedings of the International Conference on Omni-Layer Intelligent Systems*, pages 211–216, 2019.
65. Georgios Kokolakis, Athanasios Moschos, and Angelos D Keromytis. Harnessing the power of general-purpose LLMs in hardware Trojan design. In *International conference on applied cryptography and network security*, pages 176–194. Springer, 2024.
66. Gaurav Kolhe, Tyler Sheaves, Kevin Immanuel Gubbi, Soheil Salehi, Setareh Rafatirad, Sai Manoj PD, Avesta Sasan, and Houman Homayoun. Lock&roll: Deep-learning power side-channel attack mitigation using emerging reconfigurable devices and logic locking. In *Proceedings of the 59th ACM/IEEE Design Automation Conference*, pages 85–90, 2022.
67. Farinaz Koushanfar. Integrated circuits metering for piracy protection and digital rights management: An overview. In *Proceedings of the 21st edition of the great lakes symposium on Great lakes symposium on VLSI*, pages 449–454, 2011.
68. Banafsheh Saber Latibari, Sujan Ghimire, Muhtasim Alam Chowdhury, Najmeh Nazari, Kevin Immanuel Gubbi, Houman Homayoun, Avesta Sasan, and Soheil Salehi. Automated hardware logic obfuscation framework using GPT. In *2024 IEEE 17th Dallas Circuits and Systems Conference (DCAS)*, pages 1–5. IEEE, 2024.
69. Wei Liang, Dafang Zhang, Xia Lei, Mingdong Tang, Kuan-Ching Li, and Albert Y Zomaya. Circuit copyright blockchain: Blockchain-based homomorphic encryption for IP circuit protection. *IEEE Transactions on Emerging Topics in Computing*, 9(3):1410–1420, 2020.
70. Nimisha Limaye, Emmanouil Kalligeros, Nikolaos Karousos, Irene G Karybali, and Ozgur Sinanoglu. Thwarting all logic locking attacks: Dishonest oracle with truly random logic locking. *IEEE Transactions on Computer-Aided Design of Integrated Circuits and Systems*, 40(9):1740–1753, 2020.
71. Mingjie Liu, Teodor-Dumitru Ene, Robert Kirby, Chris Cheng, Nathaniel Pinckney, Rongjian Liang, Jonah Alben, Himyanshu Anand, Sanmitra Banerjee, Ismet Bayraktaroglu, et al. ChipNeMo: Domain-adapted LLMs for chip design. *arXiv preprint arXiv:2311.00176*, 2023.

72. Mingjie Liu, Nathaniel Pinckney, Brucek Khailany, and Haoxing Ren. VerilogEval: Evaluating large language models for verilog code generation. In *2023 IEEE/ACM International Conference on Computer Aided Design (ICCAD)*, pages 1–8. IEEE, 2023.
73. Shang Liu, Wenji Fang, Yao Lu, Qijun Zhang, Hongce Zhang, and Zhiyao Xie. RTL-Coder: Outperforming GPT-3.5 in design RTL generation with our open-source dataset and lightweight solution. In *2024 IEEE LLM Aided Design Workshop (LAD)*, pages 1–5. IEEE, 2024.
74. Yao Lu, Shang Liu, Qijun Zhang, and Zhiyao Xie. RTLLM: An open-source benchmark for design RTL generation with large language model. In *2024 29th Asia and South Pacific Design Automation Conference (ASP-DAC)*, pages 722–727. IEEE, 2024.
75. Yong-Qiang Lv, Qiang Zhou, Yi-Ci Cai, and Gang Qu. Trusted integrated circuits: The problem and challenges. *Journal of computer science and technology*, 29(5):918–928, 2014.
76. M. Rostami, F. Koushanfar, R. Karri. A Primer on Hardware Security: Models, Methods, and Metrics. *Proceedings of the IEEE*, 102(8):1283–1295, 2014.
77. M. Tehranipoor, C. Wang. *Introduction to hardware security and trust.* Springer Science & Business Media, 2011.
78. Jeffrey T Macher, David C Mowery, and Timothy S Simcoe. E-business and disintegration of the semiconductor industry value chain. *Industry and Innovation*, 9(3):155–181, 2002.
79. Bhabesh Mali, Karthik Maddala, Vatsal Gupta, Sweeya Reddy, Chandan Karfa, and Ramesh Karri. CHIRAAG: ChatGPT informed rapid and automated assertion generation. In *2024 IEEE Computer Society Annual Symposium on VLSI (ISVLSI)*, pages 680–683. IEEE, 2024.
80. Nowfel Mashnoor, Mohammad Akyash, Hadi Kamali, and Kimia Azar. CircuitGuard: Mitigating LLM memorization in RTL code generation against IP leakage. *2025 IEEE 43rd International Conference on Computer Design (ICCD)*, pages 790–797, 2025.
81. Nowfel Mashnoor, Mohammad Akyash, Hadi Kamali, and Kimia Azar. LLM-IFT: LLM-powered information flow tracking for secure hardware. In *2025 IEEE 43rd VLSI Test Symposium (VTS)*, pages 1–5. IEEE, 2025.
82. Prabhat Mishra, Mark Tehranipoor, and Swarup Bhunia. Security and trust vulnerabilities in third-party IPs. In *Hardware IP Security and Trust*, pages 3–14. Springer, 2017.
83. Tahoura Mosavirik, Patrick Schaumont, and Shahin Tajik. Impedanceverif: On-chip impedance sensing for system-level tampering detection. *IACR Transactions on Cryptographic Hardware and Embedded Systems*, pages 301–325, 2023.
84. Md Rafid Muttaki, Hadi M Kamali, Mark Tehranipoor, and Farimah Farahmandi. Pallet: Protecting analog devices using a last-level edit technique. In *2023 IEEE Physical Assurance and Inspection of Electronics (PAINE)*, pages 1–7. IEEE, 2023.
85. Md Rafid Muttaki, Roshanak Mohammadivojdan, Hadi Mardani Kamali, Mark Tehranipoor, and Farimah Farahmandi. Hlock+: A robust and low-overhead logic locking at the high-level language. *IEEE Transactions on Computer-Aided Design of Integrated Circuits and Systems*, 42(7):2149–2162, 2022.
86. Md Rafid Muttaki, Shuvagata Saha, Hadi M Kamali, Fahim Rahman, Mark Tehranipoor, and Farimah Farahmandi. Rtlock: IP protection using scan-aware logic locking at RTL. In *2023 Design, Automation & Test in Europe Conference & Exhibition (DATE)*, pages 1–6. IEEE, 2023.
87. Etienne Ollion, Rubing Shen, Ana Macanovic, and Arnault Chatelain. The dangers of using proprietary LLMs for research. *Nature Machine Intelligence*, 6(1):4–5, 2024.
88. Sudipta Paria, Aritra Dasgupta, and Swarup Bhunia. Divas: An LLM-based end-to-end framework for SoC security analysis and policy-based protection. *arXiv preprint arXiv:2308.06932*, 2023.
89. Shahed E Quadir, Junlin Chen, Domenic Forte, Navid Asadizanjani, Sina Shahbazmohamadi, Lei Wang, John Chandy, and Mark Tehranipoor. A survey on chip to system reverse engineering. *ACM journal on emerging technologies in computing systems (JETC)*, 13(1):1–34, 2016.
90. M Sazadur Rahman, Rui Guo, Hadi M Kamali, Fahim Rahman, Farimah Farahmandi, Mohamed Abdel-Moneum, and Mark Tehranipoor. O'clock: lock the clock via clock-gating

for SoC IP protection. In *Proceedings of the 59th ACM/IEEE Design Automation Conference*, pages 775–780, 2022.

91. M Sazadur Rahman, Henian Li, Rui Guo, Fahim Rahman, Farimah Farahmandi, and Mark Tehranipoor. LL-ATPG: logic-locking aware test using valet keys in an untrusted environment. In *2021 IEEE International Test Conference (ITC)*, pages 180–189. IEEE, 2021.
92. M Tanjidur Rahman, Shahin Tajik, M Sazadur Rahman, Mark Tehranipoor, and Navid Asadizanjani. The key is left under the mat: On the inappropriate security assumption of logic locking schemes. In *2020 IEEE international symposium on hardware oriented security and trust (HOST)*, pages 262–272. IEEE, 2020.
93. Md Tauhidur Rahman, Domenic Forte, Quihang Shi, Gustavo K Contreras, and Mohammad Tehranipoor. CSST: Preventing distribution of unlicensed and rejected ICs by untrusted foundry and assembly. In *2014 IEEE International symposium on defect and fault tolerance in VLSI and nanotechnology systems (DFT)*, pages 46–51. IEEE, 2014.
94. Mridha Md Mashahedur Rahman, M Sazadur Rahman, Rasheed Kibria, Mike Borza, Bandy Reddy, Adam Cron, Fahim Rahman, Mark Tehranipoor, and Farimah Farahmandi. CAPEC: A cellular automata guided FSM-based IP authentication scheme. In *2023 IEEE 41st VLSI Test Symposium (VTS)*, pages 1–8. IEEE, 2023.
95. Jeyavijayan Rajendran, Michael Sam, Ozgur Sinanoglu, and Ramesh Karri. Security analysis of integrated circuit camouflaging. In *Proceedings of the 2013 ACM SIGSAC conference on Computer & communications security*, pages 709–720, 2013.
96. Siladitya Ray. Samsung bans ChatGPT and other chatbots for employees after sensitive code leak. https://www.forbes.com/sites/siladityaray/2023/05/02/samsung-bans-chatgpt-and-other-chatbots-for-employees-after-sensitive-code-leak/, May 2023. Forbes.
97. Shervin Roshanisefat, Hadi Mardani Kamali, Houman Homayoun, and Avesta Sasan. Rane: An open-source formal de-obfuscation attack for reverse engineering of logic encrypted circuits. In *Proceedings of the 2021 Great Lakes Symposium on VLSI*, pages 221–228, 2021.
98. Subroto Roy and K Sivakumar. Managing intellectual property in global outsourcing for innovation generation. *Journal of Product Innovation Management*, 28(1):48–62, 2011.
99. Akashdeep Saha, Prithwish Basu Roy, Johann Knechtel, Ramesh Karri, Ozgur Sinanoglu, and Lilas Alrahis. GLLaMoR: Graph-based logic locking by large language models for enhanced robustness. In *2025 IEEE 43rd VLSI Test Symposium (VTS)*, pages 1–5. IEEE, 2025.
100. Akashdeep Saha, Zeng Wang, Prithwish Basu Roy, Johann Knechtel, Ozgur Sinanoglu, and Ramesh Karri. LockForge: Automating paper-to-code for logic locking with multi-agent reasoning LLMs. *arXiv preprint arXiv:2511.18531*, 2025.
101. Dipayan Saha, Shams Tarek, Katayoon Yahyaei, Sujan Kumar Saha, Jingbo Zhou, Mark Tehranipoor, and Farimah Farahmandi. LLM for SoC security: A paradigm shift. *IEEE Access*, 12:155498–155521, 2024.
102. Dipayan Saha, Katayoon Yahyaei, Sujan Kumar Saha, Mark Tehranipoor, and Farimah Farahmandi. Empowering hardware security with LLM: The development of a vulnerable hardware database. In *2024 IEEE International Symposium on Hardware Oriented Security and Trust (HOST)*, pages 233–243. IEEE, 2024.
103. Hassan Salmani. The global integrated circuit supply chain flow and the hardware Trojan attack. In *Trusted digital circuits: hardware Trojan vulnerabilities, prevention and detection*, pages 1–11. Springer, 2018.
104. Robert Salomon and Xavier Martin. Learning, knowledge transfer, and technology implementation performance: A study of time-to-build in the global semiconductor industry. *Management Science*, 54(7):1266–1280, 2008.
105. Md Sami Ul Islam Sami, Hadi Mardani Kamali, Farimah Farahmandi, Fahim Rahman, and Mark Tehranipoor. Enabling security of heterogeneous integration: From supply chain to in-field operations. *IEEE Design & Test*, 40(5):86–95, 2023.
106. Md Sami Ul Islam Sami, Tao Zhang, Amit Mazumder Shuvo, Md Saad Ul Haque, Paul E Calzada, Kimia Zamiri Azar, Hadi Mardani Kamali, Fahim Rahman, Farimah Farahmandi, and Mark Tehranipoor. Advancing trustworthiness in system-in-package: A novel root-of-trust hardware security module for heterogeneous integration. *IEEE Access*, 12:48081–48107, 2024.

107. Pantha Protim Sarker, Upoma Das, Mohammad Bin Monjil, Jingbo Zhou, Farimah Farahmandi, and Mark Tehranipoor. Gem-water: Generation of EM-based watermark with hidden FSM for SoC IPs to combat piracy. *IEEE Transactions on Computer-Aided Design of Integrated Circuits and Systems*, 2025.
108. Pantha Protim Sarker, Upoma Das, Nitin Varshney, Shang Shi, Akshay Kulkarni, Farimah Farahmandi, and Mark Tehranipoor. When everyday devices become weapons: A closer look at the pager and walkie-talkie attacks. *arXiv preprint arXiv:2501.17405*, 2025.
109. Anirban Sengupta. Intellectual property cores: Protection designs for CE products. *IEEE Consumer Electronics Magazine*, 5(1):83–88, 2015.
110. Kaveh Shamsi, Meng Li, Kenneth Plaks, Saverio Fazzari, David Z Pan, and Yier Jin. IP protection and supply chain security through logic obfuscation: A systematic overview. *ACM Transactions on Design Automation of Electronic Systems (TODAES)*, 24(6):1–36, 2019.
111. Kaveh Shamsi, Travis Meade, Meng Li, David Z Pan, and Yier Jin. On the approximation resiliency of logic locking and IC camouflaging schemes. *IEEE Transactions on Information Forensics and Security*, 14(2):347–359, 2018.
112. Deepak Sirone and Pramod Subramanyan. Functional analysis attacks on logic locking. *IEEE Transactions on Information Forensics and Security*, 15:2514–2527, 2020.
113. Dominik Sisejkovic, Farhad Merchant, Lennart M Reimann, and Rainer Leupers. Deceptive logic locking for hardware integrity protection against machine learning attacks. *IEEE Transactions on Computer-Aided Design of Integrated Circuits and Systems*, 41(6):1716–1729, 2021.
114. Dominik Sisejkovic, Farhad Merchant, Lennart M Reimann, Harshit Srivastava, Ahmed Hallawa, and Rainer Leupers. Challenging the security of logic locking schemes in the era of deep learning: A neuroevolutionary approach. *ACM Journal on Emerging Technologies in Computing Systems (JETC)*, 17(3):1–26, 2021.
115. Dominik Sisejkovic, Lennart M Reimann, Elmira Moussavi, Farhad Merchant, and Rainer Leupers. Logic locking at the frontiers of machine learning: A survey on developments and opportunities. In *2021 IFIP/IEEE 29th International Conference on Very Large Scale Integration (VLSI-SoC)*, pages 1–6. IEEE, 2021.
116. Andrew Stern, Huanyu Wang, Fahim Rahman, Farimah Farahmandi, and Mark Tehranipoor. ACED-IT: Assuring confidential electronic design against insider threats in a zero-trust environment. *IEEE Transactions on Computer-Aided Design of Integrated Circuits and Systems*, 41(10):3202–3215, 2021.
117. Srivatsa Subramanya and Jonathan Herald. Reverse shoulder arthroplasty with patient-specific glenoid implant positioning guides. *Techniques in Shoulder & Elbow Surgery*, 15(4):122–129, 2014.
118. Pramod Subramanyan, Sayak Ray, and Sharad Malik. Evaluating the security of logic encryption algorithms. In *2015 IEEE International Symposium on Hardware Oriented Security and Trust (HOST)*, pages 137–143. IEEE, 2015.
119. Shams Tarek, Dipayan Saha, Sujan Kumar Saha, Mark Tehranipoor, and Farimah Farahmandi. SoCureLLM: An LLM-driven approach for large-scale system-on-chip security verification and policy generation. In *2025 IEEE International Symposium on Hardware Oriented Security and Trust (HOST)*, pages 335–345. IEEE, 2025.
120. Ishraq Tashdid, Kimia Tasnia, Alexander Garcia, Jonathan Valamehr, and Sazadur Rahman. ATLAS: AI-assisted threat-to-assertion learning for system-on-chip security verification, 2026.
121. Mark Tehranipoor, Ujjwal Guin, and Domenic Forte. Counterfeit integrated circuits. In *Counterfeit Integrated Circuits: Detection and Avoidance*, pages 15–36. Springer, 2015.
122. Mark Tehranipoor, Kimia Zamiri Azar, Navid Asadizanjani, Fahim Rahman, Hadi Mardani Kamali, and Farimah Farahmandi. Rethinking hardware watermark. In *Hardware Security: A Look into the Future*, pages 143–182. Springer, 2024.
123. Mohmmad Tehranipoor, Hassan Salmani, and Xuehui Zhang. Integrated circuit authentication. *Switzerland: Springer, Cham.*, 10:978–3, 2014.

124. Shailja Thakur, Baleegh Ahmad, Hammond Pearce, Benjamin Tan, Brendan Dolan-Gavitt, Ramesh Karri, and Siddharth Garg. Verigen: A large language model for verilog code generation. *ACM Transactions on Design Automation of Electronic Systems*, 29(3):1–31, 2024.
125. Shailja Thakur, Jason Blocklove, Hammond Pearce, Benjamin Tan, Siddharth Garg, and Ramesh Karri. AutoChip: Automating HDL generation using LLM feedback. *arXiv preprint arXiv:2311.04887*, 2023.
126. Ranga Vemuri and Suyuan Chen. *Split manufacturing of integrated circuits for hardware security and trust*. Springer, 2021.
127. Chi-Tai Wang and Chui-Sheng Chiu. Competitive strategies for Taiwan's semiconductor industry in a new world economy. *Technology in Society*, 36:60–73, 2014.
128. Yiting Wang, Guoheng Sun, Wanghao Ye, Gang Qu, and Ang Li. VeriReason: Reinforcement learning with testbench feedback for reasoning-enhanced verilog generation. *arXiv preprint arXiv:2505.11849*, 2025.
129. Zeng Wang, Minghao Shao, Mohammed Thari Nabeel, Prithwish Basu Roy, Likhitha Mankali, Jitendra Bhandari, Ramesh Karri, Ozgur Sinanoglu, Muhammed Shafique, and Johann Knechtel. VeriLeaky: Navigating IP protection vs utility in fine-tuning for LLM-driven verilog coding. *2025 IEEE International Conference on LLM-Aided Design (ICLAD)*, pages 100–107, 2025.
130. Yang Xie and Ankur Srivastava. Anti-SAT: Mitigating sat attack on logic locking. *IEEE Transactions on Computer-Aided Design of Integrated Circuits and Systems*, 38(2):199–207, 2018.
131. Kangwei Xu, Ruidi Qiu, Zhuorui Zhao, Grace Li Zhang, Ulf Schlichtmann, and Bing Li. LLM-aided efficient hardware design automation. *arXiv preprint arXiv:2410.18582*, 2024.
132. Xiaolin Xu, Fahim Rahman, Bicky Shakya, Apostol Vassilev, Domenic Forte, and Mark Tehranipoor. Electronics supply chain integrity enabled by blockchain. *ACM Transactions on Design Automation of Electronic Systems (TODAES)*, 24(3):1–25, 2019.
133. Xiaolin Xu, Bicky Shakya, Mark M Tehranipoor, and Domenic Forte. Novel bypass attack and BDD-based tradeoff analysis against all known logic locking attacks. In *International conference on cryptographic hardware and embedded systems*, pages 189–210. Springer, 2017.
134. Fangfei Yang, Ming Tang, and Ozgur Sinanoglu. Stripped functionality logic locking with hamming distance-based restore unit (SFLL-HD)–unlocked. *IEEE Transactions on Information Forensics and Security*, 14(10):2778–2786, 2019.
135. YangLiu YangLiu, Xiaojun Xu, and Yuanshun Yao. Large language model unlearning. *Advances in Neural Information Processing Systems 37*, pages 105425–105475, 2024.
136. Muhammad Yasin, Bodhisatwa Mazumdar, Jeyavijayan JV Rajendran, and Ozgur Sinanoglu. SARLock: SAT attack resistant logic locking. In *2016 IEEE International Symposium on Hardware Oriented Security and Trust (HOST)*, pages 236–241. IEEE, 2016.
137. Muhammad Yasin, Bodhisatwa Mazumdar, Ozgur Sinanoglu, and Jeyavijayan Rajendran. Removal attacks on logic locking and camouflaging techniques. *IEEE Transactions on Emerging Topics in Computing*, 8(2):517–532, 2017.
138. Muhammad Yasin and Ozgur Sinanoglu. Transforming between logic locking and IC camouflaging. In *2015 10th International Design & Test Symposium (IDT)*, pages 1–4. IEEE, 2015.
139. Muhammad Yasin and Ozgur Sinanoglu. Evolution of logic locking. In *2017 IFIP/IEEE International Conference on Very Large Scale Integration (VLSI-SoC)*, pages 1–6. IEEE, 2017.
140. Henry Wai-chung Yeung. Explaining geographic shifts of chip making toward East Asia and market dynamics in semiconductor global production networks. *Economic Geography*, 98(3):272–298, 2022.
141. Kimia Zamiri Azar, Hadi Mardani Kamali, Farimah Farahmandi, and Mark Tehranipoor. IP protection: A historical view. In *Understanding Logic Locking*, pages 47–62. Springer, 2023.

142. Kimia Zamiri Azar, Hadi Mardani Kamali, Farimah Farahmandi, and Mark Tehranipoor. Making a case for logic locking. *Understanding Logic Locking*, pages 63–87, 2023.
143. Kimia Zamiri Azar, Hadi Mardani Kamali, Houman Homayoun, and Avesta Sasan. Threats on logic locking: A decade later. In *Proceedings of the 2019 Great Lakes Symposium on VLSI*, pages 471–476, 2019.
144. Kaiyuan Zhang, Siyuan Cheng, Hanxi Guo, Yuetian Chen, Zian Su, Shengwei An, Yuntao Du, Charles Fleming, Ashish Kundu, Xiangyu Zhang, and Ninghui Li. Soft: Selective data obfuscation for protecting LLM fine-tuning against membership inference attacks. *ArXiv*, abs/2506.10424, 2025.
145. Xuehui Zhang, Nicholas Tuzzio, and Mohammad Tehranipoor. Identification of recovered ICs using fingerprints from a light-weight on-chip sensor. In *Proceedings of the 49th annual design automation conference*, pages 703–708, 2012.
146. Yongan Zhang, Zhongzhi Yu, Yonggan Fu, Cheng Wan, and Yingyan Celine Lin. MG-Verilog: Multi-grained dataset towards enhanced LLM-assisted verilog generation. In *2024 IEEE LLM Aided Design Workshop (LAD)*, pages 1–5. IEEE, 2024.
147. Yang Zhao, Di Huang, Chongxiao Li, Pengwei Jin, Muxin Song, Yinan Xu, Ziyuan Nan, Mingju Gao, Tianyun Ma, Lei Qi, et al. Codev: Empowering LLMs with HDL generation through multi-level summarization. *IEEE Transactions on Computer-Aided Design of Integrated Circuits and Systems*, 2025.
148. Yujie Zhao, Hejia Zhang, Hanxian Huang, Zhongming Yu, and Jishen Zhao. Mage: A multi-agent engine for automated RTL code generation. In *2025 62nd ACM/IEEE Design Automation Conference (DAC)*, pages 1–7. IEEE, 2025.

Chapter 8
Future Research Direction

Upoma Das, Mark Tehranipoor, and Farimah Farahmandi

The integration of large language models into system-on-chip design and security has moved rapidly from theoretical exploration to practical implementation. As the industry transitions from viewing LLMs as mere code generation novelties to viewing them as foundational infrastructure, the trajectory of research and development must shift. The future of this field lies not in generating isolated toy examples, but in tackling the messy, complex, and highly constrained realities of modern silicon engineering.

8.1 High-Quality Data Collection in Gen-AI Era

The efficacy of any large language model in system-on-chip (SoC) design and security is inextricably bound to the quality of its training and fine-tuning data. Currently, the semiconductor industry relies heavily on base models trained on vast, unstructured, and often noisy open-source repositories. While sufficient for generating boilerplate register-transfer level (RTL) code, this "sheer volume" approach breaks down when applied to high-stakes, precision-critical tasks such as identifying fault-injection vulnerabilities, designing secure boot flows, or establishing a hardware root of trust [8]. Moving forward, the trajectory of LLM development for EDA must pivot toward the systematic enlargement and rigorous refinement of hardware-specific datasets [12, 13].

U. Das (✉) · M. Tehranipoor · F. Farahmandi
University of Florida, Gainesville, FL, USA
e-mail: upoma.das@ufl.edu

F. Farahmandi et al. (eds.), *LLM for SoC Design and Security*,
https://doi.org/10.1007/978-3-032-34116-7_8

8.1.1 The Challenge of Task-Specific Recollection

Different stages of the SoC lifecycle require fundamentally different data representations. A model tasked with functional verification needs to ingest simulation logs and value change dump (VCD) files, whereas a model evaluating side-channel leakage requires power profiles and layout constraints [5, 9, 14]. A primary future direction involves the deliberate recollection of data and the reorganization of labeling taxonomies to suit these highly specific EDA tasks. Treating all hardware data as generic text is a dead end. Researchers and industry practitioners must build specialized corpora where RTL is tightly coupled with its corresponding design specifications, testbenches, and verified security assertions.

While enlarging and relabeling datasets is a necessary baseline, simply aggregating more data or relying on manual curation is an incomplete and unscalable solution. In silicon engineering, a hallucinated syntax error is a minor annoyance; a hallucinated security boundary is a catastrophic liability. The emphasis must shift from the quantity of data to its verifiable quality. Manual recollection and relabeling by human domain experts are too slow to keep pace with the scale required to train next-generation foundation models [12, 13]. Therefore, future research must pioneer automated, synthetic data generation pipelines.

8.1.2 Self-Curating Data Engines

- **Forward Generation: From Specification to IP** The first pillar focuses on expanding the dataset by generating diverse hardware implementations from given specifications. Instead of relying solely on human-written code scraped from repositories, the data engine utilizes advanced LLMs to ingest formal or natural language specifications and output multiple, distinct RTL implementations. By systematically varying the prompt constraints (e.g., asking for a pipelined vs. nonpipelined architecture), the engine generates a massive, diverse corpus of (specification, IP) pairs.
- **Reverse Annotation: From Legacy IP to Specifications and Labels** Only RTL design cannot provide enough information for the LLM applied in SoC design and security verification. Hence, the second pillar flips the paradigm to address the industry's vast backlog of undocumented or poorly labeled legacy IP, which can further solve the data scarcity problem. In this phase, the data engine acts as an analyzer, digesting raw, existing RTL and automatically extracting its underlying specifications, security properties, and metadata labels [2, 4, 7]. This "reverse generation" is critical for dataset enlargement because it unlocks millions of lines of proprietary, battle-tested silicon code that currently lacks the natural language descriptions required to train an LLM. By automatically generating high-quality labels for existing IP, the engine bridges the semantic gap between raw code and human intent.

- **Holistic Validation: Beyond Functional Correctness** The third and most critical pillar is the "Check" phase, which serves as the ultimate gatekeeper for data quality. A generated (Spec, IP) pair is only valuable if it is entirely accurate. Traditionally, automated checks have focused strictly on functional verification—using bounded model checking or simulation to ensure the RTL logically matches the spec. However, in physical silicon design, functional correctness is only half the battle. A future-proof data engine must also rigorously check whether the generated IP satisfies the expected features and properties, such as power, performance (clock), and area (PPA) constraints dictated by the specification.

8.2 Brownfield over Greenfield: Upgrading Legacy IP

Much of the current academic discourse surrounding LLMs in hardware engineering centers on greenfield design—using natural language prompts to generate IP blocks entirely from scratch. However, in the semiconductor industry, completely generating an RTL design from scratch is usually rare. The overwhelming majority of engineering effort is dedicated to brownfield design: updating, upgrading, patching, and refactoring legacy IP to meet new protocol standards, tighter timing constraints, or modern security architectures [2, 3].

Transitioning an LLM from generating an isolated, textbook design, such as FIFO buffer, to safely upgrading a decade-old, highly optimized memory controller represents a massive leap in complexity. Legacy designs are frequently riddled with undocumented workarounds, deep hierarchical dependencies, and implicit timing contracts. Consequently, the future of LLM-assisted hardware design must focus on methodologies that allow these models to safely navigate and modify these fragile ecosystems without introducing subtle vulnerabilities or breaking existing functionality.

To make automated brownfield upgrades a reality, future research must pursue several specific technical directions.

8.2.1 Graph-Augmented Context Retention

Text-based retrieval-augmented generation (RAG) is insufficient for legacy refactoring because standard text chunks cannot capture the multidimensional dependencies of a hardware module. Future models must integrate RAG with structural dependency graphs—such as variable dependency graphs (VDGs) and abstract syntax trees (ASTs). By natively ingesting these graph structures alongside the RTL text, the LLM gains a topological understanding of the design. When tasked with upgrading a module (e.g., migrating an interface from AHB to AXI), a graph-augmented LLM can accurately map the proposed changes, ensuring that the LLM to comprehend how data and control signals propagate through the legacy IP.

8.2.2 Specification Reverse-Engineering and Spec-Driven Upgrades

From the past experience, it is difficult for LLM to understand RTL design logic [11, 12, 15]. Hence, a common misconception is that LLMs should be used to perform "localized" or "surgical" modifications directly on the legacy RTL. In physical hardware, modifying a local part in isolation is extremely dangerous, as it often breaks critical timing paths and introduces catastrophic state mismatches. Instead, future LLM workflows can leverage the spec generation workflow proposed in Sect. 8.1 to upgrade legacy design as follows:

- **Extract** First, the LLM leverages its understanding of the design's topology and legacy RTL to reverse-engineer the original, undocumented specification.
- **Upgrade:** Next, the LLM merges this original specification with the new engineering requirements (e.g., migrating an interface from AHB to AXI, or adding a new security encryption wrapper) to author an upgraded specification.
- **Regenerate:** Finally, the LLM generates the updated version of the RTL design holistically from this newly upgraded specification assisted with the legacy RTL code using the IP generation workflow shown in Sect. 8.1. By elevating the modification process to the specification level, the LLM ensures that all interconnected signals, state machines, and timing contracts are generated cohesively, avoiding the severe risks of localized RTL hacking.

8.2.3 Continuous Equivalence and Property Checking

The capability to generate an upgraded IP block from a new specification is only useful if the resulting RTL can be mathematically proven safe. Therefore, a critical future direction is the tight, automated integration of LLM upgrade agents with logic equivalence checking (LEC) and formal property verification (FPV) tools.

To achieve functional equivalence, the automated workflow must continuously run LEC to guarantee that the core datapath logic—outside of the newly upgraded features—remains strictly identical to the legacy design. Concurrently, rigorous security validation must be enforced. If an LLM is tasked with patching a vulnerability or introducing a new protection mechanism, the workflow must automatically generate and formally verify new systemVerilog assertions (SVA) derived directly from the upgraded specification. This dual approach ensures that the newly generated RTL not only preserves the trusted functionality of its predecessor but also mathematically guarantees that the modernized security goals have been met.

8.3 The LLM as an EDA Orchestrator

Traditionally, the EDA toolchain has been highly fragmented, requiring specialized expertise for each phase of the design and verification lifecycle. A hardware security engineer might need to use Cadence Genus for synthesis, transition to JasperGold for formal property verification, and then invoke an entirely different software suite for power or timing analysis. Mastering the unique command-line interfaces, proprietary scripting languages (such as Tcl), and dense output log formats of each distinct tool creates a steep barrier to entry. Future LLM applications will transcend raw code generation to act as intelligent orchestrators across this fragmented ecosystem, bridging the gap between high-level engineering objectives and low-level tool execution.

8.3.1 The LLM as an Orchestration Assistant

Rather than replacing the foundational EDA tools, future LLMs will serve as a high-level orchestration assistant. When presented with a complex objective, the LLM agent can analyze the requirements, determine which specific EDA tools to invoke, and formulate a cohesive, multitool workflow. For example, it can parse a failing synthesis log, suggest the appropriate formal verification check to isolate the fault, and draft the necessary tool-specific scripts to execute that check. This capability is particularly transformative for experienced engineers who possess deep domain knowledge in hardware architecture or security, but may not be syntax experts for every individual point tool on the market. The LLM translates the engineer's domain expertise into the specific terminology and execution flows required by the diverse toolchain.

8.3.2 The Critical Role of the Human-in-the-Loop

While the concept of an automated orchestrator is highly compelling, it is crucial to recognize that the AI does not operate in a vacuum, nor does it render the human engineer obsolete. Because EDA tools require exact, structured inputs and carry the risk of compounding errors if misdirected, the system relies on a mandatory "human-in-the-loop" mechanism. The LLM's role is to summarize and propose the execution chain; the human engineer's role is to review, monitor, and authorize those critical steps [3, 10].

The engineer acts as a high-level manager, evaluating the LLM's proposed workflow to ensure it aligns with the overall security constraints and design goals. The AI may not be a flawless expert on any single tool, but it excels at synthesizing the broader workflow for the engineer to review. Only once the human expert validates and approves the proposed plan does the automated chain of EDA tools actually begin running. This collaborative paradigm prevents unexpected tool behaviors from corrupting the design database, while drastically accelerating the overall engineering tempo.

8.4 Beyond Text: How Can We Leverage LLM Differently?

Currently, the semiconductor industry predominantly treats large language models as natural language and text-processing engines. The metrics for success are largely tied to how well a model can write SystemVerilog, summarize a specification, or generate a Python script. However, the underlying Transformer architecture is fundamentally a highly advanced sequence and pattern modeler. A critical future direction involves transcending text and applying the Transformer architecture directly to nontextual hardware domains. We must look beyond merely evaluating an LLM's performance on language-related tasks or using it strictly as an agentic orchestrator that calls external tools. Instead, future research must focus on the model architecture itself, exploring how Transformers can be embedded natively into other domains to fundamentally increase the algorithmic performance of the EDA tools themselves.

This paradigm shift moves the AI from being a conversational assistant to a core computational engine. For example, recent studies in hardware fuzzing demonstrate that Transformer-based models can effectively navigate complex hardware state spaces, often outperforming traditional reinforcement learning approaches [1, 6]. In this context, the Transformer is not writing the fuzzing script; it is the fuzzer. Future research should aggressively explore using these architectures to natively model execution traces, analyze power consumption profiles for side-channel leakage, and predict complex fault-injection sequences, bypassing text entirely. By treating hardware behavior as a complex sequence of signals and states, the model can develop a direct, mathematical understanding of the silicon.

Realizing this vision, however, requires overcoming massive encoding and representational challenges. While tokenizing a paragraph of human language or sequential code is relatively straightforward, tokenizing the concurrent, cycle-by-cycle state of a hardware design is not. Representing the state of a 256-bit register transitioning simultaneously alongside asynchronous interrupt signals and intricate clock domain crossings requires entirely novel embedding strategies. The research community must invent hardware-native tokenization methods that can capture this extreme concurrency and spatial-temporal complexity. Overcoming this hurdle is the necessary next step to unlock the Transformer's true potential, evolving it from a high-level coding and orchestration assistant into a foundational, native hardware modeling engine.

8.5 What to Expect in Next Era

As large language models and generative AI transition from experimental research novelties to foundational infrastructure, the landscape of semiconductor design and hardware security is on the precipice of a fundamental restructuring. This emerging era will not merely introduce new features to existing workflows; it will catalyze a paradigm shift across two critical dimensions: the commercial business model of the electronic design automation (EDA) industry, and the fundamental skill set expected of individual hardware engineers.

8.5.1 Transforming the EDA Business Model

For decades, the commercial landscape of EDA has revolved around licensing highly specialized, fragmented point tools. A semiconductor company historically purchased separate licenses for synthesis, formal property verification, power profiling, and physical layout. Each tool required deep, specialized knowledge of proprietary command-line interfaces and arcane scripting languages.

In the next era, as LLMs mature into capable orchestration assistants, this business model will evolve significantly. EDA vendors will transition from selling standalone software licenses to providing task-oriented, integrated "agents." Instead of purchasing access to a specific simulation environment, design houses will license a "Security Verification Agent" or an "IP Integration Agent." These agents will act as intelligent wrappers that lower the barrier to entry, seamlessly translating high-level engineering intent into executed, multitool workflows. The commercial value will no longer reside solely in the underlying algorithmic engines, but in the AI's ability to orchestrate an entire suite of tools to solve high-level design problems. EDA companies will effectively sell autonomous problem-solving capabilities, abstracting away the fragmented terminology and syntax that have historically siloed the industry.

8.5.2 The Evolution of the Hardware Engineer

This commercial transformation directly dictates a profound shift in what is expected of the individual hardware engineer. Historically, significant career capital was built upon mastering a single, specific EDA tool or hyper-specializing in one narrow task, such as writing constrained-random SystemVerilog testbenches or parsing complex synthesis logs.

In the next era, generative AI will commoditize these lower-level implementation details. Consequently, the industry will no longer demand—nor heavily reward—engineers who merely act as syntax translators or single-tool operators. Instead, the baseline expectation for engineers will elevate dramatically toward comprehensive, systems-level thinking.

When an LLM agent orchestrates a complex design upgrade or a multistep security verification flow, the human engineer must possess the "big picture" perspective to oversee it. Engineers will be expected to know exactly why a specific tool in the chain was called, what architectural purpose a given step serves, and, most critically, how to accurately judge whether the agent's output is mathematically and physically sound. The future hardware engineer will transition from a manual coder to a strategic director—defining rigorous security boundaries, managing autonomous agents, and applying deep domain expertise to validate that the AI-assisted silicon is secure, efficient, and ready for tape-out. The true value of the human engineer will lie not in generating the code, but in guaranteeing the integrity of the system it builds.

8.5.3 The Friction of Progress: Apprenticeship and Liability

While this transition promises unprecedented agility, it is not without profound friction. AI does not only transform the world in the ways we desire; it introduces new systemic vulnerabilities that the industry must confront.

Commercially, as EDA vendors transition from providing deterministic tools to providing probabilistic agents, the industry will face complex new legal and liability frameworks. If an automated security agent misses a subtle side-channel vulnerability, or introduces a hidden hardware Trojan that persists into taped-out silicon, the burden of liability becomes a tangled web. This commercial shift will necessitate robust, mathematically backed verification standards to shield both vendors and design houses, requiring the human-in-the-loop not just for engineering oversight, but as an indispensable legal and security firewall.

Culturally, the industry must proactively solve the looming "apprenticeship gap." Historically, junior engineers developed deep architectural intuition precisely by struggling with the low-level mechanics—writing testbenches, debugging convoluted synthesis logs, and fighting syntax errors. As generative AI abstracts away these foundational struggles, hardware teams must reinvent how they educate and onboard talent. The challenge lies in training new engineers to become the strategic systems directors of tomorrow without having them first serve as the syntax translators of yesterday. Navigating these commercial and human challenges will be just as critical to the future of SoC design and security as the algorithmic advancements themselves.

References

1. Pranav Agarwal, Aamer Abdul Rahman, Pierre-Luc St-Charles, Simon JD Prince, and Samira Ebrahimi Kahou. Transformers in reinforcement learning: a survey. *arXiv preprint arXiv:2307.05979*, 2023.
2. Bulbul Ahmed, Sujan Kumar Saha, Jingbo Zhou, Sohrab Aftabjahani, Mark Tehranipoor, and Farimah Farahmandi. Continuity in security: Leveraging LLM for translating security properties across hardware designs. In *2024 IFIP/IEEE 32nd International Conference on Very Large Scale Integration (VLSI-SoC)*, pages 1–6. IEEE, 2024.
3. Hasan Al Shaikh, Shuvagata Saha, Sujan Kumar Saha, Kimia Zamiri Azar, Farimah Farahmandi, and Mark Tehranipoor. Rethinking system-on-chip verification for secure cross-layer interactions. *IEEE Design & Test*, 2025.
4. Avinash Ayalasomayajula, Nusrat Farzana Dipu, Mark M Tehranipoor, and Farimah Farahmandi. Automatic asset identification for assertion-based SoC security verification. *IEEE Transactions on Computer-Aided Design of Integrated Circuits and Systems*, 43(10):3264–3277, 2024.
5. Md Kawser Bepary, Tao Zhang, Farimah Farahmandi, and Mark Tehranipoor. PreSCAN: A comprehensive review of pre-silicon physical side-channel vulnerability assessment methodologies. *Chips*, 3(4):311–333, 2024.
6. Lili Chen, Kevin Lu, Aravind Rajeswaran, Kimin Lee, Aditya Grover, Misha Laskin, Pieter Abbeel, Aravind Srinivas, and Igor Mordatch. Decision transformer: Reinforcement learning via sequence modeling. *Advances in neural information processing systems*, 34:15084–15097, 2021.
7. Nusrat Farzana Dipu, Avinash Ayalasomayajula, Mark M Tehranipoor, and Farimah Farahmandi. Agile: Automated assertion generation to detect information leakage vulnerabilities. *IEEE Transactions on Information Forensics and Security*, 19:1794–1809, 2023.
8. Zahin Ibnat, Paul E Calzada, Dipayan Saha, Hasan Al-Shaikh, Sujan Kumar Saha, Jingbo Zhou, Farimah Farahmandi, and Mark Tehranipoor. Trusting the machine: How secure is LLM-generated RTL code? In *2025 ACM/IEEE 7th Symposium on Machine Learning for CAD (MLCAD)*, pages 1–8. IEEE, 2025.
9. Nitin Pundir, Jungmin Park, Farimah Farahmandi, and Mark Tehranipoor. Power side-channel leakage assessment framework at register-transfer level. *IEEE Transactions on Very Large Scale Integration (VLSI) Systems*, 30(9):1207–1218, 2022.
10. Mridha Md Mashahedur Rahman, Shams Tarek, Kimia Zamiri Azar, Mark M Tehranipoor, and Farimah Farahmandi. Efficient SoC security monitoring: quality attributes and potential solutions. *IEEE Design & Test*, 41(4):26–34, 2023.
11. Dipayan Saha, Shams Tarek, Hasan Al Shaikh, Khan Thamid Hasan, Pavan Sai Nalluri, Md Ajoad Hasan, Nashmin Alam, Jingbo Zhou, Sujan Kumar Saha, Mark Tehranipoor, et al. SV-LLM: An agentic approach for SoC security verification using large language models. *arXiv preprint arXiv:2506.20415*, 2025.
12. Dipayan Saha, Shams Tarek, Katayoon Yahyaei, Sujan Kumar Saha, Jingbo Zhou, Mark Tehranipoor, and Farimah Farahmandi. LLM for SoC security: A paradigm shift. *IEEe Access*, 12:155498–155521, 2024.
13. Dipayan Saha, Katayoon Yahyaei, Sujan Kumar Saha, Mark Tehranipoor, and Farimah Farahmandi. Empowering hardware security with LLM: The development of a vulnerable hardware database. In *2024 IEEE International Symposium on Hardware Oriented Security and Trust (HOST)*, pages 233–243. IEEE, 2024.
14. Amit Mazumder Shuvo, Tao Zhang, Farimah Farahmandi, and Mark Tehranipoor. Flat: Layout-aware and security property-assisted timing fault-injection attack assessment. *IEEE Transactions on Very Large Scale Integration (VLSI) Systems*, 32(6):1150–1163, 2024.
15. Shams Tarek, Dipayan Saha, Sujan Kumar Saha, Mark Tehranipoor, and Farimah Farahmandi. SocureLLM: An LLM-driven approach for large-scale system-on-chip security verification and policy generation. In *2025 IEEE International Symposium on Hardware Oriented Security and Trust (HOST)*, pages 335–345. IEEE, 2025.

Index

F. Farahmandi et al. (eds.), *LLM for SoC Design and Security*,
https://doi.org/10.1007/978-3-032-34116-7

FSC
www.fsc.org
MIX
Papier | Fördert gute Waldnutzung
FSC® C083411

Kolibri 360 GmbH
Ferdinand-Jühlke-Straße 7
99095 Erfurt, Deutschland
produktsicherheit@kolibri360.de